ASHRAE Laboratory Design Guide

Second Edition

This publication was prepared under the auspices of
TC 9.10, Laboratory Systems.

Updates/errata for this publication
will be posted on the ASHRAE website at
www.ashrae.org/publicationupdates.

ASHRAE Laboratory Design Guide

Planning and Operation of Laboratory HVAC Systems

Second Edition

Peachtree Corners

ISBN 978-1-936504-98-5

180 Technology Parkway
Peachtree Corners, GA 30092
www.ashrae.org

Cover design by Laura Haass

Library of Congress Cataloging-in-Publication Data

ASHRAE laboratory design guide : planning and operation of laboratory HVAC systems. -- Second edition.
pages cm
Revised edition of: ASHRAE laboratory design guide / Ian B.D. McIntosh, Chad B. Dorgan, Charles E. Dorgan. 2001.
Includes bibliographical references and index.
Summary: "Reference manual for planning, design, and operation of laboratory HVAC systems to reduce the laboratory's energy footprint while ensuring safety, providing good comfort and indoor air quality, and protecting the integrity of experiments; includes online access to electronic design tools that illustrate features of laboratories and provide practical design aids"-- Provided by publisher.
ISBN 978-1-936504-98-5 (softcover : alk. paper) 1. Laboratories--Design and construction. 2. Laboratories--Safety measures. 3. Laboratories--Energy conservation. 4. Buildings--Environmental engineering. I. American Society of Heating, Refrigerating and Air-Conditioning Engineers.
TH7688.L3M35 2015
697.9--dc23
2015016133

Contents

ASHRAE Laboratory Design Guide: Planning and Operation of Laboratory HVAC Systems is accompanied by supplemental online content, which can be found at www.ashrae.org/LabDG.

Preface

ASHRAE Laboratory Design Guide: Planning and Operation of Laboratory HVAC Systems is a significant revision of the first edition of the design guide, which was published in 2001. It is intended that the information in this Guide assist engineers, owners, and system operators in determining the needs of their laboratory facility and how to best match these needs to mechanical system options.

This Guide is written in a manner that progresses from general to specific to reach a wide target audience of designers, architects, engineers, owners, operation and maintenance personnel, and others in the heating, ventilating, air-conditioning, and refrigerating (HVAC&R) industry. The first chapters of the Guide present general background information and prescribed design frameworks, whereas the later chapters are more specific, providing detailed design and application information. A primary benefit of this structure is that many types of readers can use the Guide. Whereas an owner may only read the first two chapters, an engineer may focus on the design chapters.

This new edition has been updated to reflect current standards and industry practices. Web links to these standards and to additional laboratory-related resources have been expanded. Many of the chapters have been significantly revised or rewritten.

This edition also contains two new chapters, one that covers sustainability and one that covers computer modeling of ventilation effectiveness.

It is hoped that this revised edition of the Guide will be even more useful to the laboratory designer, owner and operator, and anyone else involved in the design, construction, and operation of laboratories.

Acknowledgments

This Design Guide is the product of many members of Technical Committee (TC) 9.10, Laboratory Systems, as well as others in the laboratory and HVAC community.

Special thanks go to the chapter leaders and contributors:

- Pete Gardner was the leader and a contributor for Chapters 2 and 7.
- John Varley and Erik Eaves were the leaders and contributors to the revision of Chapter 3.
- John Castelvecchi was the leader for the significantly revised Chapter 5. In addition to John, George Sestak, Kurt Rindoks, and Bill Peters also were contributors to this chapter. John also was the leader for Chapter 11, which is a totally rewritten and expanded chapter. In addition to John, Jim Coogan and Gordon Sharp were contributors. John also was the leader and a contributor for Chapter 14.
- Carl Crow was the leader and a contributor to Chapter 6.
- Charles Henck was the leader and a contributor for Chapter 8.
- Brad Cochran was the leader and a contributor for Chapter 9.
- Roland Charneux was the leader and a contributor for Chapters 10 and 18. He authored Chapter 18, which is a new chapter to the Design Guide, with John Varley as a contributor.
- Mike Ratcliff was the leader and primary author for Chapter 17, which is a new chapter. Gordon Sharp and Bob Weidner were also contributors.
- As the leader and a contributor for Chapters 1, 4, 12, 13, 15, and the totally rewritten Chapter 16, I worked with Carl Crow as a major contributor to Chapter 1, with John Varley on Chapter 4, with Tom Smith and Jeff Traylor on Chapter 13, and with Chris Kiley, David Duthu, and Carl Crow as major contributors to Chapter 16.

Also, much thanks goes to the chapter reviewers, who in most cases reviewed multiple chapters and whose comments have greatly improved this edition of the Design Guide: Mike Brueggeroff, Hwakong Cheng, Bob Cochran, Jim Coogan, Carl Crow, Brent Eubanks, Mary Foutz, Dan Frasier, Pete Gardner, Kevin Gilkison, Traci Hanegan, Nathan Ho, Mark Hydeman, Chris Kiley, Ken Kuntz, Guy Perreault, Mike Ratcliff, David Rausch, Scott Rusk, George Sestak, Gordon Sharp, Tom Smith, John Varley, Mike Walsh, Bob Weidner, Ron Wendorski, and myself.

Shinji Sunohara did a great service in providing the SI units for the Guide. Thanks also to So-Yeng Chen for coordinating this effort.

Roland Charneux was the FTP site manager for the editing process and provided much logistical support.

David Rausch assisted the chapter leaders with obtaining permissions.

The members of the TC 9.10 Laboratory Design Guide Subcommittee also provided much support and guidance during this process.

In addition to the members already mentioned above, I would like to acknowledge the following: Leon Alevantis, Adam Bare, Patrick Carpenter, Charles Coward, Wade Conlan, Kelley Cramm, Carol Donovan, Andrew Dymek, Carl Lawson, Lloyd Le, Gaylon Richardson, Tony Rossi, and Mitch Swann.

As this second edition relied on the first edition for the basis of many of the chapters, additional thanks go to all those who contributed to the first edition of this book, which was prepared under ASHRAE Research Project 969 and authored by Ian B.D. McIntosh, Chad B. Dorgan, and Charles E. Dorgan. The Project Monitoring Subcommittee for the first edition was chaired by John Mentzer and included Pete Gardner, John Varley, Bob Weidner, and Randall Lacey. Jack Wunder performed a detailed review of the draft of the first edition, and the following key people supplied input that contributed to the success of the first edition: Todd Hardwick, Otto Van Geet, Victor Neuman, Carl Lawson, Greg De Luga, Geoffrey Bell, Luis Solarte, Daniel Ghidoni, Patrick Chudecke, Edward Fiance, Robert Haugen, Tom Begley, Bob Parsons, and Gary Butler. The authors of the first edition also recognized the following Dorgan Associates' staff members for their contributions: Svein Morner and Zachary Obert for writing various sections and reviewing background material to help ensure technical accuracy, Chad Grindle for creating and editing the graphics, Joan Dorgan for proofreading, and Suzanne Bowen for word processing and proofreading.

Many individuals worked on this Design Guide. Acknowledging those who contributed to and supported this effort involves the risk of omitting some of these individuals. I apologize if this is the case.

Henry Hays
Chair, TC 9.10 Laboratory Design Guide Subcommittee

Abbreviations and Acronyms

ACGIH	American Conference of Governmental Industrial Hygienists
ach, ACH	air changes per hour
AHU	air-handling unit
AIHA	American Industrial Hygiene Association
AMCA	Air Movement and Control Association International
APR	air pressure resistant
BMBL	*Biosafety in Microbiological and Biomedical Laboratories*
BOD	Basis of Design
BSC	biological safety cabinet
BSL	biosafety level
CDC	Centers for Disease Control and Prevention
CFD	computational fluid dynamics
cfm	cubic feet per minute
CPVC	chlorinated polyvinyl chloride
EDS	effluent decontamination system
EMCS	energy management and control system
EPA	U.S. Environmental Protection Agency
fpm	feet per minute
FRP	fiber reinforced plastic
HEPA	high-efficiency particulate air
IAQ	indoor air quality
L/s	litres per second
LCCA	life-cycle cost analysis
LEL	lower explosive limit
m/s	metres per second
NIBS	National Institute of Building Sciences
NIH	National Institutes of Health
NIOSH	National Institute for Occupational Safety and Health
O&M	operation and maintenance
OPR	Owner's Project Requirements

PVC	polyvinyl chloride
rpm, RPM	revolutions per minute
SHGC	solar heat gain coefficient
SMACNA	Sheet Metal and Air Conditioning Contractors' National Association
TAB	testing, adjusting, and balancing
TLV	threshold limit value
UPS	uninteruptible power supply
USDA	United States Department of Agriculture
USGBC	U.S. Green Building Council
VAV	variable air volume
VFD	variable-frequency drive
VSD	variable-speed drive

Introduction

1

OVERVIEW

Laboratories come in various types and configurations. A laboratory may be a few rooms in a building or it may be a large, multistory facility. Laboratories range from containing relatively low-risk materials to containing extremely high-risk biological pathogens. Each laboratory has unique requirements, especially in regard to the ventilation and exhaust systems. In addition, the realization that laboratories are energy intensive has resulted in design and control strategies for reducing the energy footprint while providing acceptable ventilation levels. This Design Guide is intended to be a comprehensive reference manual for the planning and design of laboratories.

To improve the comprehension of the material for those unfamiliar with laboratory design, this Guide is organized around a typical project, progressing through the basic steps of planning, design, construction, and operation and maintenance (O&M). The Guide's 18 chapters are discussed in detail in the following section.

ORGANIZATION

Chapter 2, "Background," provides basic background information on laboratories, including their various types and the typical equipment found in them. The intent of this chapter is to provide a basic understanding of laboratories and their importance as well as their different functions and needs.

Chapter 3, "Design Process," outlines the design process with guidance provided to designers and engineers on the key steps in meeting or exceeding the owner's design intent. Also included in this chapter is a review of special space design considerations for the different HVAC system types encountered in a laboratory.

Chapter 4, "Laboratory Planning," describes the many important issues that are addressed during the planning phase of a laboratory project. At the very core of this phase is the need to provide utmost safety for laboratory occupants via risk assessment and hazard analyses while achieving experimental integrity, good comfort and indoor air quality, capacity to meet operating loads, and successful integration of architectural and engineering systems. These various important considerations are then carefully documented in the Owner's Project Requirements (OPR) and the owner's design intent.

Chapter 5, "Exhaust Hoods," details the operability, types, applications, selection, testing and monitoring of fume exhaust hoods.

Chapter 6, "Primary Air Systems," describes the many HVAC system options available for conditioning and contaminant removal, including supply systems, room or space air diffusion, exhaust systems, duct construction, and energy efficiency.

Chapter 7, "Process Cooling," examines the different system options available to meet supplementary process cooling needs. This includes the types of water-cooled loads found in laboratories, water treatment and quality, temperature and pressure requirements, and pumping system configurations.

Chapter 8, "Air Treatment," presents several methods of treating exhaust and makeup air as it is either expelled to the outside environment or drawn into the laboratory space from the outside.

Chapter 9, "Exhaust Stack Design," deals with the final outlet through which air is discharged into the environment—exhaust stacks. Required elements for good exhaust stack design, in addition to modeling techniques available for design verification, are described.

Chapter 10, "Energy Recovery," addresses the ability to realize economic savings via the recovery of the temperature and/or humidity content of the exhaust air or other sources of energy that would otherwise be wasted. The two main energy recovery technologies discussed are air-to-air and water-to-air heat exchange. In addition, key parameters used to properly select energy recovery options are presented.

Chapter 11, "Controls," discusses controls for constant-volume fume hoods, variable-air-volume (VAV) fume hoods, and other exhaust equipment. Room control discussion includes the theory of room control, outdoor air needs, minimum ventilation air changes per hour, control stability, constant-volume and VAV strategies, control of critical spaces, and building pressurization.

Chapter 12, "Airflow Patterns and Testing Procedures," provides an understanding of airflow patterns throughout the laboratory environment and how they need to be maintained in a specific direction and velocity to protect against pollutants or contamination. This is followed by descriptions of proper air and hydronic balancing and testing procedures.

Chapter 13, "O&M for Ventilation and Exhaust Systems," details how to properly operate and maintain laboratory systems for the continual safety of occupants, integrity of laboratory experiments, and efficiency of HVAC systems and laboratory operations. To this end, this chapter discusses the training needed for O&M personnel and laboratory users. In addition, operating costs and the importance of periodic fume hood and biological safety cabinet (BSC) testing and certification are also addressed.

Chapter 14, "Laboratory Commissioning Process," introduces the quality assurance procedure known as the *commissioning process*, which commences during the planning phase and follows through to the design, construction, acceptance, turnover, and occupancy and operations phases of a building project. The commissioning process ensures that the diverse requirements for the laboratory and the owner's design intent are met.

Chapter 15, "HVAC System Economics," covers both the initial costs (costs associated with planning, design, and construction) and life-cycle costs (costs associated with operation and maintenance) of a laboratory facility. Although initial costs are usually given more emphasis because of budgetary constraints, this chapter also discusses the life-cycle costs in designing the laboratory. This is important because a substantial investment is required to effectively operate and maintain laboratory systems and equipment over a life span of 15 to 30 years.

Chapter 16, "Microbiological and Biomedical Laboratories," provides specific information about laboratories that specialize in biological containment and that house research animals. Because of the importance of containing and controlling biohazards and product protection within these special laboratories, issues such as system reliability, redundancy, proper space pressurization, envelope design, and sanitation are addressed.

Chapter 17, "CFD Modeling of Laboratory Ventilation," provides information regarding evaluating airflow patterns and contaminant concentrations using computational fluid dynamics (CFD) computer modeling. The results of the modeling can help the designer address issues including supply and exhaust device type and location, fume hood placement, control of heat or contaminant sources, and occupational safety. This chapter also presents guidelines on using CFD for analyzing laboratory ventilation.

Chapter 18, "Sustainable Design," addresses the opportunities to incorporate high-performance building design concepts into the design process. Laboratory facilities present unique challenges and opportunities for sustainability. This chapter provides an overview and presents examples of the use of these opportunities.

This Guide is also accompanied by supplemental information that can be found at www.ashrae.org/LabDG. Included are short videos illustrating concepts presented in the text, sample room data sheets in PDF format, a laboratory sustainability checklist for reviewing opportunities to enhance laboratory sustainability and a sample room schedule in Microsoft® Excel® format, PDFs of articles on laboratory HVAC published in *ASHRAE Journal* and *ASHRAE Transactions* that provide case studies and practical application information, recordings of related seminars presented at ASHRAE Conferences, and the program DilutionCalc for calculating downwind dilutions/concentrations from an exhaust stack. If the files or information at the link are not accessible, please contact the publisher.

REFERENCE SOURCES

There are many laboratory-related resources available that may be helpful to engineers, owners, and system operators. These resources include professional associations and societies, research institutions, government offices and regulators, design forums and publications, and ASHRAE committees.

PROFESSIONAL ASSOCIATIONS AND SOCIETIES

AAALAC—Association for Assessment and Accreditation of Laboratory Animal Care International

The Association for Assessment and Accreditation of Laboratory Animal Care International (AAALAC) is a private, nonprofit organization that promotes the humane treatment of animals in science through voluntary accreditation and assessment programs.

www.aaalac.org

AALAS—American Association for Laboratory Animal Science

The American Association for Laboratory Animal Science (AALAS) is a membership association of professionals employed around the world in academia, government, and private industry who are dedicated to the humane care and treatment of laboratory animals as well as the quality research that leads to scientific gains that benefit people and animals.

www.aalas.org

ABSA—American Biological Safety Association

The American Biological Safety Association (ABSA) was founded in 1984 to promote biosafety as a scientific discipline and serve the growing needs of biosafety professionals throughout the world. The Association's goals are to provide a professional association that represents the interests and needs of practitioners of biological safety and to provide a forum for the continued and timely exchange of biosafety information.

http://absa.org

ACGIH—American Conference of Governmental Industrial Hygienists

The American Conference of Governmental Industrial Hygienists (ACGIH) is a member-based organization that advances occupational and environmental health.

www.acgih.org

ACS—American Chemical Society

The American Chemical Society (ACS) is a scientific society based in the United States that supports scientific inquiry in the field of chemistry. Founded in 1876 at New York University, ACS currently has more than 163,000 members at all degree levels and in all fields of chemistry, chemical engineering, and related fields. It is the world's largest scientific society and one of the leading sources of authoritative scientific information.

www.acs.org

AESP—Association of Energy Services Professionals

The Association of Energy Services Professionals (AESP) is a member-based association dedicated to improving the delivery and implementation of energy efficiency, energy management, and distributed renewable resources.

www.aesp.org

AIA—The American Institute of Architects

The American Institute of Architects (AIA) is a professional organization for architects in the United States. Headquartered in Washington, DC, AIA offers education, government advocacy, community redevelopment, and public outreach to support the architecture profession and improve its public image. AIA also works with other members of the design and construction team to help coordinate the building industry.

www.aia.org

AIHA—American Industrial Hygiene Association

Founded in 1939, the American Industrial Hygiene Association (AIHA) is a nonprofit organization devoted to achieving and maintaining the highest professional standards for its members. AIHA administers comprehensive education programs that keep occupational and environmental health and safety (OEHS) professionals current in the field of industrial hygiene. AIHA is one of the largest international associations serving OEHS professionals practicing industrial hygiene and is a resource for those in large corporations, small businesses and who work independently as consultants.

www.aiha.org

AHRI—Air Conditioning, Heating, and Refrigeration Institute

The Air Conditioning, Heating, and Refrigeration Institute (AHRI), formed in 2008 by a merger of the Air-Conditioning and Refrigeration Institute (ARI) and the Gas Appliance Manufacturers Association (GAMA), is a North American trade association of manufacturers of air conditioning, heating, and commercial refrigeration equipment. The organization performs political advocacy on behalf of its member industries, maintains technical standards, certifies products, shares data, conducts research, and awards scholarships.

www.ari.org

ANSI—American National Standards Institute

The American National Standards Institute (ANSI) is a private nonprofit organization that oversees the development of voluntary consensus standards for products, services, processes, systems, and personnel in the United States. The organization also coordinates U.S. standards with international standards so that American products can be used worldwide.

www.ansi.org

ASCE—American Society of Civil Engineers

The American Society of Civil Engineers (ASCE) is a tax-exempt professional body founded in 1852 to represent members of the civil engineering profession worldwide. Based in Reston, Virginia, it is the oldest national engineering society in the United States.

www.asce.org

ASHE—American Society for Healthcare Engineering

Chicago-based American Society for Healthcare Engineering (ASHE) is a membership group of the American Hospital Association, with more than 10,000 members focused on optimizing the healthcare physical environment.

www.ashe.org

ASHRAE

ASHRAE, founded in 1894, is a global society advancing human well-being through sustainable technology for the built environment. The Society and its members focus on building systems, energy efficiency, indoor air quality, refrigeration and sustainability within the industry. Through research, standards writing, publishing and continuing education, ASHRAE shapes tomorrow's built environment today.

www.ashrae.org

ASME—American Society of Mechanical Engineers

The American Society of Mechanical Engineers (ASME) is a professional association that promotes the art, science, and practice of multidisciplinary engineering and allied sciences around the globe via continuing education, training and professional development, codes and standards, research, conferences and publications, government relations, and other forms of outreach.

www.asme.org

ASPE—American Society of Plumbing Engineers

The American Society of Plumbing Engineers (ASPE) is a nonprofit association for the advancement and development of plumbing engineering and related professions.

www.aspe.org

ASSE—American Society of Safety Engineers

Founded in 1911, the American Society of Safety Engineers (ASSE) sets the standards for excellence and ethics for the occupational safety, health, and environmental community. ASSE is a global association of occupational safety professionals representing more than 36,000 members worldwide who create safer work environments by preventing workplace fatalities, injuries, and illnesses. ASSE has taken over as the

Secretariat of ANSI/AIHA/ASSE Z9.5, *Laboratory Ventilation* (AIHA 2012), published by the American Industrial Hygiene Association (AIHA).

www.asse.org

CETA—Controlled Environment Testing Association

The Controlled Environment Testing Association (CETA) was established in 1992. It was developed to promote quality assurance in controlled environments through the exchange and dissemination of information on certification issues. Controlled environments include cleanrooms, laminar flow devices, sterile ovens, tunnels, dryers, biological safety cabinets (BSCs), fume hoods, containment suites, and other similar environments that require precision testing and certification.

www.cetainternational.org

CIBSE—The Chartered Institution of Building Services Engineers

The Chartered Institution of Building Services Engineers (CIBSE) received its Royal Charter in 1976. It is the professional body that exists to promote the career of building services engineers by accrediting courses of study in further and higher education, by approving work-based training programs and providing routes to full professional registration and membership, including Chartered Engineer, Incorporated Engineer, and Engineering Technician.

http://cibse.org

GBCI—Green Building Certification Institute

The Green Building Certification Institute (GBCI) is a third-party organization that provides independent oversight of professional credentialing and project certification programs related to green building. Established in 2008 to administer certifications and professional designations within the framework of the U.S. Green Building Council (USGBC) Leadership in Energy and Environmental Design® (LEED®) Green Building Rating System.

www.gbci.org

IEST—Institute of Environmental Sciences and Technology

The Institute of Environmental Sciences and Technology (IEST), founded in 1953 as a nonprofit organization, is an international society focused on contamination control and publishes recommended practices for testing cleanrooms, HEPA/ULPA filters, and clean-air filtration systems.

www.iest.org

ILAR—Institute for Laboratory Animal Research

The Institute for Laboratory Animal Research (ILAR) is a unit in the Division on Earth and Life Studies of the National Research Council of the National Academies of Science. It has been a national leader in the research, publication, and collection of information pertinent to animal care and use in the laboratory setting since 1952. Its mission is to evaluate and report on the scientific, technological, and ethical use of animals and their related biological resources and of non-animal alternatives in research, testing, education, and the production of pharmaceuticals.

www.dels.nas.edu/ilar

ICC—International Code Council

The International Code Council (ICC) is a member-focused association dedicated to helping the building safety community and construction industry provide safe, sustainable, and affordable construction through the development of codes and standards used in the design, build, and compliance process.

www.iccsafe.org

ISPE—International Society for Pharmaceutical Engineering

Founded in 1980, the International Society for Pharmaceutical Engineering (ISPE) is a nonprofit industry trade group for pharmaceutical science and manufacturing professionals. It has 25,000 members in more than 90 countries.

www.ispe.org

I^{2}SL—International Institute for Sustainable Laboratories

The International Institute for Sustainable Laboratories (I^{2}SL) is devoted to the principles of sustainable laboratories, from design to engineering to practice. Through worldwide partnerships and the exchange of technical information, I^{2}SL helps produce high-technology facilities that address the rapid pace of science, medicine, research, and development in an ever-changing and dynamic world.

www.i2sl.org

NFPA—National Fire Protection Association

The National Fire Protection Association (NFPA) is an international leader in fire prevention and public safety. This nonprofit member association was founded in 1896 to develop consensus codes and standards as well as training, research, and education in fire and other hazard protections in various buildings. NFPA has over 67,000 members and is associated with 80 national and professional organizations from across the globe.

www.nfpa.org

NIBS—National Institute of Building Sciences

The National Institute of Building Sciences (NIBS) is a nonprofit, nongovernmental organization that brings together representatives of government, the professions, industry, labor and consumer interests, and regulatory agencies to focus on the identification and resolution of problems and potential problems that hamper the construction of safe, affordable structures for housing, commerce, and industry throughout the United States. The Institute's mission to serve the public interest is accomplished by supporting advances in building sciences and technologies for the purpose of improving the performance of buildings while reducing waste and conserving energy and resources.

www.nibs.org

NSF International

NSF International, formerly the National Sanitation Foundation, was founded in 1944 and has been committed to public health, safety, and protection of the environment by developing standards, product testing, and conformity assessments. Most relevant to biosafety, NSF International has developed a biological safety cabinet (BSC) standard.

www.nsf.org

NSPE—National Society of Professional Engineers

The National Society of Professional Engineers (NSPE) is an American professional engineering organization representing the voice of licensed engineers and advocates for all related disciplines. The society concerns itself with a variety of issues relating to engineering, including licensure and ethics, the image of the engineer, the rights of engineers at the national and state levels, news of the profession, and continuing education opportunities.

www.nspe.org

SEFA—The Scientific Equipment and Furniture Association

The Scientific Equipment and Furniture Association (SEFA) was organized in 1988 by laboratory equipment manufacturers to enhance individual member company performance and improve the quality of laboratory facilities. SEFA accomplishes these goals by establishing industry standards for laboratory equipment, installation, and testing.

www.sefalabs.com

USGBC—U.S. Green Building Council

The U.S. Green Building Council (USGBC), founded in 1993, is a private, membership-based nonprofit organization that promotes sustainability in how buildings are designed, built, and operated. USGBC is best known for its development of the Leadership in Energy and Environmental Design® (LEED®) Green Building Rating System.

www.usgbc.org

RESEARCH INSTITUTIONS

CDC—Centers for Disease Control and Prevention

Centers for Disease Control and Prevention (CDC) is a federal institution whose primary goal and pledge is to promote health and quality of life by preventing and controlling disease, injury, and disability. To achieve this pledge, the CDC uses federal and private funding for research and development in several fields and sciences at its 11 different laboratories and locations.

www.cdc.gov

LBNL—Lawrence Berkeley National Laboratory

Lawrence Berkeley National Laboratory (LBNL) is a U.S. Department of Energy National Laboratory managed by the University of California. The Environmental Energy Technologies Division, established in 1973, develops technology that uses, converts, and stores energy more efficiently and with less environmental impact. To better identify and prioritize the many research activities necessary to achieve this goal, LBNL prepared research and development roadmaps for cleanrooms and laboratories as well as for data centers.

www.lbl.gov

NIH—National Institutes of Health

National Institutes of Health (NIH) is an agency of the U.S. Department of Health and Human Services and is the primary agency of the United States government responsible for biomedical and health-related research. Founded in 1887, NIH funds national and international research in numerous medical fields to accomplish this dedication to health. It comprises 27 separate institutes, centers, and offices.

www.nih.gov

NREL—National Renewable Energy Laboratory

The National Renewable Energy Laboratory (NREL) is the U.S. Department of Energy's primary national laboratory for renewable energy and energy efficiency research and development.

www.nrel.gov

NRC—National Research Council

The National Research Council (NRC) was formed by the National Academies of Science in 1916 to bridge the gap between the science and technology communities and the federal government. Today, the NRC continues this initiative by supplying reports and advice to federal and public sectors on the current issues and advances in science and technology that shape policies, inform public opinion, and advance the pursuit of science, engineering, and medicine.

www.nationalacademies.org/nrc

GOVERNMENT OFFICES AND REGULATORS

CCOHS—Canadian Centre for Occupational Health and Safety

Established in 1978, the Canadian Centre for Occupational Health and Safety (CCOHS) promotes the total well-being—physical, psychosocial and mental health—of working Canadians by providing information, training, education, management systems, and solutions that support health, safety, and wellness programs. A nonprofit federal department corporation, CCOHS is governed by a tripartite Council—representing government, employers, and labor—to ensure a balanced approach to workplace health and safety issues.

www.ccohs.ca

FDA—U.S. Food and Drug Administration

The U.S. Food and Drug Administration (FDA) is an agency of the U.S. Department of Health and Human Services, one of the United States federal executive departments. The FDA is responsible for protecting and promoting public health through the regulation and supervision of food safety, tobacco products, dietary supplements, prescription and over-the-counter pharmaceutical drugs (medications), vaccines, biopharmaceuticals, blood transfusions, medical devices, electromagnetic radiation emitting devices (EREDs), and veterinary products.

www.fda.gov

HHS—U.S. Department of Health and Human Services

The U.S. Department of Health and Human Services (HHS) is the federal administration primarily in charge of funding and developing programs, research, and education that benefit and protect the health of Americans. Some of the agencies that fall under the supervision or funding of the HHS include Centers for Disease Control and Prevention (CDC), U.S. Food and Drug Administration (FDA), and Health Resources and Services Administration (HRSA).

www.hhs.gov

Labs21—Laboratories for the 21st Century

Laboratories for the 21st Century (Labs21) was a joint initiative between U.S. Department of Energy (DOE) and U.S. Environmental Protection Agency (EPA) and was a voluntary partnership program dedicated to improving the environmental performance of U.S. laboratories. Although Labs21 is no longer an active organization, the nongovernmental organization International Institute for Sustainable Laboratories (I^2SL) has taken over many Labs21 functions and maintains online access to many of the Labs21 publications. (See the information for I^2SL in the Professional Associations and Societies section of this chapter.)

http://energy.gov/eere/femp/laboratories-21st-century

NIOSH—The National Institute for Occupational Safety and Health

The National Institute for Occupational Safety and Health (NIOSH) is part of the Centers for Disease Control and Prevention (CDC) in the U.S. Department of Health and Human Services. NIOSH is an agency established to help ensure safe and healthful working conditions for working men and women by providing research, information, education, and training in the field of occupational safety and health. Information pertaining to the specific responsibilities of NIOSH is found in Section 22 of the Occupational Safety and Health Act of 1970 (29 CFR § 671).

www.cdc.gov/niosh

NRC—U.S. Nuclear Regulatory Commission

The U.S. Nuclear Regulatory Commission (NRC) was created as an independent agency by Congress in 1974 to ensure the safe use of radioactive materials for beneficial civilian purposes while protecting people and the environment. The NRC regulates commercial nuclear power plants and other uses of nuclear materials, such as in nuclear medicine, through licensing, inspection, and enforcement of its requirements.

www.nrc.gov

OSHA—Occupational Safety and Health Administration

The Occupational Safety and Health Administration (OSHA) is a federal body whose primary purpose is to protect and save the lives of American workers through the development and implementation of various standards and regulations related to the work environment. Currently OSHA employs over 2100 inspectors at 200 nationwide locations to cover the ever-increasing workforce of over 100 million in the United States.

www.osha.gov

DESIGN FORUMS AND PUBLICATIONS

Design forums, newsletters, and trade journals such as the following often include articles covering technical and design-related issues that pertain to laboratory design.

ALN Magazine
www.alnmag.com
Applied Biosafety: Journal of the American Biological Safety Association (ABSA)
www.absa.org/pubabj.html
ASHRAE Journal
www.ashrae.org/ASHRAEjournal
Consulting-Specifying Engineer
www.csemag.com
Controlled Environments
www.cemag.us
Engineered Systems
www.esmagazine.com
FacilitiesNet
www.facilitiesnet.com
HPAC Engineering
http://hpac.com
ILAR Journal
http://ilarjournal.oxfordjournals.org
Journal of Mechanical Design (ASME)
http://mechanicaldesign.asmedigitalcollection.asme.org/journal.aspx

Journal of the American Association for Laboratory Animal Science
www.aalas.org/publications/jaalas
Laboratory Animal Science Professional
www.aalas.org/publications/las-pro
Laboratory Design Newsletter
www.labdesignnews.com
Plumbing Engineer
www.plumbingengineer.com
PM Engineer Magazine
www.pmengineer.com/publications/3
R&D Magazine
www.rdmag.com
Sustainable Laboratory Times
www.i2sl.org/resources/newsletter.html
The NEWS
www.achrnews.com/publications/3
The NEWS HVACR Directory & Source Guide
http://directory.achrnews.com/hvacrdirectory
Tradeline Industry News
www.tradelineinc.com/news

University Research Theses and Papers

Universities are a primary source of information on specific details of laboratory design. Individual university libraries and organizations can be accessed to assist in locating information on specific topics.

Guidelines for Laboratory Design and Safety

A Design Guide for Energy-Efficient Research Laboratories (LBNL 2008)
http://ateam.lbl.gov/Design-Guide

Biosafety in Microbiological and Biomedical Laboratories (*BMBL*), 5th Edition (CDC 2009)
www.cdc.gov/biosafety/publications/bmbl5

Chapter 16, "Laboratories," of *ASHRAE Handbook—HVAC Applications* (ASHRAE 2015)

Guide for the Care and Use of Laboratory Animals, 8th Edition (ILAR 2011)
http://grants.nih.gov/grants/olaw/Guide-for-the-care-and-Use-of-Laboratory-Animals.pdf

Labs21 Tool Kit
www.i2sl.org/resources/toolkit.html

NIH Design Policy and Guidelines (NIH 2013)
http://orf.od.nih.gov/PoliciesAndGuidelines/BiomedicalandAnimalResearchFacilitiesDesignPoliciesandGuidelines/Pages/policy-index.aspx

"Primary Containment for Biohazards: Selection, Installation, and Use of Biological Safety Cabinets" (Currently published as Appendix A of *BMBL* but planned as a future separate publication)
www.cdc.gov/biosafety/publications/bmbl5/BMBL5_appendixA.pdf

Section on Research Facilities in *Whole Building Design Guide* (NIBS 2010)
www.wbdg.org/design/research.php

RELATED ASHRAE TECHNICAL COMMITTEES (TCs)

TC 2.2, Plant and Animal Environment
http://tc22.ashraetcs.org
TC 4.3, Ventilation Requirements and Infiltration
http://tc43.ashraetcs.org
TC 9.6, Healthcare Facilities
http://tc96.ashraetcs.org
TC 9.10, Laboratory Systems
http://tc910.ashraetcs.org
TC 9.11, Clean Spaces
http://tc911.ashraetcs.org

REFERENCES

ASHRAE. 2015. Chapter 16, Laboratories. In ASHRAE *handbook—HVAC applications*. Atlanta: ASHRAE.

CDC. 2009. *Biosafety in microbiological and biomedical laboratories* (*BMBL*), 5th Edition. DHHS Publication No. (CDC) 21-1112. Atlanta: Centers for Disease Control and Prevention, U.S. Department of Health and Human Services. www.cdc.gov/biosafety/publications/bmbl5.

ILAR. 2011. *Guide for the care and use of laboratory animals*, 8th ed. Washington, DC: Institute for Laboratory Animal Research, Division on Earth and Life Studies, National Research Council, National Academies Press. http://grants.nih.gov/grants/olaw/Guide-for-the-care-and-Use-of-Laboratory-Animals.pdf.

LBNL. 2008. *A design guide for energy-efficient research laboratories*, Version 4.0. Berkeley, CA: Lawrence Berkeley National Laboratory. http://ateam.lbl.gov/Design-Guide.

NIBS. 2010. Research Facilities. *Whole Building Design Guide*. Washington, DC: National Institute of Building Sciences. www.wbdg.org/design/research.php.

NIH. 2013. *NIH design policy and guidelines*. Bethesda, MD: National Institutes of Health. http://orf.od.nih.gov/PoliciesAndGuidelines/BiomedicalandAnimalResearchFacilitiesDesignPoliciesandGuidelines/Pages/policy-index.aspx.

Background

2

OVERVIEW

This chapter provides a brief summary of the following information that helps form a common framework for the terminology and systems in this Guide:

- Laboratory types
- Laboratory equipment

This discussion is of laboratory types and equipment typically encountered in laboratory facilities and does not address all available types and equipment.

LABORATORY TYPES

Because there are numerous types of laboratories, there are numerous potential applications for capturing pollutants to minimize adverse impacts on occupant health. The more common laboratory types and nonlaboratory spaces are as follows:

- Biological laboratories
  - Animal laboratories
  - Biological containment laboratories using infectious agents
  - Hospital laboratories
- Chemical laboratories
  - Synthesis laboratories
  - Analytical laboratories
  - Radiochemistry laboratories
  - Teaching laboratories
- Physical laboratories
  - Cleanrooms
  - Materials testing laboratories
  - Nanotechnology/electronics/instrumentation laboratories
- Support spaces

BIOLOGICAL LABORATORIES

Biological laboratories are used to work with biologically active materials or involve chemical or physical manipulation of these materials. Laboratory types include biochemistry, microbiology, genomics, immunology, botany, pharmacology, and toxicology.

Animal Laboratories

Animal laboratories need to maintain clean and humane conditions for animals and provide safe separation of the animals from laboratory personnel. Design of animal laboratories should be performed with close interaction with veterinarians and researchers who will be working in the laboratory. There are numerous standards for animal laboratories that must be followed, such as those of the Association for Assessment and Accreditation of Laboratory Animal Care International (AAALAC), which follows National Institutes of Health (NIH) and United States Department of Agriculture (USDA) guidance. Animal research facilities are often referred to as *vivariums*. Vivariums often contain animal holding rooms, procedure rooms, and some biological laboratories.

Biological Containment Laboratories

Some biological laboratory work is with infectious agents that are characterized as being in biosafety levels (BSLs) 1 through 4, as described in *Biosafety in Microbiological and Biomedical Laboratories* (CDC 2009). This reference addresses both biological laboratory and animal laboratory design and operation.

Hospital or Clinical Laboratories

Hospital laboratories provide chemical and biological testing of specimens associated with patient care. These laboratories do not use large quantities of dangerous materials and typically do not perform general research. Although it depends on the specific type of hospital, these laboratories usually contain chemical fume hoods, local exhaust, and vented Class II biological safety cabinets (BSCs).

CHEMICAL LABORATORIES

Synthesis Laboratories

A synthesis laboratory is used for organic or inorganic reactions and analytical instrumentation for a wide variety of chemicals. Synthesis laboratories are generally not used for extremely hazardous materials such as potent compounds or large quantities of explosive materials. There are typically few specialized built-in equipment needs for synthesis laboratories beyond chemical fume hoods and limited direct equipment exhaust.

Analytical Laboratories

Analytical laboratories often contain a high density of analytical equipment such as gas or liquid chromatographs. Localized exhaust may be included to capture emissions for analytical processes. There is usually a need for sample preparation space and areas to manage the samples as they are analyzed.

Radiochemistry Laboratories

A radiochemistry laboratory is often similar to a synthesis laboratory, with additional design requirements to contain direct radiation and prevent the release of radioactive chemicals and materials. Laboratories that use radioactive materials have strict requirements that are determined by the U.S. Nuclear Regulatory Commission (NRC), Occupational Safety and Health Administration (OSHA), and the U.S. Environmental Protection Agency (EPA).

Teaching Laboratories

Teaching laboratories are designed to provide a safe learning environment for large groups (usually 30, sometimes more) of high school and undergraduate college students. Teaching laboratories typically use the same types of materials as synthesis laboratories, although in much smaller quantities. Graduate student work is typically associated with research laboratories, which are not considered teaching laboratories. The primary needs of teaching laboratories are for the instructor to maintain eye contact with all students and to have quick access to the controls of the laboratory equipment to prevent hazards.

PHYSICAL LABORATORIES

A wide range of research activities, including work with lasers, computer chip development, and nanotechnology, are performed in physical laboratories.

Cleanrooms

Cleanrooms are designed to provide a working environment with minimal airborne particulate materials and constant temperature, humidity, air pressure, and airflow patterns. Their environments require controlled atmospheres to regulate gases, negative and positive ions, and the use of robotics. The most current standards are ISO 14644-1, *Cleanrooms and associated controlled environments—Part 1: Classification of air cleanliness* (ISO 1999), and ISO 14644-2, *Cleanrooms and associated controlled environments—Part 2: Specifications for testing and monitoring to prove continued compliance with ISO 14644-1* (ISO 2000). These two standards, issued by the International Organization for Standardization (ISO), are components of a 10-part series of standards that cover more cleanroom environmental parameters and practices than the earlier Federal Standard 209E, *Airborne Particulate Cleanliness Classes in Clean Rooms and Clean Zones* (IES 1992).

Materials Testing Laboratories

Materials testing laboratories are used for physical experimentation with materials, such as wear and strength testing. The types of equipment needed for a materials laboratory depend largely on the types of testing performed. Materials laboratories usually have some process cooling requirements and may have limited chemical exhaust requirements similar to a general chemistry laboratory.

Nanotechnology/Electronics/Instrumentation Laboratories

Nanotechnology laboratories can be similar to a cleanroom in that low levels of airborne particulates and precise control of temperature, humidity, and airflow are needed. Electronics laboratories, such as those that produce electronic chips and semiconductor wafers, often need to prevent the release of hazardous chemicals by using local exhaust while still maintaining a positive room pressure. There can be toxic gases for fabrication activities that require a specialized exhaust system. There may also be scanning electron microscopes that may have special vibration and airflow velocity requirements.

SUPPORT SPACES

Laboratory support spaces contribute to research and experimentation activities by providing general laboratory maintenance, material and chemical storage, and equipment and material preparation. Support spaces may require special exhaust for chemical storage or heat rejection from equipment such as ice machines and packaged environmental chambers and laboratory exhaust equipment, such as fume hoods and BSCs, depending on the type of support provided.

LABORATORY EQUIPMENT

The design of a laboratory depends largely on the type of equipment needed to safely protect the laboratory personnel. The most common types of equipment include fume hoods, BSCs, and storage cabinets. These are discussed in further detail in Chapter 5.

FUME HOODS

Fume hoods are used in a variety of laboratory types to protect personnel performing general chemistry research or testing. They provide protection by maintaining inward airflow to the opening of a hood, which contains and exhausts the airborne materials generated during experimentation, if used properly. Fume hoods should be performance tested annually.

BIOLOGICAL SAFETY CABINETS

Biological safety cabinets (BSCs) are used to control the release of toxic particulates and infectious biological aerosols. They operate similarly to fume hoods in that they provide protection by maintaining an inward flow of air, but they also make use of various levels of filtered supply and exhaust air to protect the materials in the cabinet and to prevent the release of the materials to the atmosphere. BSCs should also be performance tested annually.

FLAMMABLE AND SOLVENT STORAGE CABINETS

Storage cabinets are used to contain fumes from large quantities of flammable materials, prevent excessive internal temperatures in the presence of fire, and contain spilled flammable liquids to prevent the spread of fire. Proper construction, correct venting (if required), conspicuous labeling, and storage of materials at or below the maximum permissible limits will ensure a cabinet's safe use and function.

REFERENCES

CDC. 2009. *Biosafety in microbiological and biomedical laboratories (BMBL)*, 5th Edition. DHHS Publication No. (CDC) 21-1112. Atlanta: Centers for Disease Control and Prevention, U.S. Department of Health and Human Services. www.cdc.gov/biosafety/publications/bmbl5.

IES. 1992. Federal Standard 209E, *Airborne particulate cleanliness classes in cleanrooms and clean zones*. Revision approved by the U.S. General Services Administration. Mount Prospect, Illinois: Institute of Environmental Sciences.

ISO. 1999. ISO 14644-1:1999, *Cleanrooms and associated controlled environments—Part 1: Classification of air cleanliness*. Geneva: International Organization for Standardization.

ISO. 2000. ISO 14644-2:2000, *Cleanrooms and associated controlled environments—Part 2: Specifications for testing and monitoring to prove continued compliance with ISO 14644-1*. Geneva: International Organization for Standardization.

Design Process

3

OVERVIEW

The goal of the design process is to create a practical model of the Owner's Project Requirements (OPR) that translates into a successful building. Essential to achieving the OPR is the creation of the Basis of Design (BOD), which stipulates the technical requirements for the project. Whereas the OPR stipulates the broad requirements that the owner seeks, such as a safe and comfortable work environment, the BOD records the technical assumptions, calculations, code review, equipment selection, and system layout assumptions that were created to satisfy the OPR (Slum 2002).

A typical project design approach schedule is as follows (Hayter and Torcellini 1999):

1. Predesign Steps
   a. Simulate a base-case building model and establish energy use targets.
   b. Complete parametric analysis.
   c. Brainstorm solutions with all design team members.
   d. Perform simulations on base-case variants considering economic criteria.
2. Design Steps
   a. Prepare preliminary architectural drawings.
   b. Design the HVAC and lighting systems.
   c. Finalize plans and specifications.
3. Construction/Occupation Steps
   a. Rerun simulations before making construction design changes.
   b. Commission all equipment and controls. Educate building operators to make sure that they operate the building as intended.

One method of discovering and documenting the OPR is to conduct a charrette with the owner and users. A charrette is a work session with all of the users of the space and the design team. This meeting would occur very early in the conceptual design phase of the project. Some of the information collected during the charrette process is provided in the following list:

- Space usage
- Work flow
- Laboratory equipment with utility requirements
- Required pressure relationships between spaces (positive, neutral, negative)

- Chemicals used or stored
- Redundancy requirements
- User protocols or process requirements
- Space temperature requirements, particularly if unusually tight temperature control is necessary
- Vibration limitations, if applicable (e.g., for optical equipment)
- Air velocity limitations and airflow uniformity requirements if applicable (e.g., for cleanrooms)
- Special control requirements, if any

Throughout the design process, the OPR is continuously refined and the changes are documented. The requirements recorded in the OPR are determined by the owner, but it is the mechanical engineer's responsibility (in conjunction with the architect and other members of the design team) to guide owners through this process and help them define their needs. This ensures that the OPR has all the necessary information to support the development of the BOD.

As the BOD is derived from the OPR, it must also be reviewed and updated, as necessary, throughout the project. Unlike the OPR, which is defined by the owner with the help of the design team, the BOD is entirely the responsibility of the mechanical engineer.

The BOD should be developed from the OPR early in the project, before the start of the predesign phase. This ensures that the BOD will guide design decisions from the very start. Also, the process of developing the BOD often exposes limitations or gaps in the OPR, which allows the designer to correct or clarify the owner's needs before misunderstood requirements are "baked into" the design. See Tables 3-1 and 3-2, which can serve as templates for collecting the information needed in the BOD.

DESIGN REQUIREMENTS

For a laboratory facility, the BOD associated with the mechanical design defines the following requirements, which are discussed in the subsections that follow and are covered more exhaustively in the chapters noted in parentheses following each item:

- Code compliance and industry standards (Chapter 4)
- Risk assessment (Chapter 4)
- Indoor air quality (IAQ) requirements
- System sizing (Chapter 6)
- System selection (Chapters 5, 6, and 7)
- Vibration, acoustical, and seismic requirements
- Air treatment (Chapter 8)
- Stack design criteria (Chapter 9)
- Energy recovery options (Chapter 10)
- Control strategies (Chapter 11)
- Testing, adjusting, and balancing (TAB) and certification requirements (Chapter 12)
- Operation and maintenance (O&M) requirements (Chapter 13)
- Commissioning integration (Chapter 14)
- Economic evaluation (Chapter 15)
- Special requirements for microbiological and biomedical laboratories (Chapter 16)
- Assessing ventilation effectiveness (Chapter 17)
- Sustainability goals and strategies (Chapter 18)

Although it is presented as a linear process, the design of a laboratory system is iterative due to the interrelationships between the different systems and their respective abilities to meet the OPR. Therefore, it is important to evaluate how the design assumptions impact the owner's design intent at each step in the design process.

CODE COMPLIANCE AND INDUSTRY STANDARDS

The foundational requirements are the legal requirements and industry standards that establish the level of care for the design. These are described in Chapter 4 and should be clearly documented in the BOD narrative.

RISK ASSESSMENT

A *hazard* is anything that can cause harm. *Risk* is the probability that the hazard will cause harm. In laboratory projects, hazards are inherent in the processes of the laboratories, but risks can be mitigated through engineering controls. Risk assessment plays a critical role in developing the design basis for managing the hazards in a laboratory. A risk assessment is performed in the conceptual design phase of the project. A good starting part in understanding the risk assessment process is ANSI/AIHA/ASSE Z9.5, *Laboratory Ventilation* (AIHA 2012). The use of the techniques discussed therein is critical in providing a systematic assessment of the hazards and risks of the project while identifying strategies to reduce them.

IAQ REQUIREMENTS

IAQ requirements are established early in the project and are defined in the OPR. These are usually identified during the risk assessment of the project, which outlines the following related to the indoor environment:

- Code requirements
- Contaminant control strategy
- Zones from which recirculated air is acceptable versus zones that must be 100% exhausted to the outdoors
- Ventilation rate
- Local capture devices

During the risk analysis for the laboratory, the hazards in the laboratory are determined along with the required exhaust airflow pressurization (positive or negative). This information is included in the BOD narrative. A sample format for this information is shown in Table 3-1. The recirculated air column in Table 3-1 is included for nonlaboratory spaces and for those laboratories where recirculation of the air is acceptable and allowable by the governing codes and standards (e.g., cleanrooms). For most laboratories handling hazardous chemicals, 100% of the air is exhausted, with none being recirculated.

SYSTEM SIZING

Using the assumptions and guidelines on occupancy rates, loads, and construction materials developed during the conceptual design phase, the sizes of the systems required for maintaining the interior environment are calculated. This process is summarized with the following steps:

1. Confirm interior and exterior design criteria.
2. Establish ventilation requirements.
3. Document room pressure relationships.
4. Define internal loads.
5. Calculate thermal loads.

Table 3-1
Sample Format for Documenting Airflow Requirements

Room Data				Exhaust, cfm (L/s)			Pressure Relationship		Supply, cfm (L/s)		Recirculated, cfm (L/s)
No.	Use	Air Exchange Rate, ach	Room Volume, ft^3 (m^3)	Required	Hood	General	+/–	%	Source*	Required	Required

**Source is either clean (e.g., hallway or anteroom) or dirty (e.g., adjacent laboratory space).*

Completion of these steps will establish the capacity requirements for the laboratory's HVAC system. A critical step in preparing for calculating loads is to gather and document the key system criteria and assumptions made in transforming the owner's requirements into a physical design. This documentation not only provides a checklist for the designer to make sure that key items are not missed but also provides the O&M staff with key information on the limitations of these systems and their intended operation. It is recommended that the assumptions be documented according to a format similar to that shown in Table 3-2, where the assumptions not only include the occupancy rates, equipment loading, and diversities but also include references to codes and standards that were used as the bases for the assumptions. Since most laboratories have multiple types of spaces, Table 3-2 can easily be modified to contain a column for each use type for information that changes from space to space.

In addition to the example shown in Table 3-2, the designer should clearly document the loads in each space, including the following:

- Envelope
- Occupants
- Lighting
- Computers
- Equipment
- Process

Calculating Loads for Comfort

Estimating the envelope, occupant, lighting, computers, and equipment load of a laboratory is not so different from estimating the load for a typical building. If the load is in the space or is transferred across the envelope, then the load is included in the calculation. However, if the load does not directly impact the space (i.e., it is contained within an exhaust hood), then it is not included as a load on the central system. Figure 3-1 details the various loads within a laboratory space. The gray items in Figure 3-1 are considered internal loads that must be handled by a central or local cooling/heating system. The other items' loads are removed from the space and thus do not impact the conditioning system for the space. The one unique load in laboratory spaces is the process load created by laboratory equipment and processes.

Figure 3-1
Laboratory Equipment Loads

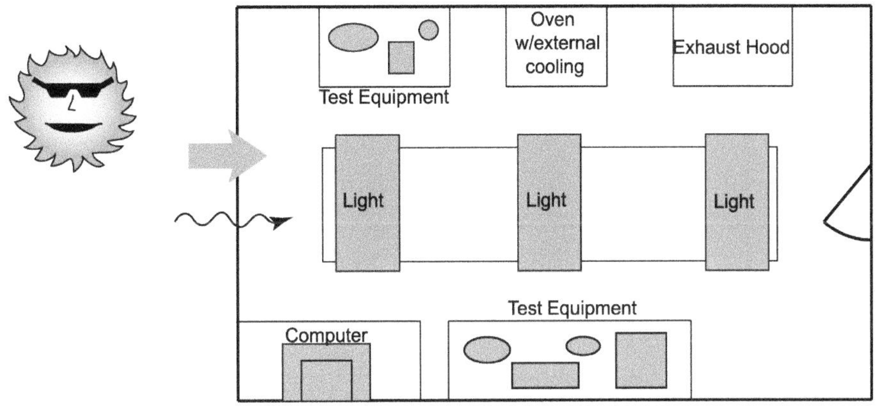

Within a laboratory, a mix of equipment is available for use by the occupants depending upon the specific experiment being conducted. Because it is rare, if not impossible, to use all of the equipment at one time, care must be taken in estimating the loads and diversity of specified equipment for the design of the systems.

One of the first tasks in estimation of equipment loads is a thorough understanding of the energy consumption of each type of equipment. Typically, this requires contacting manufacturers for the rated energy consumption for each piece of equipment or, if necessary, researching and testing the actual equipment. With these energy consumption data, standard load calculations as described in Chapter 16 of *ASHRAE Handbook—HVAC Applications* (ASHRAE 2015) can be used. Proper estimation of loads is crucial in creating acceptable conditions for personnel and experiments. In calculating the specific load for a piece of equipment, it is critical to designate where the load is being rejected to. The rejection media can be the surrounding air, domestic water, process water, or the outdoor air. Depending upon the source of rejection, the cooling load may or may not affect the central system's size requirements.

The hood use diversity factor should be estimated for the laboratory. However, caution in the use of this factor should be stressed, particularly in its use for hoods. Although the use of this factor can greatly reduce unnecessary energy consumption and first costs, careful design of the interaction of the type of controls this system uses and the main system controls must be accomplished. Also, diversity factors that lower system exhaust volumes below 75% of the sum of peak zone exhaust (without diversification) should be extensively checked and verified.

Equipment used in laboratories, including but not limited to autoclaves, cryocoolers, centrifuges, growth chambers, standard freezers, –80°C freezers, and refrigerators are available in a wide range of efficiencies. In planning for the loads in equipment rooms there are a number of energy efficiency opportunities that can save both energy and installed costs. Committing to best of class or top 30 percentile equipment can reduce the costs of electrical and cooling systems.

In planning for equipment loads many engineers use the full-load data for sizing of the electrical and HVAC equipment. Plug load studies in laboratory equipment rooms have shown that there is a lot of diversity in the equipment loads in practice. Typical equipment rooms run between 20% and 30% of nameplate ratings. Obviously the number and types of equipment should be considered in selecting the diversity.

Equipment with high heat output, such as autoclaves, should be designed with hoods over them to directly capture the steam and heat when the doors are open. An energy-efficient design for autoclaves is to design the system for a low air change for normal

Project Information		
Designer/Engineer:		
Company:		
Project:		
Date:		
Unit system: I-P SI		
Design Criteria	**Value**	**Reference/Comment**
Latitude/longitude		
Elevation		
Clearness value		
Sum.mer outdoor air design dry bulb/wet bulb		
Winter outdoor air design dry bulb		
Ground reflectance		
Cooling load methodology		
Cooling setpoint, dry bulb/relative humidity		
Heating setpoint, dry bulb/relative humidity		
Cooling/heating setback temperatures		
Roof construction type		
Wall construction type		
Ceiling U-factor		
Overall roof U-factor		
Glass U-factor		
Glass solar heat gain coefficient (SHGC)		
Overall wall U-factor		
Overall building U-factor		
People sensible/latent heat generation		
Lighting density		
Outdoor air ventilation rate		
Infiltration rate		
Cooling air change rate		
Reheat minimum value		
Supply air temperature		
Duct heat gain		
Duct leakage		
Maximum duct noise level/ceiling effect		
Air distribution system diversity		
Number of occupants at peak load		
Fan heat gain		
Equipment loads		
Expected utility costs		
Hazard types and levels		

Table 3-2
Basis of Design Documentation

conditions with a push button or spring-wound timer that is used to increase the ventilation when the doors are open. A heat or moisture sensor in the hood could also be used to automatically step up the ventilation.

Calculating Loads for Safety

The comfort loads correspond to the minimum supply air required to maintain the space at acceptable conditions. These minimum airflows must conform with the requirements for the minimum air exchange rate if it is stipulated by the BOD. Therefore, the exhaust airflow requirements for each space are compared to the supply airflow required for comfort to determine the increase in supply airflow required to maintain the required ventilation rates and/or pressure relationship. Table 3-1 is ideal for this purpose. Zones that contain chemicals and/or fume hoods frequently have process airflow requirements that exceed comfort conditioning requirements.

With the supply, return, and exhaust airflows determined for each space, the next step is the development of a pressure map for the entire facility. A pressure map graphically depicts the airflows within and between each room. The intent is to make sure that the pressure relationships documented during the planning phase are achieved.

For each room/space, the supply (primary and auxiliary), return (if any), and exhaust (general and hood) airflows are listed. If active pressure control is to be used, the required pressure differentials should also be listed. Once all room airflows are detailed, the directional airflow between the rooms is shown with arrows and values. Figure 3-2 provides an example of this process.

Though Figure 3-2 indicates air transfer between rooms occurring at discrete points, in reality the air transfer will occur at the lowest pressure loss. This can be at an open

Figure 3-2 Pressure-Mapping Process

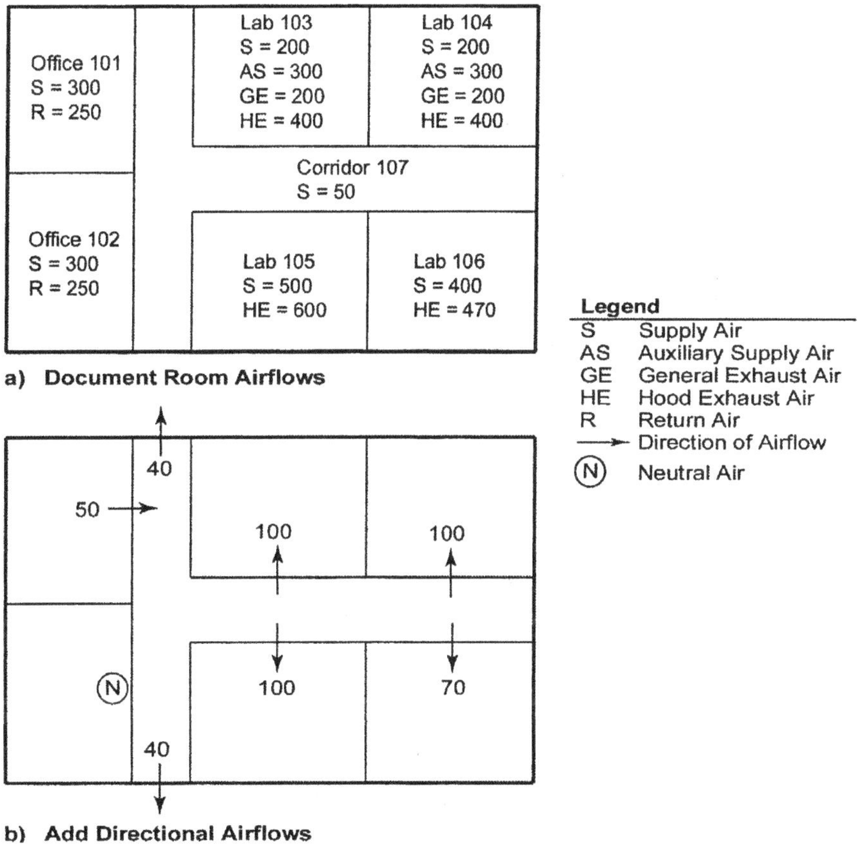

door, exterior wall/window, or interior partition. Therefore, close attention to detail on wall, window, and door construction must be taken to minimize unwanted air movement, particularly in projects that use active pressure control. It may be necessary to coordinate with the contractor and the envelope commissioning authority to make certain that the final construction can support the required pressure differentials. As the severity of hazards within a space increases, greater attention to detail is required.

SYSTEM SELECTION

With the design load and airflow values calculated for each space, the next step in the design process is to evaluate the different system options to meet these loads. Chapters 5, 6, and 7 cover the selection, design, and layout of systems in detail. Chapter 18 covers sustainability, which should be incorporated into the system design where possible. In general, the decisions to be made during this phase of the design process include the following:

- Type of exhaust system
- General exhaust/hood requirements
- Type of system
    - Constant volume
    - Constant volume (low face velocity)
    - Constant volume (ductless hoods)
    - Variable air volume
    - Auxiliary air
- Special process loads
- Space heating needs
- Division of systems/areas
- Central versus local air systems

Chapter 5 details the options available for exhaust hoods and provides guidance on their selection and application. Due to the critical nature of the exhaust systems, these typically take precedence when determining the type and location of central systems. This is followed by the design of the supply air systems. The interrelationships between the different systems and rooms makes this a critical step in the design process. The Zone Air Distribution section of Chapter 6 explains these interactions. Finally, Chapter 7 provides details on process system layout and application. Division of systems/areas is similar to a normal HVAC design process, where the division is done based on functionality of the areas, location of areas for mechanical equipment, room for ductwork, etc.

Central Supply Systems

By sizing the central air system to handle the worst-case scenario for cooling, it is possible to eliminate the need for any local systems. Variable-speed fans in conjunction with variable-air-volume (VAV) boxes are required to handle the large variability in space load. The control of this system is complicated if room pressurization is to be maintained.

Local Supply Systems (Fan-Coils)

Local air systems have their own fan, heating coil, and cooling coil. There is individual control of each unit, and the control systems are less complicated than those with one central air system and can in many cases be designed to have a higher performance than comparable central air systems. These systems typically take less building space for piping and ducting than a central air system. The disadvantages of these systems are that they require more extensive maintenance due to the many components; can cause more noise because of the relatively closer location of the fans, compressors, pumps, and motors to the conditioned space; and require many penetrations of the building envelope.

Hybrid Supply Systems

This system uses a central fan to distribute air to the zones but performs all conditioning (except preheat, if required by the climate) at the terminal units via hydronic coils. These may be either a four-pipe system that uses two coils or a changeover system in which a single coil is used for heating or cooling depending on mode. This system type is more expensive to construct due to the additional piping, but it can be substantially more energy efficient than a conventional central system by eliminating the reheat penalty. At the high airflows often required for laboratories, this reheat savings may justify the additional cost, particularly if sustainability goals impose energy use intensity (EUI) or other performance targets.

After System Selection

Upon selection of the equipment, the layout of the equipment and the distribution of services to their point of use can begin. This step in the design process, while often taken for granted, is one of the most critical. If done improperly, contaminants may not be properly contained, increasing the risk of exposure to the hazards for laboratory users; thermal loads will be unmanageable; and the system can be very costly to construct and operate. Chapter 6 addresses the many challenges in designing air-moving systems in laboratories.

VIBRATION, ACOUSTICAL, AND SEISMIC REQUIREMENTS

The evaluation of vibration, acoustical, and seismic requirements for the project should not be overlooked. Detailed design information is not provided for these issues in this Design Guide, but the reader is encouraged to consult Chapter 8 of *ASHRAE Handbook—Fundamentals* (ASHRAE 2013) for guidance on sound and vibration design and Chapters 48 and 55 of *ASHRAE Handbook—HVAC Applications* (ASHRAE 2015) for guidance on seismic design.

AIR TREATMENT

An important aspect of the air systems is their impact on the indoor and outdoor air quality. Therefore, the treatment of the supply and exhaust airstreams must be determined so that the indoor air quality is adequate for the comfort and safety of the occupants and the experiments and that the exhaust air is not a danger or nuisance to nearby people and buildings. Chapter 8 provides details on the technologies available and requirements for air treatment. In general, the following should be accomplished:

- Document acceptable ambient pollutant levels.
- Identify treatment options.
- Select and size treatment system.

It is important to recognize that, depending on the quality of the ambient outdoor air, the supply air in some instances may require extensive treatment prior to use.

STACK DESIGN CRITERIA

There is a strong correlation between the level of pollutants in the exhaust air and the design of the exhaust stack. As the concentration and hazards of the contaminants in the exhaust increase, the dilution accomplished by the exhaust discharge design must increase. Therefore, exhaust stacks are designed to reduce the concentration of pollutants at receptor points below recommended or allowable levels. This can be a challenging process because of aesthetic considerations, climatic conditions (wind), and location of stacks away from air intakes. The design of exhaust stacks is detailed in Chapter 9.

ENERGY RECOVERY OPTIONS

Laboratory facilities are high consumers of energy because of the large volume of outdoor air that must be conditioned only to be exhausted. To minimize the energy required to condition the supply air, energy can be recovered from the exhaust air and transferred to the supply air. The savings from using an energy recovery system can be as high as 80% for some conditions. However, care must be taken to avoid cross-contamination of the incoming airstream by the exhaust. Chapter 10 details the options and their application for laboratory energy recovery systems.

CONTROL STRATEGIES

The HVAC systems are sized and selected for peak design conditions. However, these conditions occur rarely. During nonpeak periods, the system must be actively controlled to efficiently maintain space comfort and safety conditions. Therefore, control strategies must be developed that maintain the following:

- Space temperature control
- Space pressurization
- Building pressurization
- Occupied/unoccupied periods
- Emergency situations

Chapter 11 provides detailed guidance on control strategies for laboratory systems.

TAB AND CERTIFICATION REQUIREMENTS

To make sure that systems can be properly started and operated, it is important to identify and ensure during the design phase that sufficient locations throughout the system can be accessed for the purpose of balancing the systems. This includes measurement ports on both sides of the valves and coils, straight runs of ductwork and piping, accessibility to dampers, and proper documentation on what is expected of the TAB firm.

Therefore, the design developed by the architect and engineer should, as a minimum, include the following:

- Typically six duct diameters of straight duct for measuring supply airflow and test instrument ports. Verify the requirements for each device.
- Typically six duct diameters of straight duct for measuring exhaust airflow for exhaust fans and ventilated workstations. Verify the requirements for each device.
- Measurement ports on both sides of valves, air-handling unit coils, energy recovery coils, and terminal reheat coils.
- Fan inlet and outlet ports for static pressure measurements.

As a minimum, the TAB and certification should

- balance all systems to design conditions,
- verify proper airflow direction/pressurization,
- verify system operation under part-load conditions (75%, 50%, 25%, and minimum),
- perform face velocity tests,

- calibrate airflow and direction monitors for all fume hoods, and
- certify all clean-air and biological safety cabinet (BSC) workstations.

Details on determining the TAB requirements for laboratories are contained in Chapter 12.

O&M REQUIREMENTS

During the development of the construction documents (drawings and specification) it is critical to identify those items necessary for the proper lifetime operation of the laboratory systems. These include the following:

- Training requirements, including a video
- System documentation, including computer-aided design (CAD) drawings and all control software
- Involvement of O&M personnel throughout the construction

Chapter 13 provides guidelines for the development of these O&M requirements, with specific emphasis on the O&M manual.

COMMISSIONING INTEGRATION

Commissioning is a quality process an owner elects to use for planning, designing, constructing, and operating the facility. The key characteristics of the commissioning process, as with any quality process, are as follows:

- Work is accomplished correctly the first time.
- The individual worker determines the level of quality.
- What constitutes a "successful" project is clearly defined.

Several commissioning tools/tasks can be integrated into an owner's current process. Specifically, the following should be completed:

- Develop owner's design intent.
- Develop commissioning plan.
- Accomplish design reviews.
- Develop Basis of Design.
- Integrate quality requirements into specifications.
- Verify installation and calibration.
- Accomplish functional performance test on installation.
- Make sure warranties are maintained.
- Have a "lessons learned" meeting.

Chapter 14 provides a detailed overview of the commissioning process relative to laboratory facilities.

ECONOMIC EVALUATION

The final step in the design process is an economic evaluation of the system. This evaluation should be based on a life-cycle cost for a typical 20- or 30-year period. Details on accomplishing a life-cycle cost analysis (LCCA) for a laboratory are in Chapter 15.

SPECIAL REQUIREMENTS FOR MICROBIOLOGICAL AND BIOMEDICAL LABORATORIES

Many laboratories work with infectious biological materials. The primary factor in designing a biocontainment laboratory is protection of laboratory workers, the environment, and the public from accidental exposure to potentially biohazardous agents. Understanding the user's protocols and requirements is essential. Information regarding biocontainment and user requirements for these laboratories is found in Chapter 16.

ASSESSING VENTILATION EFFECTIVENESS

The laboratory ventilation system is the primary means for the removal of airborne contaminants generated within the laboratory. The system's ability to remove the contaminants from the laboratory space is defined as the *ventilation effectiveness*. As designers consider reducing ventilation rates to make the laboratory more energy efficient, modeling the ventilation effectiveness becomes a critical design issue. Chapter 17 describes methods to model air flow patterns within the laboratory.

SUSTAINABILITY GOALS AND STRATEGIES

Very early in the design, the team identifies the sustainability goals. Among these goals may be certifying the building under a sustainable rating system. This Guide offers the reader many strategies that can be used to earn points for a rating system and also offers guidance that moves the designer toward a sustainable design through a proper and well-documented design process. Strategies that can be used to satisfy sustainability goals are summarized in Chapter 18.

REFERENCES

AIHA. 2012. ANSI/AIHA/ASSE Z9.5-2012, *Laboratory ventilation*. Falls Church, VA: American Industrial Hygiene Association.

ASHRAE. 2013. *ASHRAE handbook—Fundamentals*. Atlanta: ASHRAE.

ASHRAE. 2015. *ASHRAE handbook—HVAC applications*. Atlanta: ASHRAE.

Hayter, S.J., and P. Torcellini. 1999. Low-energy building design—The process and a case study: Keys to success. *ASHRAE Transactions* 105(2):802–10.

Slum, K. 2002. Design intent and Basis of Design: Clarification of terms, structure and use. *ASHRAE Transactions* 108(2):1184–85.

Laboratory Planning

4

OVERVIEW

Laboratory spaces and buildings are unique compared to other types of spaces and buildings because of the distinctive requirements to maintain the health, comfort, safety, and productivity of the various entities within. Therefore, the successful design, construction, and operation and maintenance (O&M) of a laboratory facility are achieved when sufficient planning is accomplished before undertaking the project. This chapter provides the critical information and guidance needed for the following planning considerations for a laboratory building project:

- Risk assessment
- Environmental requirements
- Appliances and occupancy
- Pressure relationships
- Ventilation and indoor air quality (IAQ)
- Laboratory codes, standards, and references
- Integration of architecture and engineered systems
- Development of planning documents

The goal of this chapter is to present and discuss the pertinent issues that must be considered and documented when planning a new or renovated laboratory facility. The fundamental concepts and principles presented in this chapter are the foundation for the remaining Guide. A good understanding of this information is critical for the success of a laboratory design and a delivered laboratory that is functional.

RISK ASSESSMENT

As mentioned in Chapter 3, a risk assessment is performed early in the design process. The purpose of a risk assessment is to address the safety issues related to the acquisition, handling, usage, storage, transportation, and waste disposal of hazardous materials. The completion of a risk assessment involves the identification and understanding of the various hazards present in each laboratory and the application of appropriate hazard analysis methods to assess the risks. In addition, the responsibilities of key individuals must be clearly defined, and adequate documentation must be available to handle all foreseeable situations.

ENVIRONMENTAL REQUIREMENTS

Laboratories have special environmental requirements, such as personnel comfort and experimental quality, that must be identified in the planning phase to ensure the laboratory can be designed and constructed to meet them. Also, the potential for changes in laboratory use should be identified so that the heating, cooling, ventilating, and exhaust systems can be designed to meet both current and future environmental requirements without substantial modifications.

Unfortunately, laboratory environmental conditions are often inhomogeneous because of the variation in the needs of the different types of occupants and functions. For example, persons that work in an animal laboratory will have very different thermal and ventilation comfort needs from the rodents being bred for a particular experiment in the same room.

To maintain the integrity of experiments, the temperature, humidity, air quality, and vibration of the environment must be considered priorities. There are two environments that should be considered: the macroenvironment, which is the actual interior space of the laboratory that surrounds the researcher, and the microenvironment, which is a localized experimental space that has its own unique requirements.

TEMPERATURE

For all laboratories, considerations must be made to avoid major swings in the room and laboratory hood temperatures. Highly sensitive experiments also typically need to avoid relatively minor temperature fluctuations, which can disrupt the controlled environment of the experiment. Exposure to direct heat or sunlight can also affect experiments conducted in the macroenvironment. For example, experiments using temperature-sensitive equipment should not be exposed to direct solar radiation. The automatic control system, laboratory layout (including the location of heat-producing equipment, windows, etc.), and air distribution methods should be properly planned so that the temperature in a laboratory is maintained to appropriate tolerances.

For experiments conducted in microenvironments, which are often subjected to extremely low temperatures, extremely high temperatures, or precisely controlled temperatures, considerations need to be made so that the required temperatures can be maintained. For example, appropriate temperature control and interlocks should be available to avoid heating or cooling experiments beyond desired limits.

HUMIDITY

As with temperature, conditions that may affect the desired humidity must be considered and counteracted to maintain experimental quality. Equipment that produces large amounts of moisture (usually due to heating/boiling of liquids) should be properly vented. Reliable automatic humidity control in addition to vapor barriers in partition walls, floors, and the ceiling that minimize loss of water vapor will help ensure a stable environment.

During planning, the necessary limits on humidity should be determined and the HVAC system planned accordingly. This may require the addition of dehumidification equipment for summer operation and the addition of a humidification system for winter operation. Cleanrooms for electronics research or manufacturing and laboratories with sensitive electronic instruments may have special minimum humidity requirements to prevent static electricity from damaging experiments and equipment.

AIR QUALITY

To maintain good air quality in laboratory spaces, planning should focus on potential sources of pollutants and contaminants and develop design intent items that minimize them. These include the following:

- Effective exhaust systems to expel and contain malicious fumes
- Proper pressure relationships between adjacent spaces
- Low-velocity diffusers that do not interfere with fume hood and biological safety cabinet (BSC) containment
- Properly sealed ductwork and envelope
- Specific pollutant sensors to maintain regulated levels

Air quality concerns related to experimental quality are often present in animal laboratories. Proper ventilation and air quality are needed to provide assurance that the results of experiments are accurate and not due to poor IAQ. Air quality can also be a concern in adjacent laboratories of different functional uses, as the contaminants from one experiment may affect the work in the adjacent laboratory.

VIBRATION

Vibration can originate from within or outside a laboratory. Interior vibration from mechanical and electrical equipment (e.g., fans, blowers, pumps, chillers, and transformers), by laboratory equipment (e.g., centrifuges and fume hoods), and pedestrians can cause interference with sensitive equipment in the laboratory. Correct balancing of mechanical equipment, using vibration isolators for both mechanical equipment and sensitive instruments, and making the building as stiff as possible can minimize these vibrations. Exterior vibrations can be caused by traffic, construction work in nearby buildings, wind, and seismic disturbances. Because this type of vibration is hard to control, selection of the location for the laboratory during the planning phase is important for minimizing possible external sources of vibration. Special vibration isolation measures may also need to be considered for exterior vibrations.

APPLIANCES AND OCCUPANCY

While the overwhelming cooling and heating load for any laboratory building is typically the conditioning of the supply air that is exhausted through the hoods, a significant amount of heat and moisture originates from appliances, lighting, and the occupants that occupy the internal laboratory environment.

APPLIANCE LOADS

The equipment used in laboratories creates sensible and latent heat gains that must be removed by the HVAC system to maintain comfortable conditions. These heat gains are released into the room and influence the thermal environment in the room. While exhaust hoods located above primary heat-generating equipment can remove most of the heat generated by that piece of equipment without influencing the temperature of that room, exhaust hoods are not always feasible. In such cases, snorkel-type exhausts, which provide localized exhaust, may be an option.

LIGHTING

In addition to the cooling load resulting from various laboratory equipment, lighting generates heat that adds to the cooling load and must be addressed by the HVAC system. The lighting levels required vary depending on the tasks performed in the area, with the lighting power depending on the types of ballasts and fixtures used and other strategies such as daylighting. Task lights are often used for specific requirements such as experimental operations that have small, detailed work. Ambient lighting ensures that the overall level of lighting satisfies the need of the most common tasks performed in that area.

OCCUPANTS

The occupants (people and animals) contribute to the cooling load of the space. Occupant densities, diversities, activity levels, and scheduling are the key criteria in determining the occupant load requirements.

PRESSURE RELATIONSHIPS

Maintaining proper pressure relationships between adjacent spaces in a laboratory building is critical to ensure airflow is in the proper direction, from clean areas to dirty areas. Therefore, pressure relationships must be determined during the planning phase. Planning includes the identification of individual pressure zones and the development of a preliminary pressure map. A primary problem with the design, construction, and operation of laboratory facilities is that air moves from space to space based on simple pressure relationships—from higher-pressure areas to lower-pressure areas. Since there is no such thing as "smart air" that follows the arrows on drawings, special attention must be paid to the layout and design of the HVAC systems and to wall construction.

There are three primary problem areas that are often overlooked when developing pressure maps. First, the outdoor wind direction places a positive pressure on the upstream wall and a negative pressure on the roof and downstream wall. Second, due to stack effect, as one moves upward in a building, the relative pressure from indoors to outdoors goes from negative to positive, resulting in air entering the lower floors and exiting the top floors. Finally, all buildings are composed of compartments (rooms) and chases (elevators and stairs), which allow air to quickly move from one location to another location in a building. Therefore, proper sealing of spaces and accounting for external forces are required in the proper layout of a system in laboratory facilities. To simplify the planning process, the need for negative, positive, and neutral pressure rooms; anterooms (air locks); and transfer air should be discussed. Figure 4-1 details each of these graphically.

NEGATIVE PRESSURE ROOM

A negative pressure room is at a lower pressure compared to adjacent spaces and, therefore, the net flow of air is into the room. For this to happen, the volume flow rate of the supply air must be less than that of the combined exhaust flow rates from all general and equipment exhausts. Laboratories that use hazardous chemicals and materials need to

Figure 4-1
Pressure Relationship Terminology

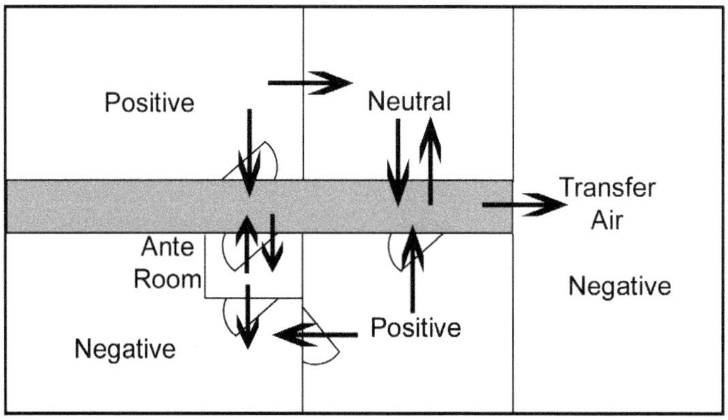

be maintained at a negative pressure so that fumes, particulates, odors, and other by-products of experiments do not migrate from the experiment to adjacent occupied spaces.

POSITIVE PRESSURE ROOM

A positive pressure room is at a higher pressure compared to adjacent spaces and, therefore, the net airflow is out of the room. The volume flow rate of the supply air must be greater than that of the combined exhaust flow rates from all general and equipment exhausts for a room to maintain a positive pressure. Cleanrooms and sterile facilities are examples of laboratories that need positive pressurization to prevent infiltration that could compromise experimental requirements for cleanliness. If hazardous materials must be used in a positively pressurized laboratory, such as an electronics manufacturing laboratory, exhaust equipment must be designed and used carefully so that it removes all hazardous fumes. More commonly, an air lock or other buffer zone can provide a negative containment area around a positive pressure room that contains hazardous material.

NEUTRAL PRESSURE ROOM

A neutral pressure room has no specific pressurization requirements. Typically, non-laboratory spaces such as office areas and corridors are maintained at a neutral pressure. Some rooms that have no pressurization requirements of their own must be maintained at either positive or negative pressure so that the pressure requirements for adjacent rooms are satisfied and airflow from clean to dirty is maintained. These rooms are identified when creating a pressure map.

ANTEROOM

An anteroom is a transition room between areas of substantially different pressures or that is used to gain access to a room that must maintain its pressure even during disturbances such as a door opening. The use of anterooms provides assurance that pressure relationships are constantly maintained and air remains flowing from clean to dirty, and they reduce the need for the HVAC control system to respond to large disturbances.

TRANSFER AIR

Transfer air moves from one space to an adjacent space through a transfer grille or other air distribution outlet device. Transferring air should only be done between spaces that have a similar level of cleanliness or from clean areas to those of lesser cleanliness. Typically, transfer air is used between subdivided rooms in a common area to prevent large pressure differences as the supply and exhaust airflows on either side change.

VENTILATION AND IAQ

During planning, careful consideration of ventilation and IAQ is needed because of the hazardous nature of the materials used in laboratories and the need to provide a safe working environment for laboratory personnel. Various codes and standards identify the ventilation requirements for laboratories. The differences in ventilation requirements from occupied to unoccupied modes should also be considered during planning, along with determining how to satisfy both the cooling load and the air change requirements for a laboratory.

SUPPLY AIR/EXHAUST AIR TREATMENT REQUIREMENTS

Because most laboratory environments are inherently hazardous, pollutants and their sources cannot be completely eradicated. Therefore, air treatment may be needed to provide the desired air quality by either preventing contaminants from leaving the laboratory and entering the environment (e.g., biological laboratories) or preventing pollutants in the environment from entering the laboratory (e.g., cleanrooms). During planning, the work to be performed in the laboratory should be evaluated for supply or exhaust air treatment needs, including the use of filtration, scrubbing, condensing, and oxidation technologies. Details on the various air treatment applications and technologies available can be found in Chapter 8.

LABORATORY CODES, STANDARDS, AND REFERENCES

In additional to the usual federal, state, and local requirements, the codes, standards, and references listed in Table 4-1 provide a starting point for laboratories. The owner's design standards, funding sources, research program, or certification requirements may result in additional requirements.

For more information on these organizations and additional laboratory-related organizations and resources, refer to Chapter 1.

INTEGRATION OF ARCHITECTURE AND ENGINEERED SYSTEMS

For the organization of the internal laboratory spaces, the architect and engineer must generate and coordinate various concepts to efficiently integrate architecture and engineered systems. This coordination effort has the objective of developing systems that complement each other while providing a safe and healthy environment at a reasonable cost. The main items that require integration between architects and engineers are building concepts, utility distribution, laboratory layout approaches, specific layout issues such as fume hood location, and budgeting. The locations of positively and negatively pressurized areas and levels of cleanliness should be planned to minimize controls and ducting.

BUILDING CONCEPT

Building concept issues include basic choices about the laboratory building layout that will affect later design and system opportunities. These issues include choosing the air intake locations to avoid contamination, selecting dedicated or manifolded supply and exhaust air systems, and considering O&M requirements.

Air Intake Location

The location of outdoor air intakes needs to be coordinated between the architects and engineers to minimize the possibility of reentrainment of contaminants from the laboratory exhaust discharge, loading docks, cooling tower discharge, vehicular traffic, and adjacent property pollution into the outdoor air intakes. Selection of air intake locations should consider issues such as the prevailing wind direction and seasonal events such as trees losing their leaves or snow, which may clog the air intake. The choice of air intake locations will affect later planning decisions, such as the location of mechanical rooms, the location and layout of utility corridors, and the exterior aesthetics of the laboratory building.

Adequate exit velocities, height, placement, and types of exhaust stacks and fans must also be considered along with the selection of outdoor air intake locations. State and local codes may specify a minimum separation distance between exhaust discharge and supply air intakes. For additional details on air intake locations relative to exhaust systems, see Chapter 9.

Table 4-1 Summary of Codes, Standards, and References Pertaining to Laboratories

Name	Reference	Summary of Purpose
ANSI/AIHA/ASSE Z9.5, *Laboratory Ventilation*	(AIHA 2012)	Establishes minimum requirements and procedures for the design and operation of laboratory ventilation systems used to protect personnel from overexposure to harmful or potentially harmful contaminants.
ANSI/ASHRAE Standard 62.1, *Ventilation for Acceptable Indoor Air Quality*	(ASHRAE 2013)	Forms the basis for the minimum outdoor air requirements for most spaces and stipulates when treatment of outdoor air and exhaust air is necessary.
ASHRAE Handbook—HVAC Applications, Chapter 16, Laboratories	(ASHRAE 2015)	Provides a condensed version of the information provided in the chapters of this Guide.
Biosafety in Microbiological and Biomedical Laboratories (BMBL)	(CDC 2009)	Provides biosafety level (BSL) 1 to 4 procedures and guidelines for the manipulations of etiologic agents in laboratory settings and animal facilities. The four levels of control that are defined range from safely dealing with microorganisms that pose no risk of disease for normal healthy individuals to dealing with the high risk of life-threatening diseases.
Guide for the Care and Use of Laboratory Animals	(ILAR 2011)	Assists institutions in caring for and using animals in ways judged to be scientifically, technically, and humanely appropriate.
NFPA 101: Life Safety Code®	(NFPA 2015a)	Addresses all the construction, protection, and occupancy features needed to minimize danger to life from fire, smoke, and panic. Forms the basis for law in many national jurisdictions.
NFPA 30: Flammable and Combustible Liquids Code	(NFPA 2015b)	Provides the most up-to-date requirements for dealing with flammable and combustible liquids and is therefore useful to design engineers, enforcing officials, insurers, and laboratory workers.
NFPA 45: Standard on Fire Protection for Laboratories Using Chemicals	(NFPA 2015c)	Provides the minimum fire protection requirements for fire-safe design and operation in educational and industrial laboratories using chemicals.
NFPA 801: Facilities Handling Radioactive Materials	(NFPA 2014)	Identifies guidelines for decreasing the risk of explosion or fire and the severity of contamination from a fire or explosion at facilities that handle materials that are radioactive (except nuclear reactors).
NSF/ANSI 49, *Biosafety Cabinetry: Design, Construction, Performance, and Field Certification*	(NSF 2014)	Provides comprehensive information and guidance on the principles and applications of air filtration, which supplies the level of particulate cleanliness required by HVAC systems.
Prudent Practices in the Laboratory: Handling and Management of Chemical Hazards	(NRC 2011)	Recommends several prudent practices that stimulate a culture of safety for chemical laboratory operations. Provides information and cross-references on how to handle compounds that pose special hazardous risks.

Dedicated and Manifolded Air Systems

The supply air to the laboratories, adjoining office spaces, storage areas, and corridors can be provided by a central system that serves all these areas or via separate (distributed) systems dedicated exclusively for either laboratory or nonlaboratory use.

Similarly, exhaust air from laboratory and other spaces can be expelled through either dedicated or manifolded exhaust systems, although separate exhaust systems for laboratory and nonlaboratory systems are almost always used. The choice of either dedicated or manifolded air systems will determine later planning and design choices, such as the space needed for utility corridors and shafts and their locations.

There are advantages and disadvantages to using either dedicated or manifolded systems, all of which must be weighed carefully before a selection is made. Dedicated exhaust systems should be considered where different exhaust streams may be incompatible. Perchloric acid exhaust, for example, must have a separate exhaust system. Also, dedicated exhaust systems may have simpler control and balancing requirements than manifolded exhaust systems. Manifolded systems, however, have lower initial cost and maintenance requirements because of the reduced number of fans and stacks that are needed, and they have better atmospheric dispersion due to the momentum of the large air mass. They also are better for energy recovery applications and easier for installing backup fans.

O&M Issues

O&M issues, such as required clearances and access locations for HVAC and laboratory equipment, must be thoroughly planned. Laboratories may need to sustain design performance conditions for extended periods of time to protect the laboratory personnel and experiments and therefore need systems and equipment that can be sufficiently maintained. The continued reliability of the equipment and systems in a laboratory can be increased if they are selected and installed with consideration of accessibility and maintainability. Coordination between architects and engineers is needed to ensure that sufficient space is allowed for various HVAC and laboratory equipment and other building systems.

UTILITY DISTRIBUTION

Various infrastructure distribution components, such as supply and exhaust ductwork, piping (HVAC, plumbing, gas), mechanical equipment technology, fire protection, and electrical conduits, must be distributed throughout the building to various laboratories, offices, and other spaces. During the planning phase, architects and engineers must coordinate the distribution of these services to fit within the available space yet remain maintainable. Methods of utility distribution for architects and engineers include utility corridors, multiple interior shafts, multiple exterior shafts, corridor ceiling distribution, and interstitial space.

Utility Corridors

A utility corridor is a separate hallway that provides space for piping, ductwork, conduit, and other components to enable access to the systems between equipment rooms. Utility corridors are a horizontal layout method, with very few vertical shafts used for multistory laboratory buildings. The use of utility corridors has the advantage of allowing O&M staff to maintain laboratory equipment from an adjacent service corridor rather than routinely having to enter the laboratory. This also allows the utilities to be placed away from public access hallways so that they can be maintained without disrupting building occupants and when building occupant access to the critical systems that serve laboratories is limited.

Multiple Interior Shafts

Multiple interior shafts are a second method of utility distribution that uses a primarily vertical layout with horizontal runs to individual equipment. Typically, multiple interior shafts only require a short horizontal run of ductwork, which is desirable for exhaust from fume hoods and other laboratory equipment. Depending on the building layout, interior shafts can be accessed either through a service corridor or a public corridor or from within the laboratory. A possible downside of this layout is that multiple shafts can occupy a significant portion of the floor area in high-rise laboratories.

Multiple Exterior Shafts

Multiple exterior shafts are similar to interior shafts except that they are located on the exterior wall of the building rather than in the interior of a building. As with interior shafts, exterior shafts typically only require short horizontal runs of ductwork. Exterior utility shafts are typically accessed from the laboratory or from the building exterior.

Corridor Ceiling Distribution

Corridor ceiling distribution uses the space above a dropped ceiling (sometimes no dropped ceiling) to contain ductwork and piping in a corridor just outside of a laboratory. This method of utility distribution requires a higher floor-to-floor height, as 3 to 4 ft (0.91 to 1.22 m) above the dropped ceiling is required for the various utilities. While this type of system can be more difficult to maintain because of the need to move ceiling panels around to find equipment, it does provide increased usable floor area compared to multiple interior shafts or a separate corridor for utilities. Typically, ceiling distribution is confined to corridors and is not used above laboratory spaces, as the need to remove ceiling panels to maintain equipment could interfere with laboratory activities (e.g., cause dust to fall on sensitive laboratory equipment or experiments).

Interstitial Space

Interstitial spaces are dedicated mechanical and electrical equipment floors located between each occupied floor. Ductwork, piping, and conduit are routed through this space to connect to the floors above and/or below. The benefits of placing systems in interstitial spaces are ease of accessibility and the ability to modify the layout of the various systems quickly without disturbing laboratory operations. However, this type of utility distribution is inherently expensive due to the need for additional floors in the laboratory building.

Some laboratories and medical research facilities have been laid out with vertical utility spaces that are 4 to 8 ft (1.22 to 2.44 m) wide. This is not a conventional layout and requires careful planning and coordination by the owner and the design team. It does provide sidewall utility access (versus ceiling and floor access), which is especially useful in laboratories that require frequent utility changes to meet use changes.

LABORATORY LAYOUT APPROACHES

Architects and engineers must coordinate to develop a plan for the internal organization of a laboratory and adjacent building spaces. The approaches to laboratory layouts include modular designs and workstation layouts based on specific functions.

Modular Design

A laboratory module is a basic work area, usually for one or two persons, that is used as a general building block in planning the complete floor plan. Each module can be contained in a separate room or combined into groups and used in larger rooms. The layout of modules should be carefully coordinated with the utility distribution layout. For laboratory buildings that house different types of work, multiple modules, one for each distinct type of laboratory work, may be chosen and used as appropriate.

Workstation Layouts Based on Specific Functions

The processes, materials, and waste products that are to be contained at a particular workstation need to be considered with regard to how they affect the function or pose a hazard to the functions of other workstations. As a result, their placement can be critical in order to avoid potential conflicts. Workstations should be located in low-traffic areas of the laboratory.

SPECIFIC LAYOUT ISSUES

Once the general building concepts, method of utility distribution, and approach to laboratory layout have been planned, the specific layout issues for laboratory areas should be determined. Specific layout issues for laboratories include the separation of laboratory and nonlaboratory areas, primary and secondary barriers, directional airflow, and workstation placement.

Separation of Laboratory and Nonlaboratory Areas

For the layout of support offices, storage rooms, and mechanical rooms associated with laboratory buildings, the following physical separation issues must be addressed during planning:

- The separation distance needed between laboratory areas and nonlaboratory areas
- The location of nonlaboratory areas, such as offices for laboratory workers, to determine if they should be placed within laboratories, across the hall, or in another part of the building
- Whether the location of administrative offices should reflect the organization's hierarchical structure, if one is in place
- The materials and construction methods required for the envelope that separates laboratory and nonlaboratory areas

Primary and Secondary Containment Barriers

Laboratories typically use a primary barrier and a secondary barrier to control the spread of contaminants. The primary barrier is usually a piece of laboratory equipment, such as a fume hood or BSC, which is intended to capture the hazardous materials used in the laboratory most of the time. A secondary barrier typically consists of a more passive measure, such as a negative pressure differential for a room or a laboratory that is completely sealed off by an air lock. The secondary barrier captures general laboratory emissions, spills, and contaminants pulled out of the workstation by moving into and out of an open-face station.

Directional Airflow

Maintaining pressure differentials between areas of different uses provides directional airflow from clean to dirty to ensure that contaminants from a laboratory are not spread throughout the building. Therefore, selection of pressure differentials (i.e., creating a pressure map) is needed, which requires coordination between mechanical and architectural systems to ensure that the chosen pressure differential can be maintained. Also, the requirements for the materials necessary to maintain the proper pressure differentials (and thus directional airflows) must be coordinated, such as any needs for tight-sealing doors, anterooms, sealed plumbing penetrations, and electrical conduits.

Workstation Placement

As one moves within a laboratory space from the exterior wall toward the primary access door and public corridor, equipment should be placed according to its decreasing hazard potential. Therefore, fume hoods and BSCs for conducting hazardous experiments should be located away from the primary access doors and pathways. In addition, locating hoods and cabinets in this manner reduces occupant traffic past them, which can negatively affect the performance. Coordination of functionality and aesthetics is needed in selecting supply air diffusers because laboratory areas need diffusers that can introduce large volumes of air at low speeds to prevent turbulence, which can disrupt airflow to open workstations, such as those with fume hoods and BSCs, as well as clean air stations.

DEVELOPMENT OF PLANNING DOCUMENTS

The Owner's Project Requirements (OPR) and the Basis of Design (BOD) are the two main planning documents used in the procurement of the laboratory facility. This section describes these programming documents and offers guidance for their development.

The OPR is not as detailed as the BOD and is used to record specific room types; their quantities and areas; the number and type of personnel who will occupy the laboratory; the types of research, teaching, or industrial functions; and an estimate of construction costs.

The BOD provides a more detailed explanation of the information provided by the OPR by describing the laboratory's functional needs; intended levels and quality of environmental requirements and control; and basic mechanical, electrical, and plumbing systems. Producing a clear design intent is very critical to defining a benchmark to be used to judge the true success of the project of constructing a laboratory facility. This document is dynamic in nature, and any changes to the design intent should be documented, reviewed, and approved by the owner. The following is a basic outline of the BOD:

- General project description
- Objectives
- Functional uses
- General quality of materials and construction
- Occupancy requirements
- Indoor environmental quality (IEQ) requirements
- Performance criteria

REFERENCES

AIHA. 2012. ANSI/AIHA/ASSE Z9.5-2012, *Laboratory ventilation.* Falls Church, VA: American Industrial Hygiene Association.

ASHRAE. 2013. ANSI/ASHRAE Standard 62.1-2013, *Ventilation for acceptable indoor air quality.* Atlanta: ASHRAE.

ASHRAE. 2015. Chapter 16, Laboratories. In *ASHRAE handbook—HVAC applications.* Atlanta: ASHRAE.

CDC. 2009. *Biosafety in microbiological and biomedical laboratories* (*BMBL*), 5th Edition. DHHS Publication No. (CDC) 21-1112. Atlanta: Centers for Disease Control and Prevention, U.S. Department of Health and Human Services. www.cdc.gov/biosafety/publications/bmbl5.

ILAR. 2011. *Guide for the care and use of laboratory animals*, 8th ed. Washington, DC: Institute for Laboratory Animal Research, Division on Earth and Life Studies, National Research Council, National Academies Press. http://grants.nih.gov/grants/olaw/Guide-for-the-care-and-Use-of-Laboratory-Animals.pdf.

NFPA. 2014. *NFPA 801: Standard for fire protection for facilities handling radioactive materials*. Quincy, MA: National Fire Protection Association.

NFPA. 2015a. *NFPA 101: Life safety code®*. Quincy, MA: National Fire Protection Association.

NFPA. 2015b. *NFPA 30: Flammable and combustible liquids code*. Quincy, MA: National Fire Protection Association.

NFPA. 2015c. *NFPA 45: Standard on fire protection for laboratories using chemicals*. Quincy, MA: National Fire Protection Association.

NRC. 2011. *Prudent practices in the laboratory: Handling and management of chemical hazards*. 2011. Washington, DC: National Research Council, National Academies Press. www.nap.edu/catalog.php?record_id=12654.

NSF. 2014. NSF/ANSI 49-2014, *Biosafety cabinetry: Design, construction, performance, and field certification*. Ann Arbor, MI: NSF International.

Exhaust Hoods

5

OVERVIEW

Detailed knowledge of exhaust hoods, including the types of hoods available and their features, limitations, and design and operational requirements, is crucial to ensure the proper design and application of laboratory HVAC systems. While safety is the primary objective, energy conservation should be a secondary objective. Achieving both safety and energy conservation requires an understanding of the different types and purposes of the exhaust hoods. To provide information for the proper application of the hoods (described in the later chapters of this Guide), this chapter discusses the types of hoods as well as hood selection and proper certification and monitoring:

- Types of exhaust hoods
  - Chemical fume hoods
  - Biological safety cabinets
  - Other exhaust devices
- Selection of exhaust hoods
- Hood performance verification and continuous monitoring

Exhaust hoods are the most visible component of a laboratory HVAC system. They are the primary safety devices used in laboratories to capture chemical fumes. An understanding of how to optimize the use of these devices with the ventilation system for proper temperature control and dilution is needed for managing energy conservation. Oversimplification leads to inefficiency. It may be easy to design a laboratory to operate with 6 to 12 air changes per hour (ach) and a constant-volume fume hood designed for a face velocity of 100 fpm (0.51 m/s), but however safe this has proven to be, this method may waste energy.

Historically, the general consensus has been to operate a fume hood at a face velocity of 100 fpm (0.51 m/s). Although face velocity is a not a direct measure of hood performance, it does serve as a good diagnostic tool to observe air patterns and help to resolve containment failures. The tradition of operating a hood at 100 fpm (0.51 m/s) has been challenged by some manufacturers who have developed fume hood products that operate at face velocities as low as 60 fpm (0.30 m/s). The norm has been to target 100 fpm (0.51 m/s) because at this face velocity most hoods seem to provide adequate containment and minimize turbulence. Too high a velocity can result in greater turbulence and lead to containment failure. Too low a face velocity could fail to overcome disturbances generated outside of the hood (such as cross drafts due to personnel walking past the hood or supply air from ceiling diffusers). A challenge to this paradigm requires knowledge of fume hood systems, control of the laboratory environment, and the test of time.

DEFINING EXHAUST HOODS

For a more detailed discussion of fume hoods, see the latest edition of SEFA 1, *Recommended Practices for Laboratory Fume Hoods*, from The Scientific Equipment and Furniture Association (SEFA) (2010a). SEFA is an international trade association composed of manufacturers of laboratory furniture, casework, and fume hoods, as well as design and installation professionals. The purpose of SEFA 1 is to provide architects, engineers, planners, specifiers, manufacturers, and end users with industry standard practices. These recommended practices cover the design, construction, installation, testing, maintenance, and safe use of laboratory fume hoods, including benchtop, distillation, floor-mounted, auxiliary air, radioisotope, and perchloric acid hoods. SEFA 1 defines a laboratory fume hood as "a safety device specifically designed to carry undesirable effluents (generated within the hood during a laboratory procedure) away from laboratory personnel and out of the building, when connected to a properly designed laboratory ventilation system" (SEFA 2010a, p. 48).

From this definition, all other exhaust hoods can be segregated. Other hoods may not function as well as a laboratory fume hood for general chemistry but may provide excellent containment for their specific purposes. Examples of other devices include demonstration hoods, California hoods, ventilated enclosures, local exhaust ventilation, laminar flow hoods, biological safety cabinets, and ductless hoods. These hoods may be appropriate for specific laboratories, and readers should refer to SEFA 1 for more information in such cases.

SEFA 1 covers three categories: as manufactured, as installed, and as used. "As manufactured" includes hood components and recommend testing at the factory. "As installed" discusses locating the hoods, safety considerations, field testing, and troubleshooting. "As used" describes safe work practices, plans for conducting experiments, personal protection, work practices and responsibilities, and providing proper hood performance.

TYPES OF EXHAUST HOODS

CHEMICAL FUME HOODS

A common type of exhaust hood found in almost all laboratory facilities is the chemical fume hood. These hoods are designed to contain general odorous, toxic, or otherwise harmful chemical substances that are used in a wide range of laboratory activities, including research, demonstration, and numerous laboratory processes and experiments. In laboratories where the primary work of the occupants involves handling chemicals, it is recommended by the American Chemical Society (ACS) and others that there be at least one fume hood per two workers and that each worker should have at least 2.5 linear feet (0.76 m) of working space at the face of the fume hood (ACS 1991). During the laboratory planning stage, the number of hoods required should be determined. Energy savings is achieved through reducing the number of hoods in the laboratory space, but the number should not be reduced below what is needed for safe operations.

The type of chemical fume hood chosen depends upon the process being accomplished within the hood, safety requirements for the intended processes, usage patterns, and energy efficiency.

In general, chemical fume hoods have several components in common, including the following:

- Work surface
- Light
- Services (sink cup, sink, water, gas, vacuum, etc.)
- Monitor
- Exterior shell
- Interior liner
- Hood baffle
- Hood exhaust collar
- Hood bypass
- Hood sash

Figure 5-1 illustrates some of these typical components of a fume hood.

The hood work surface is the horizontal surface used for setting up experiments. Typically a chemical- and heat-resistant material is used for the work surface. It is typically recessed slightly in the center to contain minor spills within the hood.

Hoods usually provide some type of light within the hood to illuminate the work surface, with a safety glass panel typically separating the lights from the hood interior.

Hood services can include a variety of utilities that may be needed in the hood or at the entrance to the hood. Typical services include water, air, vacuum, natural or propane gas, specialty gases, fire suppression systems, and electricity.

Figure 5-1
Typical Components of a Fume Hood

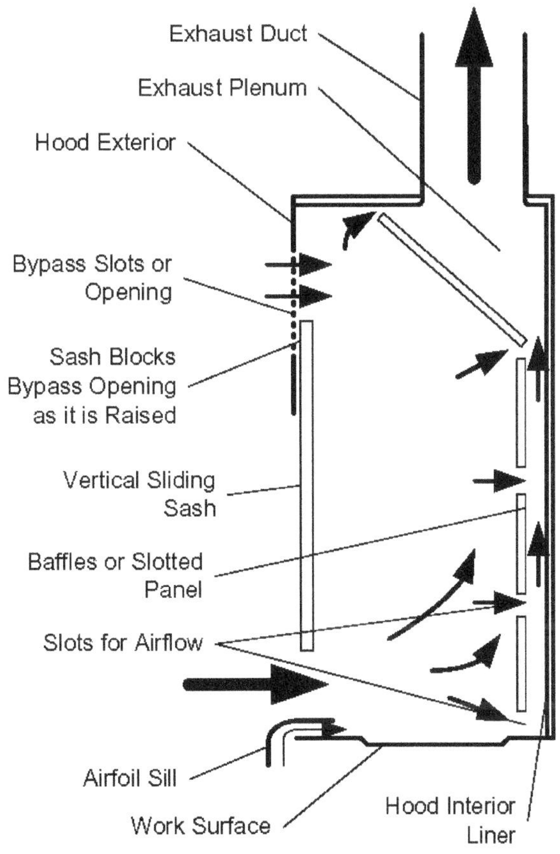

A fume hood monitor must be provided for all hoods. The monitor verifies either face velocity or exhaust flow and provides a clear indication of proper hood operation.

The hood exterior can be constructed from a variety of materials, but it is usually coated steel. The exterior forms the shell to hold the remainder of the components. The front of the hood is typically contoured with an airfoil sill to allow smooth airflow patterns entering the hood through the sash.

The interior of the hood is constructed of materials that are resistant to the chemicals being used in the hood. Typically, the interior is made from fiber-reinforced epoxy and polyester composite, phenolic resins, stainless steel, or other corrosion and/or chemically resistant materials.

The hood baffle at the rear of the hood is designed to control the airflow pattern within the hood. Slots are typically provided at the bottom and middle of the back and at the middle of the top of the hood to evenly distribute the exhaust flow throughout the hood.

The exhaust collar is typically located at the top of the hood behind the baffle. This provides the connection from the hood to the exhaust duct serving the hood. The collar shape and size can affect the performance of the hood. Bell-mouth collars provide a smoother transition and reduce noise and turbulence.

A hood bypass may be located above the sash and allows air to bypass the sash opening as the hood sash is closed. This prevents high face velocities on constant-volume hoods when the hood is partially closed. The sash typically covers the bypass opening as it is raised, forcing the air to come through the sash opening instead of the bypass. The bypass can be either an open bypass or a restricted bypass. Open bypasses are designed to limit the face velocity to no more than three times the design face velocity, when the sash is open from 6 in. (152 mm) to fully open. A restricted bypass is smaller and is usually shorter. A restricted bypass is used with variable-air-volume (VAV) hoods to reduce energy usage.

Hood sashes are the movable panel(s) on the front of the hood. They are typically transparent and made of some type of safety glass. They provide a barrier between the worker and the experiments. Sashes can be vertical moving, horizontal moving, or a combination of vertical and horizontal sashes. Figure 5-2 shows some of the typical sash arrangements.

Because of the range of requirements, many configurations of chemical fume hoods have been developed, including the following:

- Benchtop hood
- Distillation hood
- Floor-mounted hood
- Auxiliary air hood
- Radioisotope hood
- Perchloric acid hood

Benchtop hoods are designed to sit on top of a work bench or storage cabinet. The hoods come in a variety of widths and sash types. They typically are used for small to moderate quantities of low to highly toxic chemicals.

Distillation hoods are similar to benchtop hoods but have a taller working area for the taller distillation apparatus. The hoods typically have two vertical sashes or horizontal panels.

Floor-mounted hoods are similar to distillation hoods. A floor-mounted hood extends to the floor and typically has two vertical sashes or multiple horizontal panels. With floor-mounted fume hoods, the worker is only inside the hood to set up the experiment and not

Figure 5-2
Typical Sash Configurations

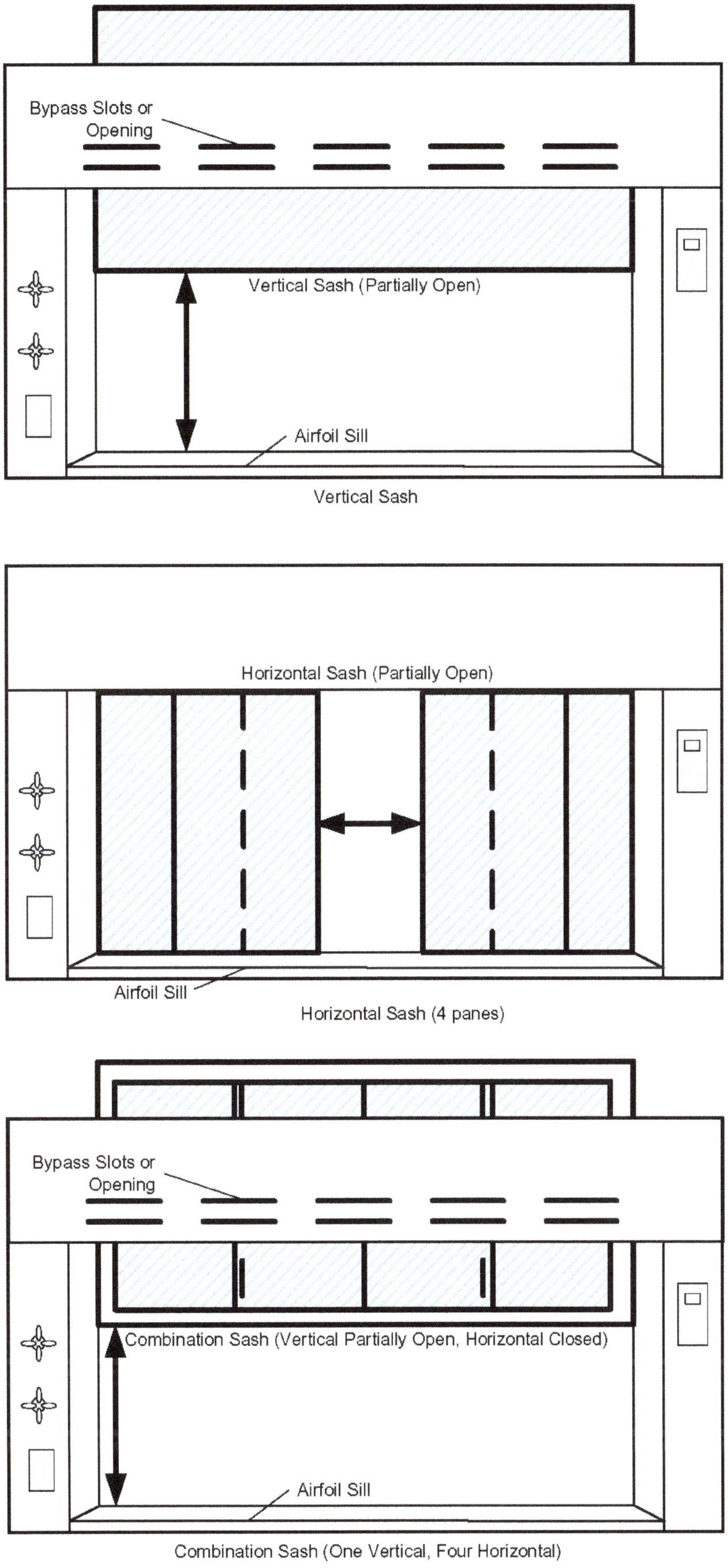

actually in the hood while the experiment is being conducted or hazardous materials are being released.

Auxiliary air hoods are similar to benchtop hoods except that makeup air is introduced at the exterior of the hood to be drawn into the hood when the sashes are open. Makeup air should be tempered to prevent cold or hot drafts at the face of the hood. Unconditioned air entering the hood may affect the results of the experiments. The auxiliary air can also affect the room temperature and humidity. Auxiliary air hoods are not commonly used in today's market.

Radioisotope hoods are similar to benchtop hoods. The hoods are designed for use with beta and gamma radiation. The interior of the hood is constructed from Type 304L stainless steel with welded seams for cleaning and decontamination. These hoods are typically provided with vertical sashes only. The work surface must be able to support the shielded containers in which the radioactive material is transported. The recommended face velocity for radioisotope fume hoods is between 100 and 125 fpm (0.51 and 0.64 m/s), since radioisotopes usually have a low permissible exposure limit (PEL). It is recommended that the exhaust duct material be 18 gage (1.27 mm) 316 stainless steel (DiBerardinis et al. 2013) welded. In some instances, the U.S. Nuclear Regulatory Commission (NRC) regulations require further safety features, such as high-efficiency particulate air (HEPA) filtration, activated charcoal absorption for exhaust air, and continuous radioactivity monitoring of the exhaust airstream. VAV control is not recommended for radioisotope fume hoods due to the potential need for air filtration and because emissions from radioisotope hoods are based on the concentration of radioactive materials, which will increase when airflow is reduced by a VAV fume hood.

Perchloric acid hoods are designed specifically for use with perchloric acid. They are similar to benchtop hoods with a seamless liner of stainless steel, chlorinated polyvinyl chloride (CPVC), or polypropylene. The hoods, exhaust system, and fans must have a washdown system to remove perchlorates from the surfaces after use. Perchlorate salts are explosive. The hoods should have a dedicated exhaust system. These hoods are typically constant-volume hoods (benchtop with full bypass).

BIOLOGICAL SAFETY CABINETS

In biological laboratories and clinics, the typical exhaust hood, which simply exhausts its contents to the atmosphere for dilution, is not sufficient to properly handle the materials used within the laboratory. In these instances, biological safety cabinets (BSCs) are needed to provide a more secure environment. These cabinets are designed to specifically handle organisms, infectious agents, processes, and particles that can potentially contaminate and harm other projects, personnel, or the environment. They also protect the product from cross-contamination. Therefore, all BSCs require filtration of exhaust air, and some require filtration of supply air, depending on the classification type of the hood. Because HEPA filters and prescribed airflows provide the protective envelope, these devices should be annually certified according to field test protocols described in NSF International Standard NSF/ANSI 49 (NSF 2014).

BSCs are divided into Classes I, II, and III, with Class II cabinets being divided further into four types depending on the percentage of exhaust air recirculation within the hood and typical face velocities used. The classification system is specified by NSF/ANSI 49 (NSF 2014). The three classes of BSCs should not be confused with the four biosafety levels (BSLs), which deal with the general level of infection and risk associated with a laboratory and are described in *Biosafety in Microbiological and Biomedical Laboratories (BMBL)* (CDC 2009). Selection of a BSC should be made by a qualified safety professional by performing a risk assessment of the BSC use. The results should then be applied to BSC classification type, exhaust requirements, and installation site assessment for proper location within the laboratory.

Class I

The Class I BSC provides personnel and environmental protection only. Class I cabinets do not use supply filtration or recirculation of supply air, but all of the exhaust air is HEPA filtered. The Class I BSC has the unfiltered room airflow drawn inward past the operator and across the work surface. Due to this airflow arrangement, Class I BSCs do not provide protection for the products used in the cabinet, as the room airflow can contaminate work materials on the work surface. The typical minimum face velocity for a Class I cabinet is between 75 and 100 fpm (0.38 and 0.51 m/s). Figure 5-3 is a schematic of a Class I BSC. Class I BSCs are typically vented out of doors with negative-pressure ductwork inside the building.

Class II

The Class II BSC provides personnel, product, and environmental protection. It also protects laboratory workers by maintaining inward unfiltered room airflow with the addition of filtered vertical unidirectional airflow using HEPA filters for the supply and exhaust airflow. Because of the vertical unidirectional flow pattern in the cabinet's work area, this class of BSC protects work materials from the inward unfiltered room airflows as well as work-area cross-contamination.

The requirement for HEPA filtration of both the supply and exhaust airflows is due to the processes intended for use in the BSCs requiring clean and controlled environments. Further, Class II cabinets contain type differences within this classification according to face velocities and exhaust recirculation percentages. The subtypes of Class II cabinets are Type A1, Type A2, Type B1, and Type B2.

Figure 5-3
Class I BSC

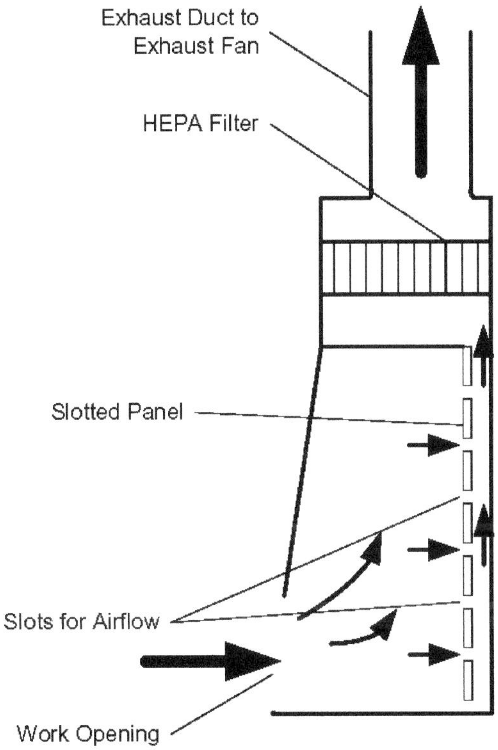

All Class II cabinets are designed for work involving procedures assigned to BSL-1, BSL-2, or BSL-3. Class II BSCs may be used with procedures requiring BSL-4 containment if used in a BSL-4 suit laboratory by a worker wearing a positive protective suit. All work procedures involving volatile chemicals and radionuclides should be reviewed by a certified industrial hygienist and a chemical hygiene plan developed for both service and use.

Type A1

The Class II Type A1 BSC provides HEPA-filtered vertical unidirectional airflow within the work area from a common plenum, where a portion of the air is exhausted and a portion is supplied to the work area. Class II Type A1 BSCs use approximately 70% recirculated air from this common plenum and exhaust the remaining 30% equal to the cabinet's inflow volume. Class II Type A1 BSCs have a minimum required face velocity of 75 fpm (0.38 m/s) and are not intended for experiments using volatile chemicals or radionuclide. The exhaust air may be relieved to the space or connected to an exhaust system through a thimble connection. Figure 5-4 is a schematic of a Class II Type A1 BSC.

Type A2

The Class II Type A2 BSC provides HEPA-filtered vertical unidirectional airflow within the work area from a common plenum, where a portion of the air is exhausted and a portion is supplied to the work area of the cabinet. Class II Type A2 BSCs use approxi-

Figure 5-4
Class II
Type A1
BSC

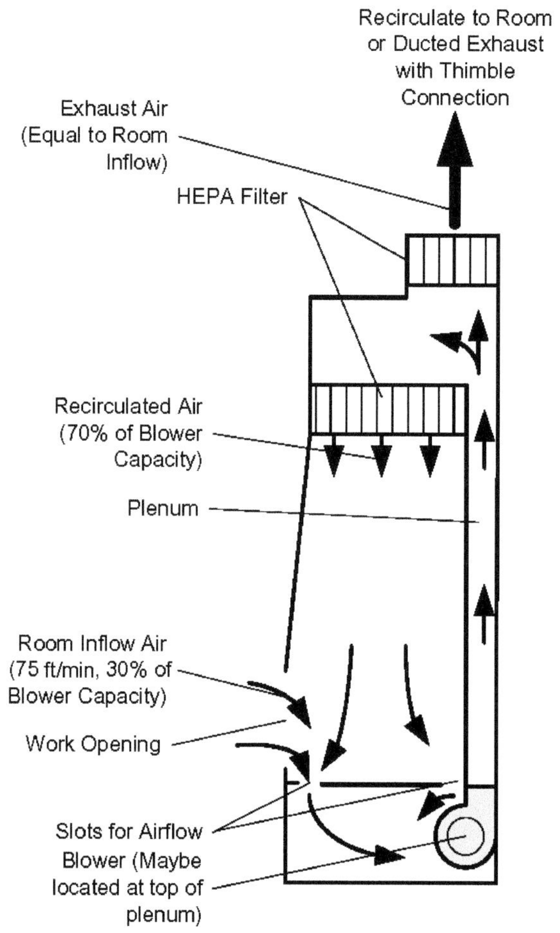

mately 70% recirculated air from the common plenum and exhaust the remaining 30% equal to the cabinet's inflow volume. Class II Type A2 BSCs have minimum inward velocity of 100 fpm (0.51 m/s) and can be used for work with volatile chemicals and radionuclide required as an adjunct to microbiological studies if exhausted through a properly functioning exhaust canopy with an exhaust monitor. Figure 5.5 is a schematic of a Class II Type A2 BSC.

Class II Type A1 and Type A2 BSCs may recirculate the air into the laboratory space since the air is particulate free, having passed through a HEPA filter. When recirculating the air to the space, volatile chemicals cannot be used. These BSCs can be connected to a ducted exhaust system through a thimble or canopy arrangement. They should not be hard ducted, as the pressure fluctuations will affect the airflow in the cabinet. The flow through the canopy or thimble connection should be 10% to 20% greater than the exhaust from the BSC. Figure 5-6 is a schematic of a typical thimble connection.

Type B1

The Class II Type B1 BSC provides HEPA-filtered vertical unidirectional airflow within the work area from the unfiltered inflow volume and 30% of the recirculation airflow. The remaining 70% of the HEPA-filtered vertical unidirectional airflow is exhausted through the work area rear grille to a direct-connected exhaust system. This type of cabinet requires that all exhaust ductwork and plenums be maintained at negative pressure or be contained in a negatively pressurized duct or plenum.

Figure 5-5
Class II
Type A2
BSC

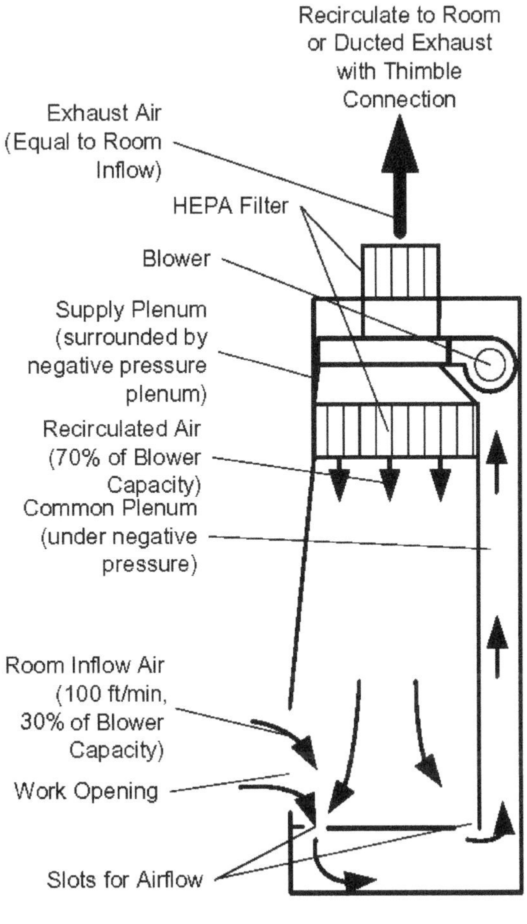

Figure 5-6
Thimble Connection for Class II Type A1 and A2 BSCs

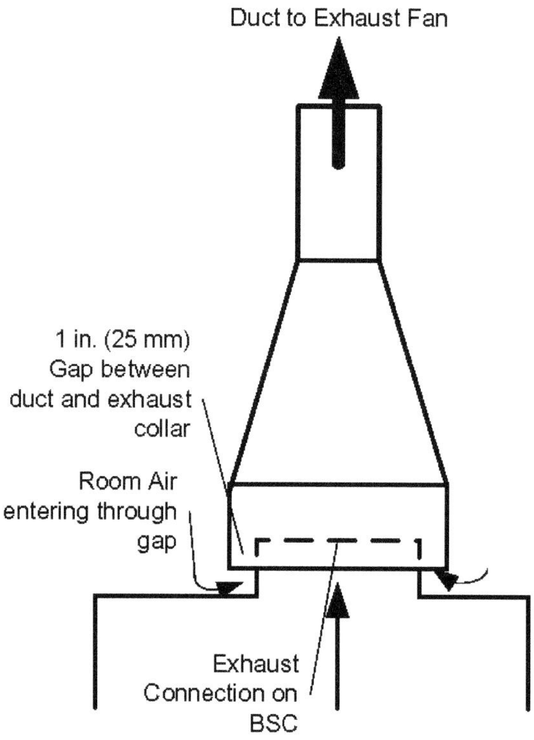

Additional features and requirements of Type B1 BSCs are that they

- provide a dedicated exhaust through the work area rear grille,
- can be used with volatile materials as determined by a risk analysis if working behind a work-area split line,
- require an exhaust system designed to maintain cabinet exhaust airflow under increased static pressure of up to 50%, and
- require an airflow monitor interlocked to the supply fan.

Figure 5-7 is a schematic of a Class II Type B1 BSC.

Type B2

The Class II Type B2 BSC provides HEPA-filtered vertical unidirectional airflow within the work area taken directly from the room as well as the unfiltered room inflow volume. For this reason, Class II Type B2 BSCs are referred to as “total exhaust” cabinets, since they exhaust 100% of both the supply air and inflow air. This type of cabinet requires that all exhaust ductwork and plenums be maintained at a negative pressure or be contained in a negatively pressurized duct or plenum.

Additional features and requirements of Type B2 BSCs are that they

- provide total exhaust of the work area,
- can use small amounts of volatile materials,
- require their exhaust system to be designed for a 50% increase in static pressure for less than a 5% decrease in airflow due to the high and variable pressure drop through the BSC, and
- require an exhaust airflow monitor interlocked to the supply fan.

Figure 5-8 is a schematic of a Class II Type B2 BSC.

Figure 5-7
Class II Type B1
BSC

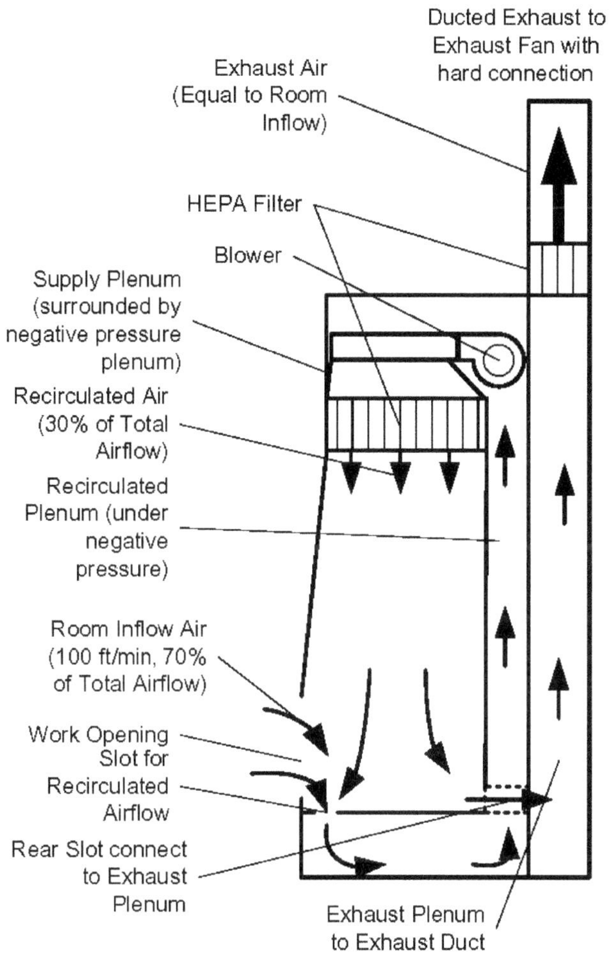

Class III

The Class III BSC is used for processes and biological agents that pose the greatest hazard risk, namely, for those designated as BSL-3 and BSL-4. These cabinets are of air-tight construction and have attached rubber gloves for handling the materials within the hood. Double-door autoclaves and chemical dunk tanks may be used, depending on the owner's requirements, to pass materials in and out of a Class III cabinet. To contain all contaminates within the cabinet, a minimum negative pressure of 0.5 in. w.g. (120 Pa) must be maintained at the required airflow rate. These BSCs require pressure monitors, decontamination ports, and gastight shutoff dampers.

Class III cabinets use single HEPA filters for supply air, use double HEPA filters in series or single HEPA filtration and incineration for all exhaust air from the cabinet, and do not recirculate any air within the cabinet. This class of cabinet requires that exhaust ductwork be under negative pressure, with the exhaust typically being separate from other exhaust systems serving the laboratory. Class III safety cabinets are only required in very specialized laboratories and are not commonly used, compared to the other classes and types of BSCs. However, this type of cabinet is the preferred method of containment when a high level of personnel and community protection is needed. Figure 5-9 shows an example of a Class III BSC.

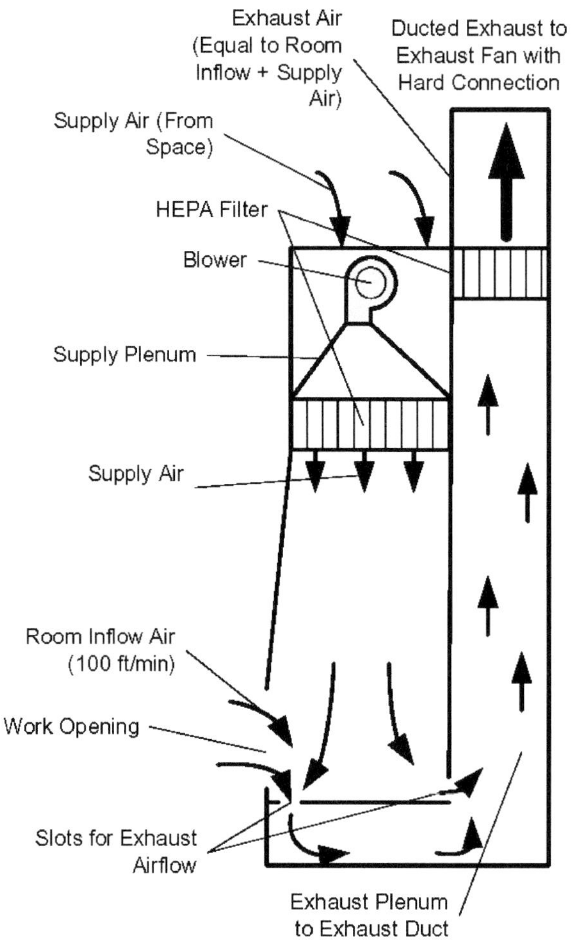

Figure 5-8
Class II Type B2 BSC

OTHER HOOD TYPES

In addition to chemical fume hoods and biological safety cabinets, there are numerous other types of primary containment devices that are available to control airborne contaminants in laboratories, including the following:

- Ventilated balance enclosure (weighing station)
- Slot hood
- Canopy hood and dedicated equipment exhaust
- Snorkels and spot exhausts
- Laminar flow clean air station
- Glove box
- Student demonstration hood
- Ventilated enclosure

Ventilated Balance Enclosure (Weighing Station)

A ventilated balance enclosure (VBE) is designed to house instrumentation and contain and exhaust hazardous materials (potent compounds) during measurement and weighing operations. Due to the hazardous nature of some materials, it is necessary to protect people from exposure to the materials when measuring and weighing the materials. However, typical fume hoods can make accurate measurements difficult due to turbulence from relatively high airflow velocities within the fume hood. Therefore, VBEs are

Figure 5-9
Class III BSC

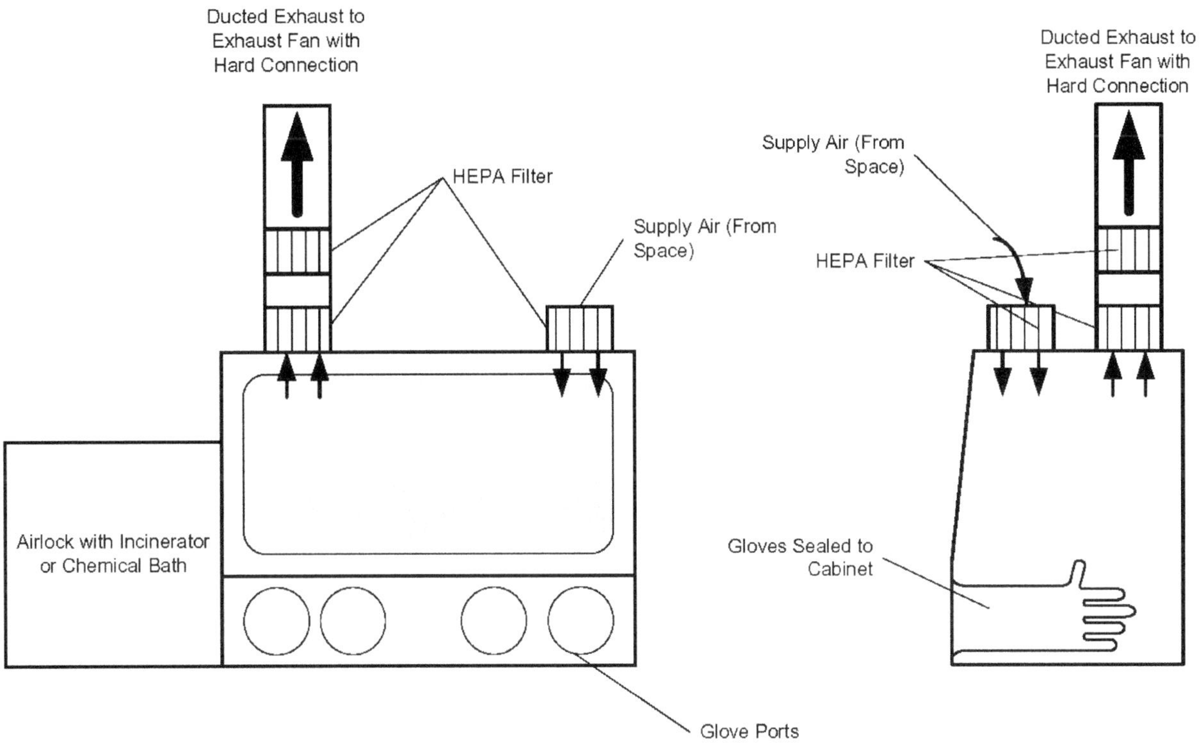

used to provide a ventilated space with low airflow velocities for measuring and weighing hazardous materials. The containment requirements are determined and verified by appropriate test methods.

Slot Hood

Slot hoods are used to capture hazardous materials close to their source when the use of chemical fume hoods is not practical or necessary. The primary advantage of a slot hood is that it does not enclose the process but relies only on capture velocity to remove materials away from the researcher.

A slot hood works by drawing materials through small, high-velocity exhaust slots in a plenum located near the contaminant source. Slot hoods are not as effective in protecting laboratory personnel as chemical fume hoods and should not be used as a substitute when materials need to be handled in an enclosed or partially enclosed fume hood. Typically the laboratory operations where slot hoods are applicable include specimen preparation, mixing, and weighing operations. The orientation of the slot hood with respect to the user can be critical to reduce potential for exposure. The aspect ratio of the slot and the slot velocity are typically in the range of 500 to 2000 fpm (2.5 to 10.2 m/s), with the slots and plenum being located not more than 12 in. (305 mm) from the generation source. Figure 5-10 shows a slot hood.

Canopy Hood and Dedicated Equipment Exhaust

Canopy hoods are defined as "receiving hoods" and are used for removal of heat, steam, or exhaust of materials of negligible hazard generated from equipment such as ovens, autoclaves, and sterilizers. This type of hood is usually used in cases where effluents are at high temperatures or are directed upward; thus, the hoods are suspended over

Figure 5-10
Slot Hood

(Courtesy of Monoxivent)

the equipment for which they are intended to provide exhaust. Exhaust velocities for canopy hoods should be based on the velocity necessary to receive and capture the rising plume. The flow, inflow velocity, and location must be carefully planned and appropriate for the generation source. The effectiveness of a canopy hood is dependent largely on the contour and structure of the opening used. For more information on the contour and structure impacts of canopy hoods, and exhaust hoods in general, consult Chapter 3, Local Exhaust Hoods, of *Industrial Ventilation: A Manual of Recommended Practice for Design* from the American Conference of Governmental Industrial Hygienists (ACGIH 2013).

As with slot hoods, canopy hoods should be placed as close to the source as possible and should not be used as a substitute when laboratory processes require chemical fume hoods. Figure 5-11 shows a canopy hood.

Snorkels and Spot Exhausts

Local exhaust snorkels and spot exhausts perform similarly. Snorkels are typically movable, whereas spot exhausts are typically fixed, small exhaust capture hoods, but both are intended to provide exhaust for small point sources. Snorkels and spot exhausts are designed to contain low-hazard airborne contaminants produced by specific processes and equipment not suited for locating within an enclosure. Both are used to remove heat or nontoxic particles that may be generated from benchtop research equipment.

Snorkels are usually funnel-shaped inlet cones connected to 3 to 6 in. (75 to 150 mm) diameter flexible or semiflexible ductwork extending from the ceiling to above the benchtop level. Snorkel exhaust dampers are typically closed when not in use. Figure 5-12 shows a typical snorkel.

Laminar Flow Clean Air Station

Laminar flow clean air stations provide product protection by design. Unidirectional air can be introduced vertically or horizontally depending on design. Some stations are

Figure 5-11
Canopy Hood

(Courtesy of Kewaunee Scientific Corporation)

Figure 5-12
Snorkel

(Courtesy of Monoxivent)

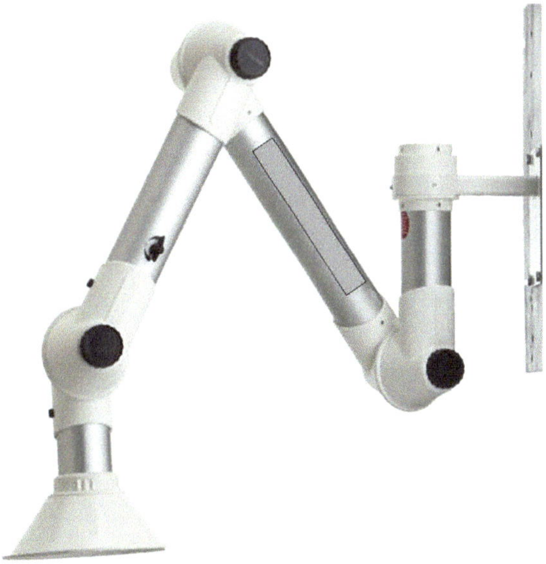

designed to be vented either partially or 100% depending on the requirements of the experimental protocol. These units are designed to meet class 100 (ISO 1999) cleanliness standards.

Glove Box

A glove box is used in cases where the toxicity or radioactivity of the airborne contaminants is too high to be handled by a chemical fume hood and in cases where the substance being handled reacts with air. In these instances, the glove box completely isolates the experiment from the surroundings of the laboratory. To accomplish this, the glove box is usually of airtight construction and is made of stainless steel and safety glass with smooth finishes. Also, as the name suggests, full-length rubber gloves are attached to the side of the box and are used for handling the materials within the box. The glove box depth should be 24 in. (610 mm) or less to prevent glove damage.

Glove boxes are maintained under a relatively high negative pressure of 0.25 to 0.5 in. w.g. (0.06 to 0.12 kPa). Depending on the material handling requirements of the glove box, the following may also be used: HEPA filtration of supply air, HEPA filtration of the exhaust air for toxic and infectious materials, activated charcoal filter absorption for volatile chemicals, or air locks for introduction of items into the box. When HEPA or activated charcoal filtration is used, a prefilter for removing particulate matter is recommended.

Glove boxes that are used to contain substances that react with air use an inert gas or one that does not react with the substance. When inert gas is used, the bypass opening for the glove box is replaced with a connection to an inert gas supply. Inert gas glove boxes may or may not be ducted to the outdoors, depending on the hazards to laboratory personnel associated with the materials that will be used in the glove box.

Student Demonstration Hood

Demonstration hoods, or pass-through hoods, are special fume hoods with viewing glass or sashes on multiple sides. They are used to allow students to view experiments being conducted in the hood. Demonstration hoods with sashes or viewing windows on the side opposite the main sash often eliminate the baffle that helps keep the inflow of air uniform across the sash opening. The hoods are designed to operate with only one sash open at a time. Typically these are constant-volume bypass hoods, but they can be operated as VAV.

Ventilated Enclosure

A ventilated enclosure is intended to contain equipment that may generate small concentrations of effluent. A ventilated enclosure typically has an opening to allow access and is sometimes equipped with a baffle at the back of the hood. Such enclosures are used to enclose low-risk applications. The front and sides are typically glass or clear plastic.

SELECTION OF EXHAUST HOODS

The selection of exhaust hoods is an important step in the design of a laboratory and can greatly impact the health and safety of the laboratory workers, the results of the laboratory experiments, and the life-cycle cost of owning and operating the laboratory. To begin the selection process, a clear understanding of the current and future types of work that are to be conducted in the laboratory needs to be researched and documented. The hazards associated with the current and future work in the laboratory are then identified, and the applicable codes, standards, and laws that regulate the anticipated types of work and associated hazards need to be located and followed. Once the hazards and applicable regulations have been identified, the appropriate types of exhaust hoods for the laboratory processes can be chosen. Figure 5-13 shows a matrix of the exhaust hood selection process.

In some cases, there are no options for the type of hood to use. If perchloric acid is to be used, a perchloric acid fume hood is required and other types of hoods cannot be substituted. The same is true for regular work with radioisotopes—only a radioisotope fume hood should be used. Biological safety cabinets allow some flexibility in the type to use, as long as the appropriate level of protection is achieved. Refer to Figure 5-13 for a diagram of laboratory hoods and their potential applications.

Figure 5-13
Exhaust Hood Selection Matrix

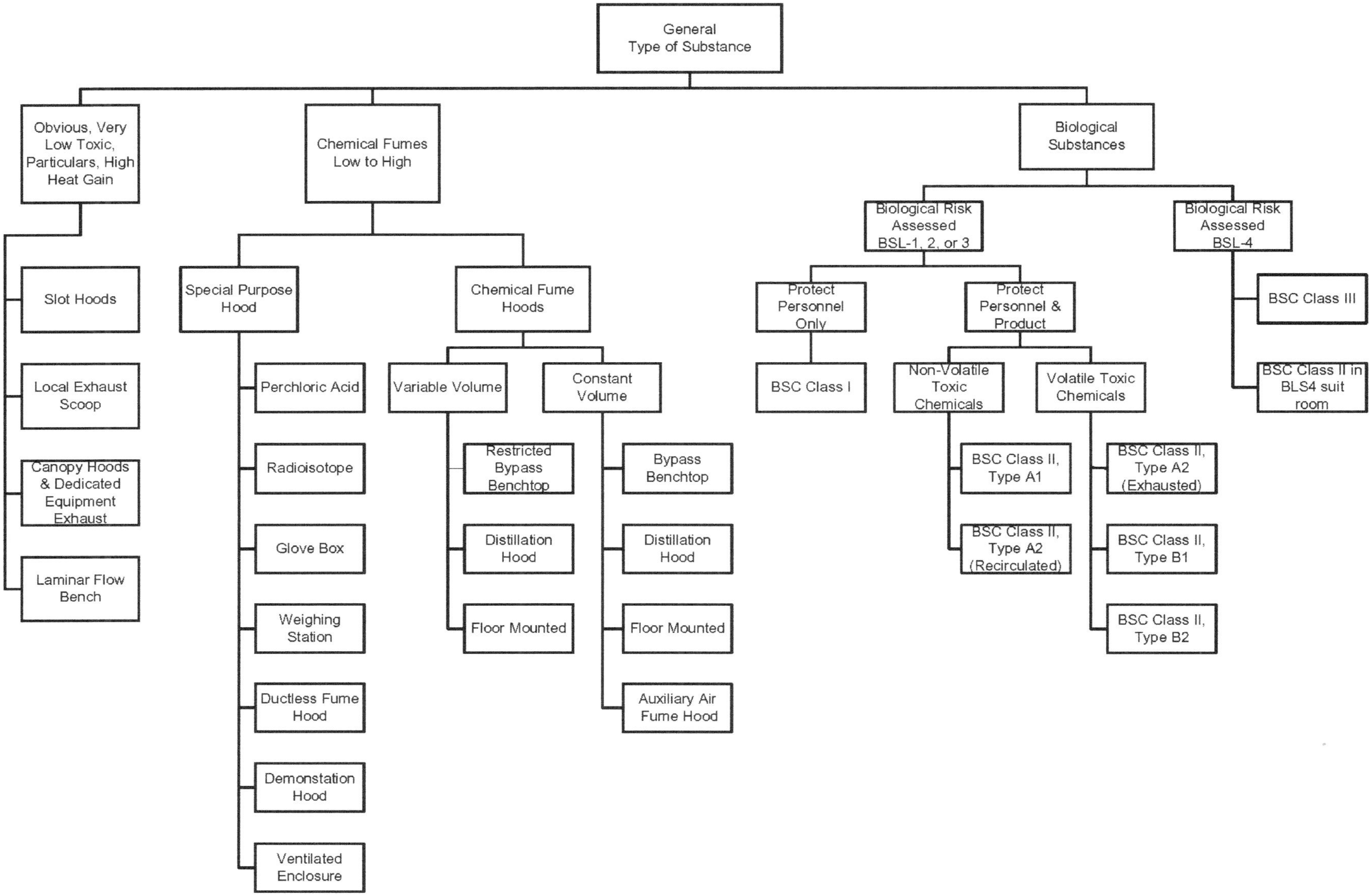

CONSTANT VOLUME VERSUS VARIABLE AIR VOLUME

Chemical fume hoods can be either constant volume or variable air volume (VAV). Constant-volume fume hoods have the benefit of being less expensive initially, require a less complex supply and exhaust system, and are generally easier to maintain. However, constant-volume fume hoods can waste substantial amounts of energy if the laboratory is occupied infrequently. VAV fume hoods have the benefit of saving substantial amounts of energy over their lifespan if the laboratory is used periodically, and they can provide continuous monitoring to document that the required face velocity of the hood is being met. VAV hoods require a more complex supply and exhaust system, require additional controls, and can be more difficult to maintain, as it is essential that their performance be checked regularly.

The choice of a constant-volume or VAV fume hood is made in conjunction with the selection of the type of primary air system that is to be used in the laboratory. (Primary air systems are discussed in Chapter 6.) Also made in conjunction with the choice of constant-volume or VAV is the choice of dedicated or manifolded exhaust air systems. Generally, manifolded exhaust systems should be grouped into areas of similar hazards.

The compatibility of materials used and the layout of the laboratory building influence choosing dedicated or manifolded exhausts systems. The choice of dedicated or manifolded exhaust systems is also closely tied to whether constant-volume or VAV fume hoods are selected. VAV systems are generally more economical when a manifolded exhaust system is used, as a dedicated system would require a separate, costly variable-frequency drive (VFD) for each fume hood fan motor. A manifolded system would only require a less expensive air control device for each fume hood and one VFD per manifolded system. The choice of dedicated or manifolded exhaust for constant-volume fume hoods is generally determined by the compatibility of the materials used and the cost difference between numerous individual runs of ductwork for a dedicated system and a few long runs of ductwork to create a manifolded system.

The choice of constant-volume or VAV fume hoods and dedicated or manifolded laboratory exhaust systems can be an iterative process. Also, a laboratory building does not need to be exclusively constant volume or exclusively VAV, exclusively dedicated exhaust or exclusively manifolded exhaust. Similar areas of work and associated hazards can be grouped together to use one or the other type of system without the whole building necessarily being the same.

VAV laboratories and the hoods designed to operate in these laboratories (constant-volume fume hoods with partially restricted bypass) can have substantial energy savings over constant-volume laboratories with constant-volume fume hoods. VAV fume hoods can reduce energy consumption (regardless of the fume hood velocity setpoint as described in the Overview section of this chapter) by reducing the volume of exhaust flow and laboratory supply air when the hood sash is closed or the hood is not in use, depending on the design of the VAV controls. The amount of energy savings depends on the reduction in fan power and reduction of conditioned air required for makeup.

VAV fume exhaust hoods can be benchtop, distillation, or floor-mounted types. What makes these hoods VAV versus constant volume is the size of the bypass opening. A restricted bypass or no bypass allows the exhaust air volume to vary according to the sash position to maintain a constant face velocity. The energy savings are a result of reduced energy for conditioning the supply air as well as reduced fan energy for both the supply and exhaust air when the fume hood sash is partially or fully closed. To achieve energy savings with VAV fume hoods, there must be times when either the laboratory is unoccupied or the fume hoods are not being used, and the laboratory occupants must be educated to keep fume hood sashes closed when the hoods are not in use. VAV fume hoods typically make use of a small bypass opening (airfoil sill) to ensure that a minimum amount of air continues to enter the fume hood even when the sash is completely closed. This bypass makes sure that any contaminant concentrations generated inside the hood are diluted and exhausted when the sash is closed. Chemicals should not be stored in the fume hoods but in chemical storage rooms or cabinets.

While there are significant operating cost savings with VAV hoods, they generally have higher initial costs due to the control systems cost for the hood and primary air system that modulate flow and achieve desired space pressurization. VAV fume hoods, like other fume hoods in general, require periodic maintenance and testing to ensure that the fume hood is operating properly and maintaining the required flow and face velocity. The operation and maintenance (O&M) personnel should be knowledgeable about the uniqueness of VAV systems compared to their constant-volume counterparts. Therefore, the ability to provide maintenance to a VAV fume hood through adequate training should be an important hood selection criterion.

ENERGY EFFICIENCY

A low-exhaust-volume (LEV) fume hood can operate safely at lower volumetric air requirements. This is achieved in one of two ways: restricted opening (low flow) or lower face velocity (low velocity). Sash restriction is achieved by reducing the sash opening (either vertically or horizontally) while maintaining a face velocity to the setpoint (usually 100 fpm [0.51 m/s]). The reduced opening requires less air overall and the energy associated with it. Another solution requires a fume hood designed specifically to operate at a lower face velocity (as low as 60 fpm [0.30 m/s]). The lower velocity of air equates to less volumetric air per hood and therefore results in less energy lost. No energy will be saved, however, if the room air change rate determined for the laboratory requires more cubic feet per minute (litres per second) than the hood exhausts. Energy savings through a reduction of airflow is only sensible when the hood demands more exhausted air than the room necessitates.

Even more savings can be achieved when combining a VAV system with a LEV fume hood. In either case (VAV with a modified constant-volume fume hood or VAV with an LEV fume hood), the energy savings from a VAV system only occurs if the sash is lowered or closed whenever possible. This can be enhanced with an automated sash-closing device that lowers the sash when an operator is not present.

Another solution for energy savings in the laboratory is to use a ductless fume hood. In certain applications, this special-purpose hood can provide a safe working environment while exhausting no air from the laboratory at all. In the case of a ductless hood, all the air exhausted from the room is through the normal room ventilation system (and should be part of the energy analysis). The hood captures the dangerous fumes in the activated carbon filter system and returns cleaned air to the room. The most thorough standard explaining the application of a ductless fume hood is SEFA 9, *Recommended Practices for Ductless Enclosures* (SEFA 2010b). Ductless hoods require that the user have more knowledge of the chemicals being used, be aware of the flammability of the filter media and the implications associated with a fire in a ductless hood, and caution the use of evaporating excessive amounts of liquid (even water vapor will consume filter media quickly). Ductless hoods require continual monitoring, filter maintenance, and user knowledge. Ductless hoods have a high first cost (although this is offset by reducing the cost of the entire ventilation system needed for each fume hood) and require a plan for filter replacement and disposal.

On the supply side energy savings are possible, as well. Traditionally, laboratories have used a guideline of 6 to 12 ach to provide adequate air turn to reduce exposure to chemical vapors while also managing heat loads and minimizing viral exposure. This air change rate was meant to accommodate the equipment and number of people working in one room for comfort and personnel health. Like the tradition that a fume hood should operate at 100 fpm (0.51 m/s), the 6 to 12 ach tradition is also being challenged. Lower air change rates are possible if there are fewer or no people in the room or if the room has fewer chemicals present.

HOOD PERFORMANCE VERIFICATION AND CONTINUOUS MONITORING

One of the key issues in ensuring proper protection for laboratory workers is the testing and monitoring of all types of exhaust hoods. Fume hoods are typically tested under ASHRAE Standard 110 (ASHRAE 1995). Usually BSCs are initially certified by NSF International and include proof of certification, then, upon installation, the cabinets are field-certified according to NSF/ANSI 49 (NSF 2014). This certification should be repeated on a regular, ongoing basis (typically yearly), whenever the hood is relocated, and if occupant complaints arise about the performance of the hood.

The following sections cover the sources of information for testing and monitoring exhaust hoods.

ASHRAE STANDARD 110

ASHRAE Standard 110, *Method of Testing Performance of Laboratory Fume Hoods* (ASHRAE 1995), specifies the test methodology that qualifies the performance of a laboratory fume hood. There are many factors to consider when evaluating a laboratory fume hood. ASHRAE Standard 110 consists of three tests:

- Flow visualization
- Face velocity measurement
- Tracer gas containment

There are many important factors (cross drafts, work procedures, obstructions, etc.) in the safe operation of a laboratory fume hood, and not all of them are covered by the test. It is the responsibility of the user, the chemical hygiene officer, and the engineer to understand the hood performance. The test standard defines three alternative ratings based on the condition of the test: as manufactured, as installed, and as used.

The standard also defines the test equipment, calibration, test conditions, data reported, and hood conditions. It provides information regarding smoke visualization tests, what to do when a hood is installed with a VAV system, and sash movements. The standard does not define acceptable criteria for passing the test. Acceptable criteria for passing the test should be established with the assistance of the user, chemical hygiene officer, and engineer.

ANSI/AIHA/ASSE Z9.5

Through the American Industrial Hygiene Association (AIHA), the ANSI/AIHA Z9.5 Subcommittee published ANSI/AIHA/ASSE Z9.5, *Laboratory Ventilation* (AIHA 2012), and the American Society of Safety Engineers (ASSE) has taken over as the Secretariat of this standard. The purpose of the standard is to establish minimum requirements and best practices for laboratory ventilation systems to protect personnel from physical harm and overexposure to harmful or potentially harmful airborne contaminants generated within the laboratory. ANSI/AIHA/ASSE Z9.5 does not cover ventilation of all types of laboratories; exceptions are listed in the standard.

Where ASHRAE Standard 110 establishes the test method, ANSI/AIHA/ASSE Z9.5 establishes acceptance criteria for fume hood containment. These criteria are widely accepted in the industry. A control level for a five-minute average test at each location conducted at a generation rate of 4 L/m is to be no greater than 0.05 ppm for an "as manufactured" test (AM 0.05) and 0.10 ppm for "as installed" (AI 0.1). Additionally, the standard requires that the laboratory establish a chemical hygiene plan, identifies responsible parties, and recommends that, among other things, a tracer gas containment "pass" criteria (e.g., AI 0.5, AI 0.1, AI 0.05) be determined.

Once installed, it should be verified that the operation and performance of the fume hood meets the acceptable containment performance criteria defined in ANSI/AIHA/ASSE Z9.5. Once the fume hood meets the performance criteria, the operation of the hood system can be verified during periodic routine tests. Routine tests may involve measurement of air velocity for a range of sash openings, cross-draft velocities, VAV response and stability, and airflow visualization as necessary to evaluate and verify proper operation and performance.

ANSI/AIHA/ASSE Z9.5 recommends a control limit of 0.05 ppm (for the manufacturer's ideal conditions) to 0.1 ppm (for an operating laboratory) for tracer gas containment tests conducted at a generation rate of 4 L/min from the ASHRAE Standard 110 ejector located at prescribed locations inside the fume hood (AIHA 2012). However, the control limits for allowable tracer gas escape are to be determined by the laboratory with respect to the processes and materials used in the hood.

Another safety requirement for fume hoods is the use of appropriate continuous monitoring devices. Monitors should be installed on all laboratory hoods used for protecting people from overexposure while working with potentially hazardous materials. Monitors can include measurement of flow, face velocity, or other parameters such as pressure that indicate proper operation. The monitor can also include audible and visual alarms to alert the user to suboptimal operation.

ANSI/AIHA/ASSE Z9.5 also provides some recommendations for energy efficiency as it relates to laboratory ventilation, identifying acceptable solutions such as lowering hood face velocity, using a VAV hood system, and using ductless hoods—although these systems may require additional commissioning, testing, application analysis, and filter management. Regarding alternative face velocities, the standard recognizes face velocities as low as 60 fpm (0.30 m/s) in some instances. A fume hood can achieve more energy efficiency by exhausting less air in two ways: by reducing the face velocity or by restricting the sash opening. To make use of these adjustments, it is important to determine the minimum allowable air changes per hour (ACH) of the fume hood, and ANSI/AIHA/ASSE Z9.5 addresses this issue. While attempting to save energy in typical higher-hood-density installations, minimum fume hood flow rates in the range of 150 to 375 ach have been used to control vapor concentrations inside hood interiors. Careful consideration must be made to determine the appropriate exhaust flow rate for any individual fume hood.

VAV fume hoods (as described in the Chemical Fume Hoods section of this chapter) provide the opportunity to reduce energy consumption by reducing airflow when the sash is closed. Exhaust flow for a fume hood can range from a minimum with the sash closed to a maximum with the sash full open (100%). The flow (Q) when the sash is open is equal to the design face velocity (V_f) multiplied by the opening area (A_h). The exhaust flow is reduced to a minimum flow and held constant when the sash is closed. The minimum flow must be sufficient for satisfactory hood containment and must sufficiently dilute contaminants to prevent accumulation of unsafe concentrations within the fume hood and exhaust duct.

Establishing the minimum flow for a fume hood is more complicated than using the formula $Q = V_f \times A_h$ and achieving a minimum inflow face velocity. ANSI/AIHA/ASSE Z9.5 recommends that minimum flow should be based on internal volume of the fume hood and internal ACH, where a range of 375 ach to as low as 150 ach has been found to be adequate in most cases (AIHA 2012). ANSI/AIHA/ASSE Z9.5 requires users to consider the following when selecting an appropriate minimum air change rate and minimum fume hood flow:

- The processes and materials generated within the hoods (maximum generation rate)
- The lower explosive limit (LEL) and the safety factor (most facilities use 10% to 25% of the LEL)
- Hood containment and dilution of hazardous concentrations within the hood
- Potential for increased corrosion
- The ability to measure and control flow
- Effect on duct transport and stack discharge velocities

Using less than 375 ach requires some extra effort. A specialized fume hood dilution investigation should be conducted to determine a safe minimum flow through VAV fume hoods. The analysis should include the following tasks as a minimum:

- Collect and evaluate information about the fume hoods, VAV control systems, and current operating specifications.
- Collect information about processes and materials used in VAV fume hoods to consider the potential to reduce flow below current design standards.
- Establish minimum flow specifications by testing a representative sample of fume hoods. The tests should include the following tasks:
  - Determine the ability of the fume hoods to maintain containment of smoke and tracer gas with the sash in the closed position.
  - Maintain containment during VAV response and stability tests (i.e., sash movement effect tests).
  - Evaluate the capability of the VAV systems to provide accurate and stable control of fume hood exhaust and laboratory supply at the reduced flow setpoint.
  - Determine the effects on duct transport velocity.
  - Conduct tests to assess the potential for accumulation of unsafe airborne concentrations within the fume hoods and exhaust duct.

The results of investigations on numerous fume hood types in various facilities indicate that the specifications are hood, laboratory, and system dependent. Establishment and application of a prescriptive specification for fume hood minimum flow without proper due diligence is risky (Smith 2013).

Although ANSI/AIHA/ASSE Z9.5 recognizes energy savings by reducing the airflow rate of the laboratory itself, this standard makes no recommendation for allowable room air change rates. Instead, the standard states that the laboratory ventilation system is to be designed to remove and dilute air contaminants in accordance with the Laboratory Ventilation Management Plan (AIHA 2012). Paradoxically, energy savings for a laboratory cannot be calculated without this information. It is incorrect to propose that energy is saved in the fume hood by reducing the air exhausted from the hood, when it may be necessary to remove the same air to meet a safe air change rate for the laboratory.

OTHER SOURCES OF INFORMATION

International Mechanical Code (Section 510)

The *International Mechanical Code*® (*IMC*®) is published by the International Code Council (ICC). The ICC was founded in 1994 from the merger of the regional code councils at the time to provide a common model code throughout the United States. The *IMC* is used in a majority of the states as the basis of their codes and is updated every three years.

Section 510 of the *2015 IMC* deals with hazardous exhaust systems. In laboratories this applies when concentrations of flammable vapors, gases, fumes, etc. exceed 25% of the lower flammability limit at room temperature or have a health hazard rating of 1, 2, 3, or 4 exceeding 1% of the median lethal concentration for acute inhalation toxicity (ICC 2015).

NFPA 45: Standard on Fire Protection for Laboratories Using Chemicals

Established in 1896, the international nonprofit National Fire Protection Association (NFPA) is the world's leading advocate of fire prevention and an authoritative source on public safety. NFPA publishes *NFPA 45: Standard on Fire Protection for Laboratories Using Chemicals*, which provides basic requirements for the protection of life and property through prevention and control of fires and explosions involving the use of chemicals in laboratory-scale operations (NFPA 2015b). The standard covers such topics as laboratory unit hazard classifications, laboratory unit design and construction, fire protection, explosion hazard protection, laboratory ventilating systems and hood requirements, chemical storage, and more.

The section covering fume hoods provides safety guidelines for hood location in the room, exhaust air discharge, duct construction, manifolds, duct velocity, and discharge. Some information also addresses fume hood interiors, sash closures, and electrical devices. *NFPA 45* specifically recognizes that a fume hood is not designed to contain explosions, and since it is common to have an open flame in a laboratory fume hood, the materials that construct a fume hood are to have a flame spread of less than 25 per ASTM E84, *Standard Test Method for Surface Burning Characteristics of Building Materials* (ASTM 2015). *NFPA 45* also requires that minimum exhaust flow be sufficient to prevent concentrations from approaching LELs inside the fume hood and exhaust duct (NFPA 2015b). The maximum allowable concentration is typically 25% of the LEL, though it is lower in some cases (e.g., 0% of the LEL) (Smith 2013).

Industrial Specialists

The goal of industrial hygienists is to keep workers, their families, and the community healthy and safe. They play a vital part in ensuring that federal, state, and local laws and regulations are followed in the work environment (AIHA 2013). They and industrial ventilation experts, laboratory safety engineers, and the company's in-house experts are other sources of information for guidance on achieving as safe and efficient a working environment as possible in the potentially hazardous field of working in a chemical laboratory. These individuals are armed with their experience and education and information pertinent to their own field, such as that found in *Industrial Ventilation: A Manual of Recommended Practice for Design* (ACGIH. 2013), ASHRAE Handbooks (ASHRAE 2015), and *Prudent Practices in the Laboratory: Handling and Management of Chemical Hazards* (NRC 2011). The skill to apply this knowledge and the appropriate application of standards is critical to the safety and reliability of the product and application while consideration is made to the responsibility to reduce energy consumption.

Decisions regarding the safety of an application and the tools used to conduct research must be analyzed continuously. Researchers must consider the safety of each application and call for help when necessary.

A properly trained laboratory safety engineer or industrial hygienist can provide knowledge on subjects from chemical exposure limits, regional and local codes, safety protocols, special handling and waste removal requirements, ventilation strategies, room air balancing, filter capacities and capabilities, and maintenance issues. When it comes time to select a new exhaust hood, these individuals provide the most valuable insight.

REFERENCES

ACS. 1991. *Design of safe chemical laboratories: Suggested references*, 2nd ed. Washington, DC: American Chemical Society, Committee on Chemical Safety.

ACGIH. 2013. *Industrial ventilation: A manual of recommended practice for design*, 28th ed. Cincinnati, Ohio: American Conference of Governmental Industrial Hygienists.

AIHA. 2012. ANSI/AIHA/ASSE Z9.5-2012, *Laboratory ventilation*. Falls Church, VA: American Industrial Hygiene Association.

AIHA. 2013. What is an industrial hygienist? Falls Church, VA: American Industrial Hygiene Association. www.aiha.org/about-aiha/Press/Documents/What%20is%20IH.pdf.

ASHRAE. 1995. ASHRAE Standard 110-1995, *Method of testing performance of laboratory fume hoods*. Atlanta: ASHRAE.

ASHRAE. 2015. Chapter 16, Laboratories. In *ASHRAE handbook—HVAC applications*. Atlanta: ASHRAE.

ASTM. 2015. ASTM E84-15, *Standard test method for surface burning characteristics of building materials*. West Conshohocken, PA: ASTM International.

CDC. 2009. *Biosafety in microbiological and biomedical laboratories* (*BMBL*), 5th Edition. DHHS Publication No. (CDC) 21-1112. Atlanta: Centers for Disease Control and Prevention, U.S. Department of Health and Human Services. www.cdc.gov/bio safety/publications/bmbl5.

DiBerardinis, L.J., J.S. Baum, M.W. First, G.T. Gatwood, and A.K. Seth. 2013. *Guidelines for laboratory design: Health, safety, and environmental considerations*, 4th ed. New York: John Wiley & Sons, Inc. http://onlinelibrary.wiley.com/book/10.1002/9781118633816.

ICC. 2015. *2015 International mechanical code*® (*IMC*®). Washington, DC: International Code Council.

ISO. 1999. ISO 14644-1:1999, *Cleanrooms and associated controlled environments Part 1: Classification of air cleanliness*. International Organization for Standardization.

NFPA. 2015a. About NFPA. Quincy, MA: National Fire Protection Association. www.nfpa.org/about-nfpa.

NFPA. 2015b. *NFPA 45: Standard on fire protection for laboratories using chemicals*. Quincy, MA: National Fire Protection Association.

NRC. 2011. *Prudent practices in the laboratory: Handling and management of chemical hazards*. 2011. Washington, DC: National Research Council, National Academies Press. www.nap.edu/catalog.php?record_id=12654.

NSF. 2014. NSF/ANSI 49-2014, *Biosafety cabinetry: Design, construction, performance, and field certification*. Ann Arbor: NSF International.

SEFA. 2010a. SEFA 1-2010, *Recommended practices for laboratory fume hoods*. Garden City, NY: Scientific Equipment Furniture Association.

SEFA. 2010b. SEFA 9-2010, *Recommended practices for ductless enclosures*. Garden City, NY: Scientific Equipment Furniture Association.

Smith, T. 2013. Minimum flow specifications for fume hoods. Seminar 60, presented at the 2013 ASHRAE Winter Conference, Dallas, TX, January 26–30.

Primary Air Systems

6

OVERVIEW

Laboratories present a unique set of problems from an air distribution perspective. Most laboratories require large volumes of supply air but must also maintain low room air velocities for fume hoods to operate correctly and provide a comfortable environment for laboratory occupants. The total airflow requirements for a laboratory are determined by the most critical of the following criteria:

- Total amount of exhaust from containment and exhaust devices
- Cooling required to offset internal heat gains
- Minimum ventilation requirements

In the development of laboratory systems, the proper layout and design of the primary supply and exhaust air systems is critical in ensuring the performance and safety of the laboratory environment. When designing the primary air systems, complex and often conflicting system options must be evaluated. These include the type of fume exhaust hoods used, special requirements for exhaust and supply duct systems, the locations and types of air distribution devices, and the first and operating costs of any proposed system. A key step in the layout and design of the laboratory system is the selection and sizing of the primary air systems. However, prior to the design of these components (air-handling units, etc.), the zone air distribution and heating system must be determined.

The main reason the design of the primary air systems is started at the zone level is that for these systems to perform properly and efficiently, the conditions in the occupied space must first be met. Therefore, for the selection and design of the primary air systems, the following should be evaluated in this order:

1. Zone air distribution
2. Zone heating
3. Exhaust air system
4. Supply air system
5. Duct construction
6. Energy efficiency

ZONE AIR DISTRIBUTION

The distribution (supply) of air into a laboratory, relative to the exhaust air requirements (based on fume hoods, snorkels, etc.), determines the integrity of hood containment and ensures proper pressurization control (direction of airflow) with respect to adjacent nonlaboratory (administrative) zones and spaces differing in cleanliness. Therefore, the locations of diffusers, the discharge velocities, and the volume control of the exhaust and supply air for a room are all critical to maintaining a safe and comfortable work environment. The quantities of supply and exhaust air, as well as the integrity of the room envelope and the influence of outside forces (wind and stack effect), determine the air movement between spaces.

Therefore, with the total supply (including auxiliary air) and exhaust air volumes known for a space, the next question is how to properly supply and exhaust the air to maintain safe and comfortable conditions within the space.

Special techniques and diffusers are often needed to introduce the large air quantities required for a laboratory without creating disturbances on the laboratory benchtops and at exhaust devices. The key to maintaining proper air movement within a laboratory is the proper selection and location of supply, return, and exhaust air devices. When designing laboratory air systems, it is important to make sure that the following are evaluated and their interactions with one another are thoroughly understood:

- Room air velocities
- Relative device locations
- Device types
- Off-peak loads and balancing

ROOM AIR VELOCITIES

Air supplied to a laboratory must be distributed properly to keep temperature gradients and air currents to a minimum. Air outlets (preferably nonaspirating diffusers) must not discharge into the face of a fume hood, a biological safety cabinet (BSC), an exhaust device, or sensitive laboratory equipment. To ensure proper containment of pollutants within a laboratory hood, it is necessary to maintain proper velocities at the entrance (face) of the hood. Disturbances in the room air motion near the face of a hood can result in entraining pollutants out of the laboratory hood into the occupied space. Therefore, it is crucial that the supply, return (if any), and general exhaust systems do not adversely affect the air velocities at laboratory hood faces.

The greatest adverse influence on air motion near a laboratory fume hood is from the supply air system through the supply air diffusers. Because of the typical requirement of 100 fpm (0.51 m/s) face velocity for a laboratory fume hood, it is recommended that the maximum supply air velocity within the occupied space, below approximately 7 ft. (2.1 m) above the floor, be 50 fpm (0.25 m/s) (DiBerardinis et al. 2013). In general, a supply air velocity of 50% of the required hood face velocity should be maintained within the occupied space. However, some experts recommend that within the frontal area of a fume hood or BSC, the room supply air velocity should be no more than 20% of the required hood velocity to maintain containment in the hood (Wunder 2000).

RELATIVE DEVICE LOCATIONS

The placement of supply air diffusers relative to laboratory fume hoods and to other diffusers determines the airflow within the room. Typically, it is best practice to place diffusers as far from hoods as possible. There are two primary reasons for this general rule.

First, when a diffuser is located too close to a hood, it is likely that there will be insufficient distance between the diffuser discharge and hood face for the air velocity to reduce to an acceptable value, thus creating drafts and compromising the containment of the hoods. Second, if diffusers are located too near a laboratory fume hood, it is possible that the ventilation of the general laboratory space will be hampered, as the supply air could short-circuit and be exhausted out of the hood.

Air currents external to the fume hood can jeopardize the hood's effectiveness and expose the researcher to materials used in the hood. Detrimental air currents can be produced by

- air supply distribution patterns in the laboratory,
- movements of the researcher,
- people walking past the fume hood,
- thermal convection, and
- opening of doors and windows.

Caplan and Knutson (1977, 1978a, 1978b) conducted tests to determine the interactions between room air motion and fume hood capture velocities with respect to the spillage of contaminants into the room. Their tests indicate that the effect of room air currents is significant and of the same order of magnitude as the effect of the hood face velocity. Consequently, improper design and/or installation of the replacement air supply can lower the performance of the fume hood. Disturbance velocities at the face of the hood should be no more than one-half and preferably one-fifth of the face velocity of the hood. This is an especially critical factor in designs that use low face velocities. For example, a fume hood with a face velocity of 100 fpm (0.51 m/s) could tolerate a maximum disturbance velocity of 50 fpm (0.25 m/s). If the design face velocity were 60 fpm (0.30 m/s), the maximum disturbance velocity would be 30 fpm (0.15 m/s). To the extent possible, the fume hood should be located so that traffic flow past the hood is minimal. Also, the fume hood should be placed to avoid any air currents generated from the opening of windows and doors. To ensure optimum placement of fume hoods, the HVAC system designer must take an active role early in the design process.

Test results (Amon et al. 2007) indicate that good hood design, selection of appropriate diffusers, and proper laboratory design are imperative for proper performance. The results also indicate that numerous variables affecting performance are beyond our current understanding. The data suggest that locating supply air diffusers to the side, rather than directly in front of, fume hoods has the least impact on hood performance, confirming that cross drafts greater than 50% of the average face velocity can impair hood performance.

DEVICE TYPES

The space velocity and diffuser location requirements make the selection and application of the room supply air system the key to the safety and comfort of the occupants. Due to the large volume of supply air being introduced, conventional outlets that typically discharge the supply air at higher velocities and high throw may not be the best selection. When air patterns from adjacent diffusers meet due to the relatively long throws at high air volume, the pattern is deflected down into the occupied zone, causing draft and discomfort. To understand what type of air distribution systems (diffusers) are commonly used in laboratory systems, a basic knowledge of air distribution fundamentals is required.

With the introduction of a jet of air into a large, relatively stagnant space, the disturbance of the air jet is well defined. As shown in Figure 6-1, the jet enters at a high velocity and dissipates with distance. This is a result of the induction of room air into the jet.

The higher the jet velocity, the greater the induced air and turbulence in the space. Conversely, the lower the velocity, the less the induced air and turbulence. However, there is still the concern of occupant comfort. If the air velocity is too high, the occupants may complain of drafts, and if too low, they may complain of stuffiness. Therefore, the proper selection of supply air diffusers must address the velocity near hood faces, the general mixing of room air and supply air, and the local velocity of room air. It is the combination of these requirements with the number and type of diffusers that determines the final room air motion.

From Figure 6-1 it is obvious that a diffuser producing the represented characteristic of nondirectional flow is poor for laboratories because of its relatively turbulent nature. To minimize the turbulence within the space, the diffuser supply velocity needs to be reduced. This can be accomplished through the addition of multiple diffusers or the use of a diffuser with multiple orifices. While each orifice has the same general performance characteristics, having multiple orifices reduces the jet velocity and a more unidirectional (laminar) flow is achieved. Figure 6-2 illustrates this principle.

To achieve the low-velocity, high-volume air requirements typical in most laboratories, several laminar-flow air distribution systems have been developed:

- Perforated duct
- Perforated diffusers
- Perforated ceiling panels
- High-capacity radial

Figure 6-1
Generic Air Jet Characteristics

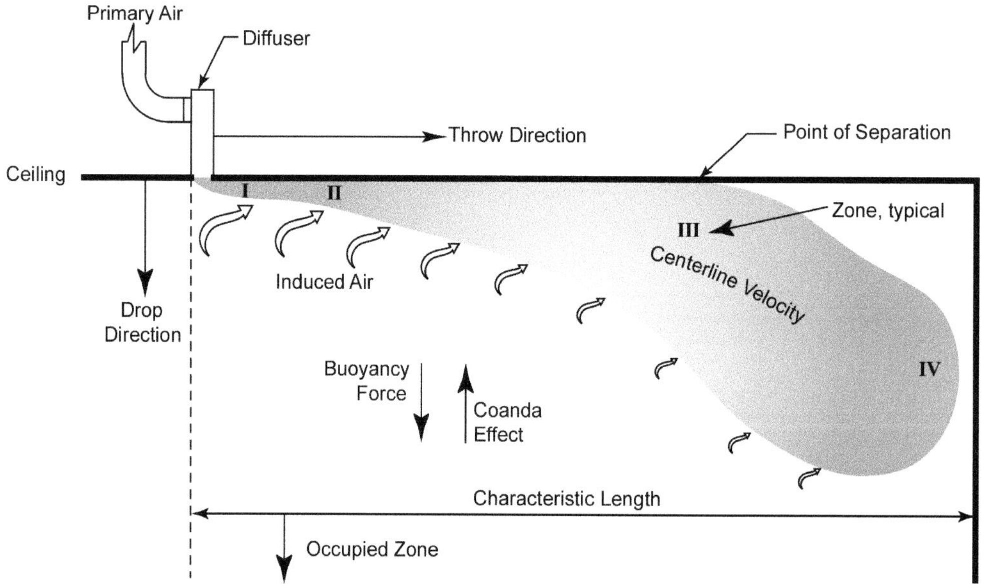

$$\text{ADPI} = f\left(\frac{T_n}{L}\right)$$

where:

T_n = Throw at centerline velocity n, ft (m)
n = Centerline velocity, throw typically 200, 150, 100, and 50 fpm (1, 0.75, 0.5, and 0.25 m)
L = Characteristic length, ft (m)

Note: Figure is for illustrative purposes only

These options are shown in Figure 6-3. It is important to understand that the systems shown in Figure 6-3 are specialized for critical environments and are separate product lines from those of diffuser manufacturers. Though these types of supply systems offer the best characteristics for safety and comfort in laboratories, they are not always practical due to cost limitations.

To select a system under these circumstances requires returning to basic air distribution principles. Figure 6-4 shows a typical sectional view of a laboratory, and Figure 6-5 presents a plan view with airflow patterns indicated.

The key to diffuser selection is that the throw of the diffuser where the velocity is above 50 fpm (0.25 m/s), designated as T_{50}, should not enter the area near the fume hoods. Since most air distribution systems are laid out symmetrically, the T_{50} envelope should not enter the occupied space. Although this criterion appears simple to meet, it is complicated by the fact that sufficient velocities are required to thoroughly mix the room air for comfort. Table 6-1 lists the T_{50} throw divided by diffuser characteristic length, *L*, required for comfort.

Figure 6-2
Nondirectional versus Laminar Flow

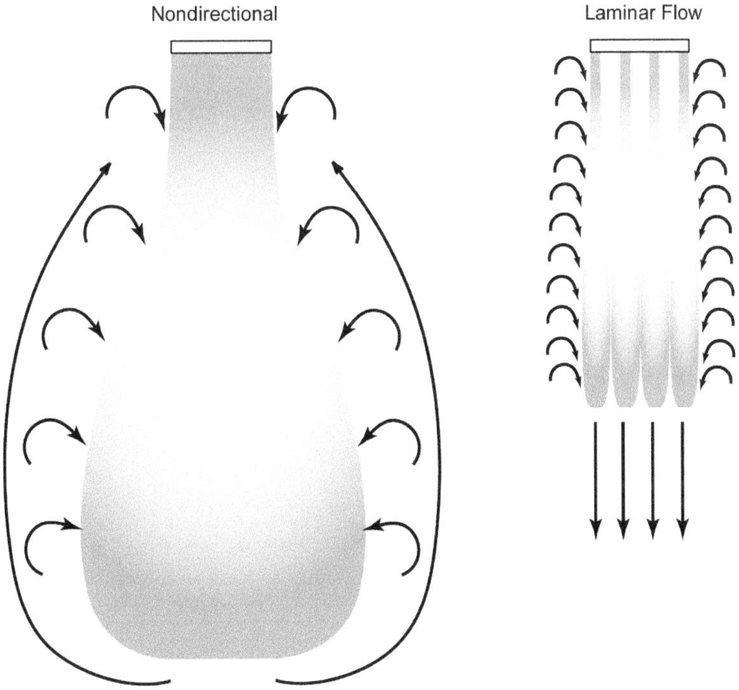

Figure 6-3
Perforated Supply Systems

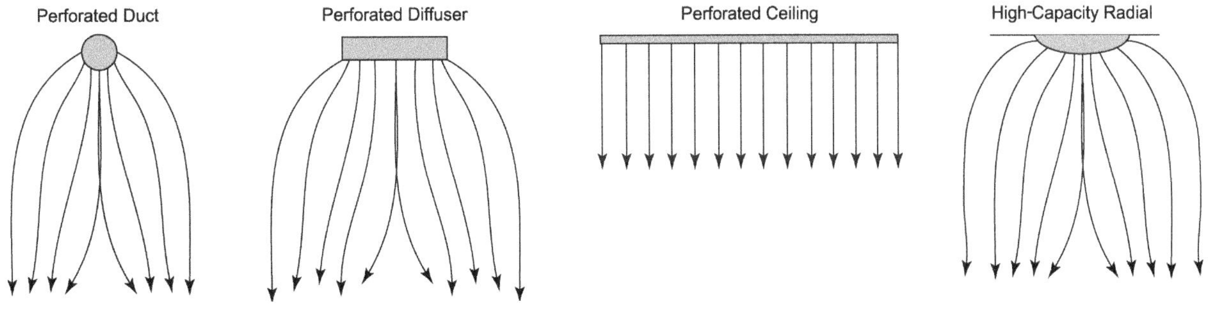

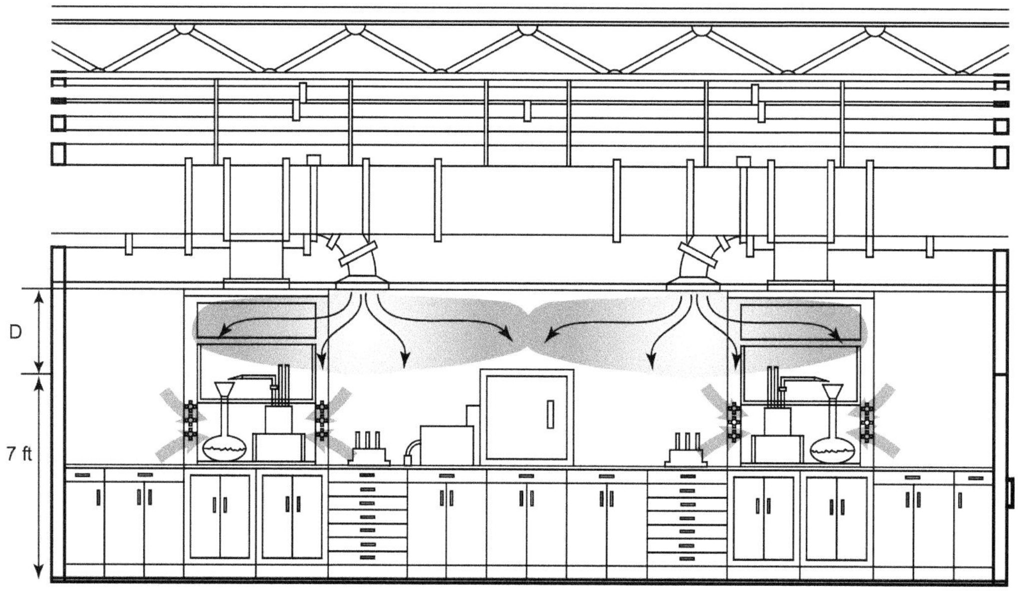

Figure 6-4
Sectional View

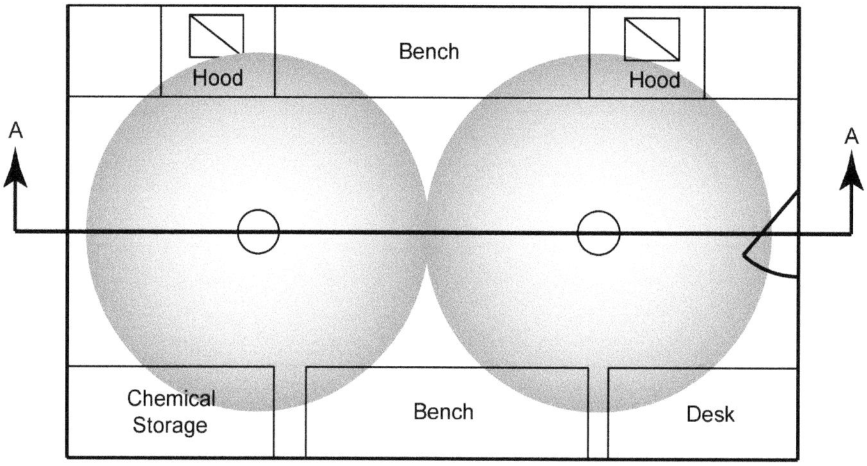

Figure 6-5
Plan View

Table 6-1
T_{50}/L Ranges

Diffuser Type	T_{50}/L Range
Sidewall	1.3 to 2.0
Ceiling—Round Pattern	0.6 to 1.2
Ceiling—Cross Pattern	1.0 to 2.0
Slot	0.5 to 3.3

For any diffuser, a T_{50}/L greater than unity will result in the throw being greater than 50 fpm (0.25 m/s) at walls and within the occupied space. Therefore, the use of sidewall and ceiling cross-pattern diffusers must be considered and evaluated carefully. Detailed data on the specific diffusers to be used should be analyzed.

While T_{50}/L addresses the horizontal direction, of more concern is how far down the 50 fpm (0.25 m/s) velocity profile extends into the occupied space. Unfortunately, most manufacturers' catalogs do not provide drop information for diffusers, which must either be obtained from the manufacturer directly or be physically verified by testing prior to selection of a diffuser. However, even if T_{50} falls within the occupied space, the diffuser

Figure 6-6
Layout of Laboratory and Diffuser Selection

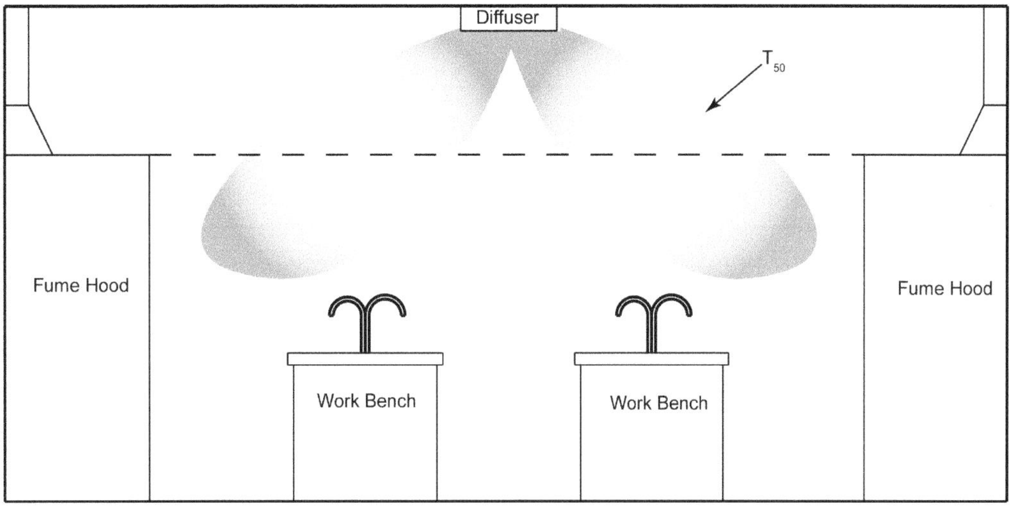

Figure 6-7
Misapplication of Diffuser Layout

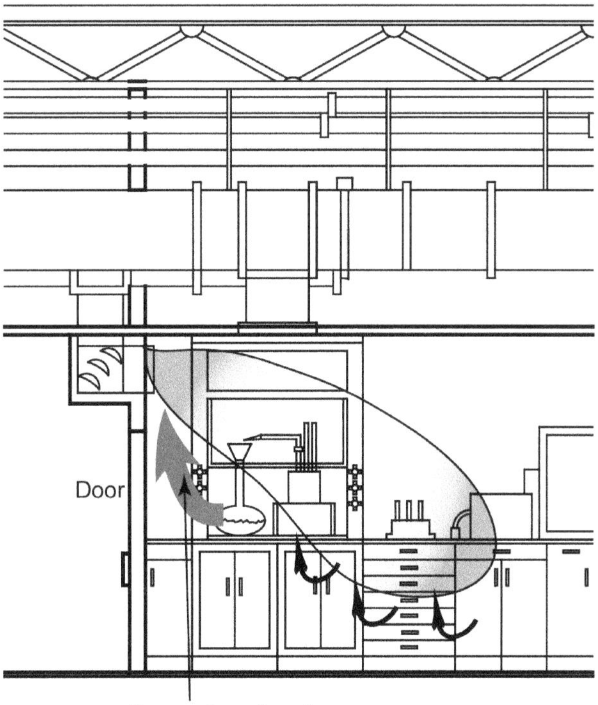

may still work. Figure 6-6 shows an example of one such instance. As Figure 6-6 illustrates, the diffuser can be used if the T_{50} line does not reach the hood face. Care must be taken in these instances.

Other situations, such as that shown in Figure 6-7, must be avoided. In this instance, the sidewall diffuser application results in higher-than-acceptable velocities and turbulence near the hood face due to the vane position and the characteristics of the sidewall diffuser.

One appropriate method for distributing air within laboratories uses radial-flow diffusers, supplying air in multiple directions and flushing the room with large volumes of low-velocity, clean, conditioned air. The low velocity and high volume of these diffusers allows them to be located closer to fume hoods and exhaust devices than conventional diffusers. The low velocity and short throw of the discharge from these diffusers greatly reduces the chance of the supply air disrupting the flow of air into the fume hoods and exhaust devices in the laboratory. The short throw of the radial-flow diffuser also reduces the possibility of creating drafts within the occupied zone.

Regardless of the type of diffuser or air distribution system chosen, substitutions must NOT be allowed during construction without thorough analysis and approval of the change. This must be accomplished for every laboratory space and every unique diffuser.

Finally, it is recommended that nonlocking adjustable louvers not be allowed on diffusers. When designed properly, the louvers typically only provide proper air distribution and safety in one position. However, if used, the positions of the adjustable louvers may change due to occupant or operation and maintenance (O&M) personnel intervention, usually resulting in more problems, not fewer.

Advanced laboratory modeling techniques, such as computational fluid dynamics (CFD) simulations, are increasingly being used to help determine the laboratory's airflow characteristics in order to optimize the ventilation rate. CFD can be used to analyze various dynamic scenarios within laboratories, including the modeling of spills and their impact on fume hood containment, comparing the performance of different ventilation schemes, and analyzing the performance of energy-efficient technologies such as active or passive beams.

OFF-PEAK LOADS AND BALANCING

The final step in the design of a zone distribution system is to ensure that the room air motion is maintained under all system conditions. The primary concern is with variable-air-volume (VAV) systems. However, there are also fluctuations in constant-volume systems because of changes in processes and equipment. Therefore, the supply air diffuser performance (throw and drop) must be verified for minimum flows.

The last consideration is to make sure adequate dampers and test ports are available for balancing of the system. Of critical importance is the requirement for straight exhaust duct from each laboratory hood at least 10 duct diameters downstream and 5 duct diameters upstream of any obstruction. This is required so that accurate exhaust flow measurements are possible during balancing.

Once installation of all air distribution systems has been completed, the system must be properly balanced. This work should be accomplished by an independent contractor with the aid of a commissioning authority to ensure that the balancing is done properly and that any errors in the system are resolved. Air balancing must be completed to ensure that the volume difference between supply and exhaust airflows is maintained at peak and nonpeak conditions, that the specified airflow from diffusers is within acceptable volumes and velocities, and that comfort conditions are maintained.

For more information on air distribution and balancing in laboratories, refer to Chapter 12.

ZONE HEATING

Because the exhaust requirement in a laboratory often exceeds the volume of supply air required for comfort, heating is often required all year. However, the need for heat focuses on avoiding comfort problems from cold exterior walls during winter and not overcooling the space during cooling periods.

It is important to understand two different heating loads. The first one, from heat loss through the building envelope, is identical to any other building and is not unique to laboratories. It is important to recognize that this heating load only affects the perimeter and not the interior zones. To meet this load, sufficient heat must be provided to the room to offset the loss through the building envelope. This heat can be added to the supply airstream or provided through a separate system.

The second heating load (termed *reheat*), found in many VAV systems because excess supply air is required, is magnified in laboratory systems due to higher air exchange rates. Although the heat can be provided via a second system to meet this load, heating the supply system is often most economical and sensible.

The three main options for providing heat to individual spaces are as follows:

- Baseboard heating
- Radiant panel heating
- Supply air heating

The sources of heat for all of these options are hot water, steam, or electricity. Figure 6-8 details these systems.

BASEBOARD HEATING

Baseboard heating is typically applied along exterior walls to offset the heat loss to the outdoors during cooler periods. The primary advantages of baseboard heating are that it is easily zoned and it has few moving parts to maintain. However, there are several disadvantages of using perimeter baseboard heaters within a laboratory. First, control of space conditions may not be suitable for laboratory applications because of humidification and temperature control issues. For instance, many laboratories require that the space environment be maintained in a consistent state for experimentation consistency. Unfortunately, the response time of perimeter baseboard heating is typically much slower than is acceptable for laboratories, making it difficult to maintain the room temperature setpoint within specified limits. Finally, maintenance and cleaning of these systems can be difficult and expensive when decontamination is necessary.

One consideration of perimeter baseboard heating systems is the relationship they must have to laboratory casework and equipment. Generally, baseboard heating protrudes approximately 2 in. (50.8 mm) from the wall and cannot be blocked by any objects to ensure proper heating performance. Thus, use of this type of system requires careful planning and involves loss of perimeter space, which is very valuable in many laboratories.

Figure 6-8
Zone Heating Options

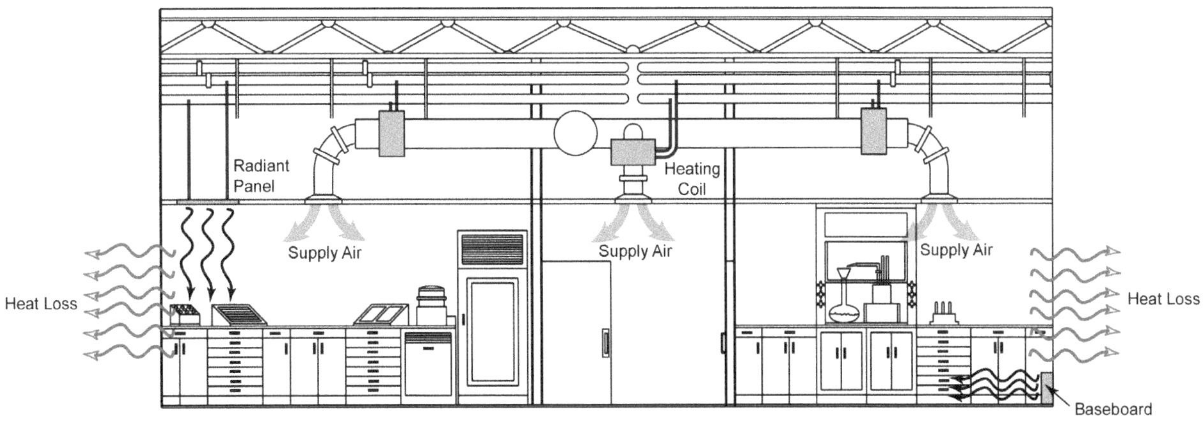

RADIANT PANEL HEATING

The second option for zone heating is the use of radiant panels. Radiant panels generally come in five configurations:

- Metal ceiling panels
- Embedded piping
- Electric ceiling panels
- Electrically heated ceiling or floor
- Air-heated floor

Similar to perimeter baseboard systems, all of these configurations use the principle of free convection, with the addition of radiant heating. These systems do not require any laboratory floor space since they are integrated into the infrastructure of the laboratory. The radiant panels are made of various materials with steam, hot-water piping, or electrical resistance elements passed through them. These panels can then be located within floors, walls, or ceilings. Unfortunately, with radiant panel systems, temperature control is often slow when heavier materials such as concrete are used, due to the storage effect. Another temperature-related problem for radiant panels is the uneven distribution of heat around objects.

SUPPLY AIR HEATING

The final heating option involves the use of a central air distribution system to heat the zones. Typically this option is the better choice in laboratory applications because of its quick response times in temperature control and its even heating distribution.

In using electric heat for terminal reheat in the supply air, care must be used in selecting the heater size and the minimum turndown on the supply air quantity prior to activating the heat source. The heater must be sized to reheat all of the supply air when the exhaust devices are at maximum flow in order to maintain the proper room pressure relative to the surrounding nonlaboratory space. Electric resistance heaters typically have a minimum air velocity that must be maintained for a given power density in the duct to avoid overheating, which would create a safety hazard. With the high reheat loads plus heating loads to offset building envelope losses, the velocity requirements may limit how low the supply can be turned down when the laboratory fume hoods are closed. Proper lengths of straight duct before and after electric duct heaters must also be maintained to provide uniform air distribution over the heating coil. Locating a duct branch take-off too close to the heater can cause turbulent airflow at the heater and trip protective safeties. Hot-water heating provides more flexibility in lowering the amount of supply air that needs to be reheated and therefore saves energy.

EXHAUST AIR SYSTEM

With the room layouts completed and the exhaust airflow requirements known and the fume exhaust hoods chosen (see Chapter 5), the primary exhaust air system can be selected and sized. Because the exhaust system in a laboratory facility is often the most critical system for ensuring the safety and health of the occupants, its design is accomplished first. Further, because most laboratories do not allow recirculation of air (100% ventilation with outdoor air), the focus of this section is on 100% exhausted spaces. However, in cases where recirculation is allowed (e.g., adjacent nonlaboratory spaces), the guidance for general exhaust can be followed.

A plan view of a typical laboratory module is shown in Figure 6-9, and Figure 6-10 shows the supply and exhaust systems for a typical laboratory module. The air systems within the typical laboratory module include the following:

- Dedicated hood exhaust
- Benchtop localized snorkel exhaust
- General exhaust
- Room supply
- Auxiliary supply

Exhaust systems are addressed in this section and supply systems in the following section.

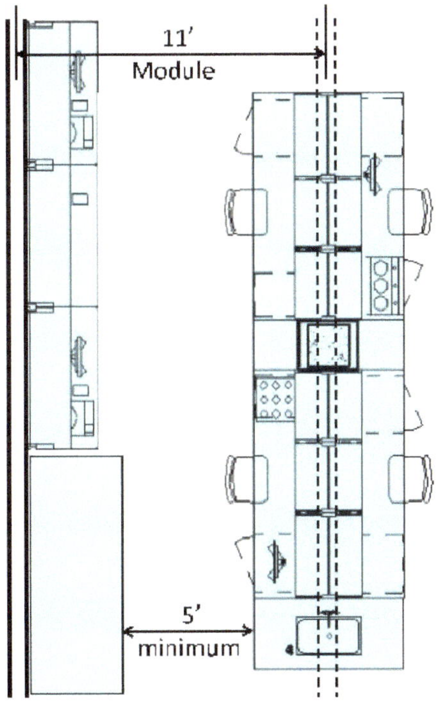

Figure 6-9
Typical Laboratory Module

(NIH 2013)

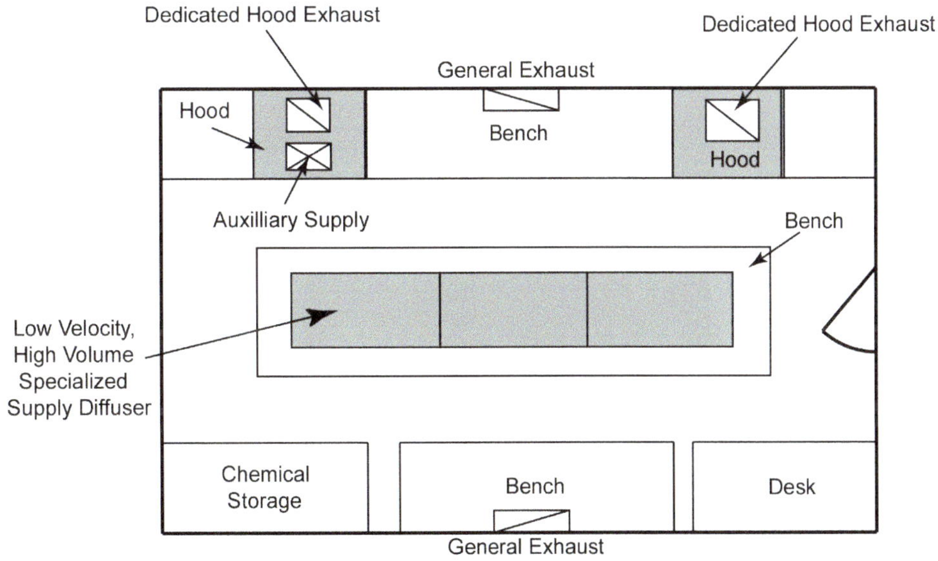

Figure 6-10
Air Systems in a Typical Laboratory

The exhaust systems must remove air from the various containment devices and from the laboratory itself. The exhaust systems must be controlled and operate in conjunction with the supply system to maintain the required room pressure relationships with the surrounding spaces and minimum air change rates within the laboratory. In a typical laboratory, seldom are all the exhaust devices in full operation at the same time. Energy conservation and lower equipment capacities may be possible if the system can be sized for the maximum simultaneous expected usage, allowing for some diversity. This can be determined through consultations with the owner and research staff.

A laboratory space can have any combination of general (room and uncontaminated exhaust sources) and contaminated air (hood, BSC, snorkel, glove box, etc.) exhaust:

- Only general exhaust
- Only contaminated air exhaust
- General and contaminated air exhaust

Application of the various exhaust points within a laboratory is dictated by the safety and health codes and standards detailed in Chapter 4. As a review, fume hood exhaust is typically required when dangerous, toxic, or carcinogenic materials are being handled or used in such a manner as would be dangerous for human contact. BSCs are used for safely working with materials contaminated with (or potentially contaminated with) pathogens requiring a defined biosafety level (BSL) (BSLs are described in *Biosafety in Microbiological and Biomedical Laboratories* [CDC 2009]). Snorkels are used to remove heat or nontoxic particles that may be generated from benchtop research equipment. Unlike a chemical hood, a glove box is fully enclosed and is under negative or positive pressure. Glove boxes are usually small units that have multiple openings in which arm-length rubber gloves are mounted. The operator works inside the box by using these gloves.

General exhaust is used for less toxic, less odorous substances and procedures, such as chemical storage, experiment preparation, and cleanup. General exhaust is also used to minimize the impact that heat-producing equipment (e.g., ovens) has on the occupant by removing the heat prior to its introduction into the occupied space. Typically, canopy hoods are used to remove heat or moisture generated by a specific research apparatus (e.g., steam sterilizer) or process. Canopy hoods cannot contain hazardous fumes adequately to protect the researcher.

The design and layout of an exhaust system for a laboratory facility is accomplished through a series of steps:

1. Determine type of exhaust system.
2. Determine separation of exhaust system.
3. Accomplish duct layout and design.
4. Accomplish fume exhaust stack design.
5. Select fume exhaust fans.

DETERMINE TYPE OF EXHAUST SYSTEM

Criteria for fume hood control differ depending on the type of hood. The exhaust volumetric flow is kept constant for standard, auxiliary air, and air-bypass fume hoods. In VAV fume hoods, the exhaust flow is varied to maintain a constant face velocity as the sash position changes. Selection of the fume hood control method should be made in consultation with the health and safety officer.

Constant air volume is required for specific systems (perchloric acid fume hoods, radioisotope fume hoods, and Class 2 Type B BSCs) and is typically used in small installations with few hoods to simplify the installation and operation and to minimize the required capital investment. These exhausts present unique challenges due to contamination control issues or large pressure drops across filters. It is standard practice to provide separate hoods and ductwork for perchloric acid and radioisotope hoods. Biological contaminants are normally provided with a separate exhaust system due to high-efficiency particulate air (HEPA) filtration requirements.

In the constant-volume exhaust system (see Figure 6-11), the volume of air exhausted from a laboratory room or hood is constant regardless of sash position or activity within the room. Typically, the constant airflow is an inherent characteristic of the exhaust hood.

Constant-volume fume hoods can be classified as either pressure-dependent or pressure-independent systems. Although simple in configuration, the pressure-dependent system is unable to adjust the damper position in response to any fluctuation in system pressure across the exhaust damper. In a pressure-independent system, each exhaust device is provided with an airflow control valve, or *air valve*. Figure 6-12 details the key components of a generic airflow controller. See Chapter 11 for more information on control systems and components.

In larger systems and where operating costs need to be minimized, VAV systems are applied (see Figure 6-13). Because the power required to exhaust the air is proportional to

Figure 6-11
Constant-Volume Laboratory Ventilation Control

(Courtesy of Siemens Building Technologies)

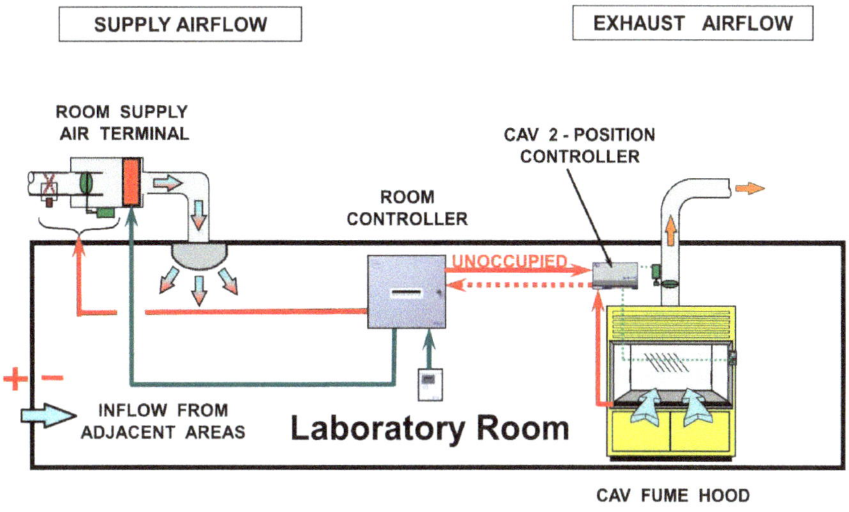

Figure 6-12
Generic Airflow Controller

(Courtesy of Siemens Building Technologies)

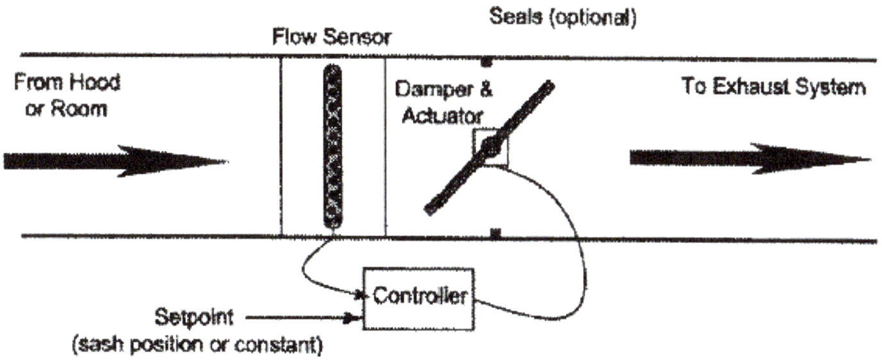

Figure 6-13
VAV Laboratory Ventilation Control

(Courtesy of Siemens Building Technologies)

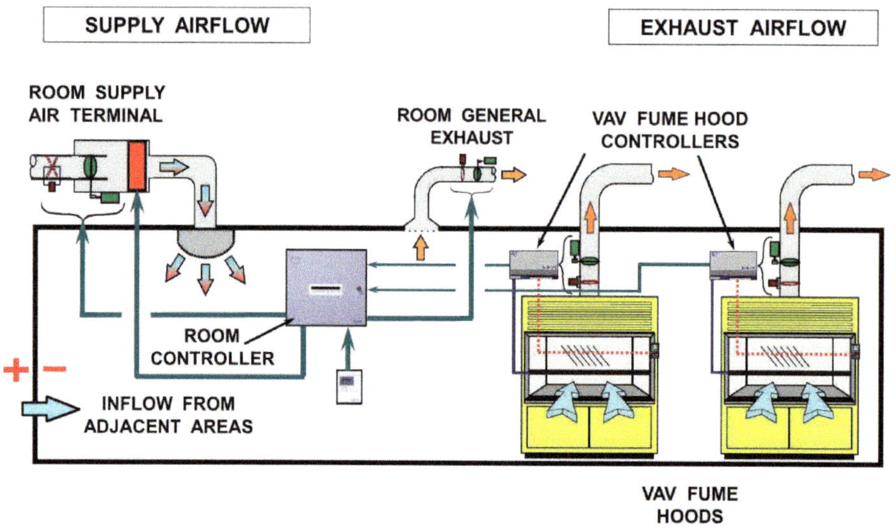

the cube of the airflow (see Equation 6-1), it is often very economical to operate a VAV exhaust system.

$$P_{new} = P_{old}\left(\frac{Q_{new}}{Q_{old}}\right)^3 \quad (6\text{-}1)$$

where

P	=	fan power, hp (kW)
Q	=	fan flow rate, cfm (L/s)
new	=	new operating point
old	=	old operating point

The primary difference between constant-volume and VAV systems is that in the VAV system each hood and space must have a flow controller attached to it and the fan must have some form of flow control. Ideally, the fan flow control would be through the use of a variable-speed drive (VSD) that is controlled to maintain a constant negative pressure at some point in the exhaust ductwork (typically two-thirds of the way down the farthest run). However, fans also have been installed to just "ride the fan curve." While there will be energy savings, the actual savings are approximately a third of what a VSD system attains. To maintain required exhaust stack discharge velocities and effective stack height, the controller may need to limit the speed reduction on the fan and open bypass dampers at the fan intake plenum that allow outdoor air to mix with the building exhaust in order to maintain.

DETERMINE SEPARATION OF EXHAUST SYSTEM

Exhaust systems within a building are separated for a number of reasons, including the following:

- Incompatible or hazardous substances in the exhaust air
- Type of exhaust (constant versus variable)
- Location of exhaust requirements

The key criterion for the separation of the exhaust systems is the types of substances found in the exhaust air. Several exhaust sources can be combined if they are from similar processes and classes of exhaust. For example, if all hoods within a facility are used for the purpose of teaching general chemistry with similar levels and types of chemicals, then the exhaust streams can be combined. However, the exhausts from BSL-1 and BSL-3 laboratories, for example, should not be combined due to the different treatment and handling procedures of the exhausts. A further consideration on whether two exhaust airstreams can be mixed is if chemicals in the two streams will react with one another. One such example is perchloric acid, a strong mineral acid commonly used as a laboratory reagent. Organic materials are highly likely to spontaneously combust if mixed with or contacted by perchloric acid. Under some circumstances, perchloric acid vapors form perchlorate salts in ductwork that can be flammable and highly explosive when subjected to heat, impact, or reaction with other specific chemicals. If hot perchloric acid is used in a vent system that does not have internal washdown capabilities, the salts can build up over time in the hood, baffles, filters, fans, ducts, and exhaust stacks.

The second criterion, constant volume versus VAV, typically dictates the separation of systems. The only time constant-volume and VAV exhaust streams are mixed is when the constant-volume source has a flow controller installed in it. It is also common in larger systems that the mixing of constant and variable airstreams is undesirable due to the use of one or both of the streams in heat recovery or treatment prior to discharging from the building.

The final criterion in determining the separation of exhaust systems is the physical location of the exhaust points. As discussed in Chapter 4, the layout of service corridors, or shafts, and mechanical rooms dictates which spaces and hoods can be physically grouped together.

Fire codes typically require compartmentation of the laboratory into multiple "laboratory units" or "control areas" with fire separations between units to avoid exceeding the limits on chemical storage and dispensing in excess of code limits on substances posing physical and/or health hazards. Exhaust ducts from each laboratory unit must be separately ducted to a point outside the building via fire-rated shafts. However, exhaust ducts from laboratory hoods and other exhaust systems within the same laboratory unit may be allowed to be combined within that laboratory unit.

ACCOMPLISH DUCT LAYOUT AND DESIGN

Typical Layouts

Fume hoods and BSCs can either be supplied with individual (dedicated) exhaust fans or have their exhausts manifolded and connected to a central exhaust fan. Based on the level of separation of systems, there are two typical layouts of exhaust systems in laboratory facilities:

- Dedicated
- Manifolded

Dedicated Exhaust System

In a dedicated exhaust system, each hood is provided with a separate duct, exhaust fan, and discharge stack (see Chapter 9). This type of exhaust system may be chosen due to a requirement for special exhaust filtration or treatment systems for specific applications, special ductwork and exhaust fan corrosion control needs, provision for selected emergency power backup, and simpler initial balancing.

Several disadvantages associated with dedicated exhaust systems are notable. Maintaining the correct flow at each exhaust fan requires periodic maintenance and balancing, as well as consideration of the flow rates with the fume hood sash in various positions. This arrangement also requires numerous roof penetrations and exhaust stacks, which can result in higher initial and operating costs and may complicate design and space availability in multistory applications. Also, a dedicated system can make the addition of hoods and alteration of laboratories costly, both in initial installation and later renovations.

It is less complicated to apply dedicated exhaust systems to single-story laboratory buildings. Exhaust fans dedicated to individual hoods are used to convey contaminated exhaust directly to an exhaust stack on the roof of the laboratory building. Figure 6-14 shows an example of a single-story laboratory facility with multiple dedicated exhaust systems.

In multistory buildings, the use of dedicated exhaust systems is more complicated since each duct must end up discharging at the roof level. Therefore, mechanical chases (shafts) must be provided to route the ductwork to the roof. Figure 6-15 details such a system.

Manifolded Exhaust System

The other option for the arrangement of hood exhaust is the use of manifolding. This type of system works by joining several hood or BSC exhausts into a common exhaust manifold, which is then routed to a single fan (or fan array) and discharge stack(s) on the roof. This type of system typically has several advantages, including the following:

- Lower capital and operational costs
- Fewer exhaust stacks
- Lower redundancy costs
- Greater adaptability of design
- Simpler effluent treatment
- Dilution and momentum
- Reduction in total exhaust requirement due to diversity of fume hood use

Using the information previously presented on which hoods can be combined, the layout of the exhaust system is simply a matter of convenience and location. Using the

Figure 6-14
Dedicated Exhaust Systems—Single-Story Building

(Reprinted with permission from Price Industries [Price 2006])

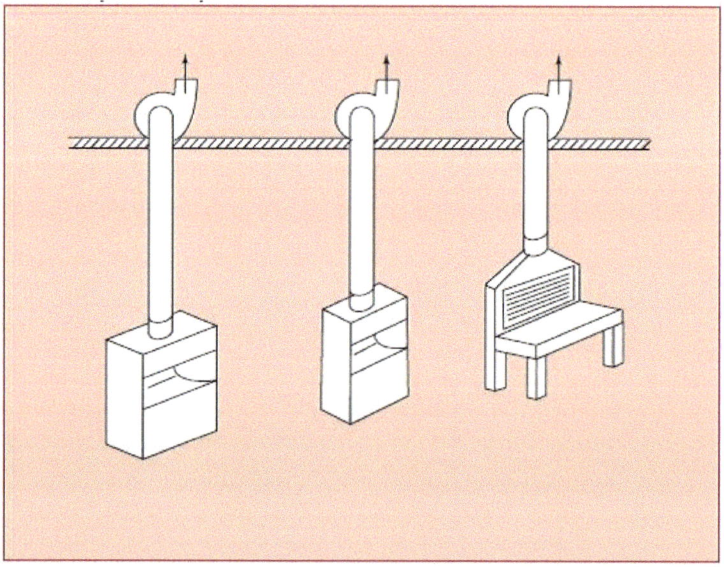

single-story and multistory examples from the dedicated system, Figures 6-16 and 6-17 detail the single-story and multistory layouts of a manifolded system.

A manifolded exhaust system can be classified as either pressure-dependent or pressure-independent. Pressure-dependent systems (see Figure 6-18) must maintain a constant exhaust volume by setting manual balancing dampers provided in the exhaust duct connection to each exhaust device. A major disadvantage of this type of system is the requirement to rebalance the entire system whenever an additional exhaust device is added to the system.

The pressure-independent system (see Figure 6-19) may be constant volume, VAV, or some combination of the two. Not only do pressure-independent systems have the flexibility to add additional exhaust devices without rebalancing the entire system, but also VAV control of the exhaust devices can be used and the exhaust fan size can be reduced by accounting for some diversity of hood use.

Figure 6-15
Dedicated Exhaust Systems—Multistory Building

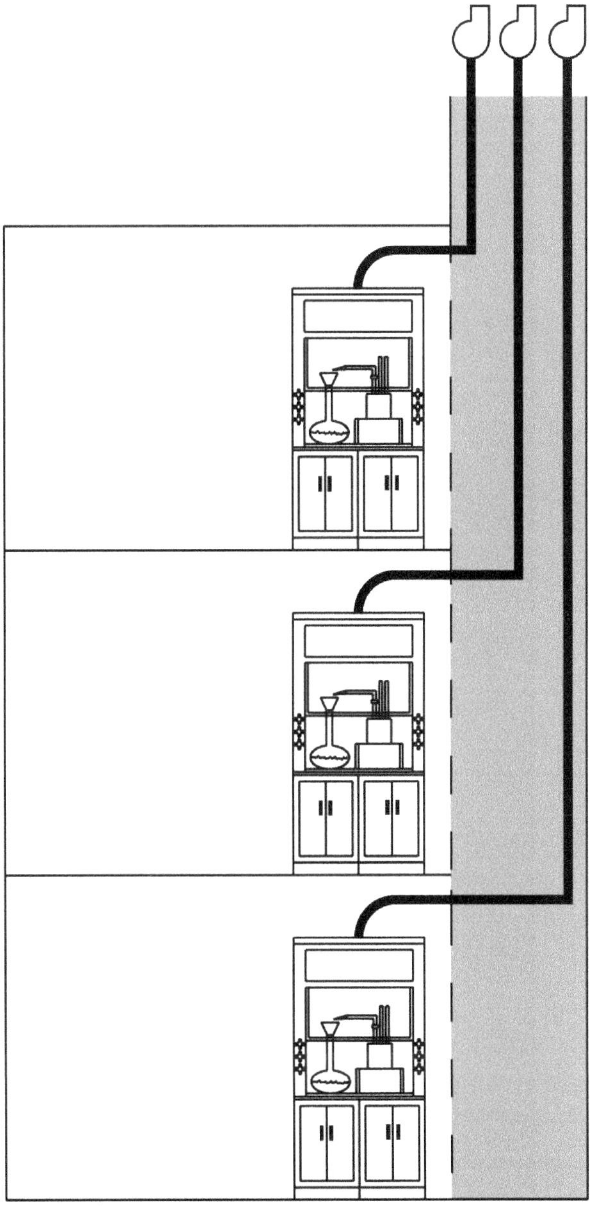

Figure 6-16
Manifolded Exhaust System—Pressure Dependent

(Reprinted with permission from Price Industries [Price 2006])

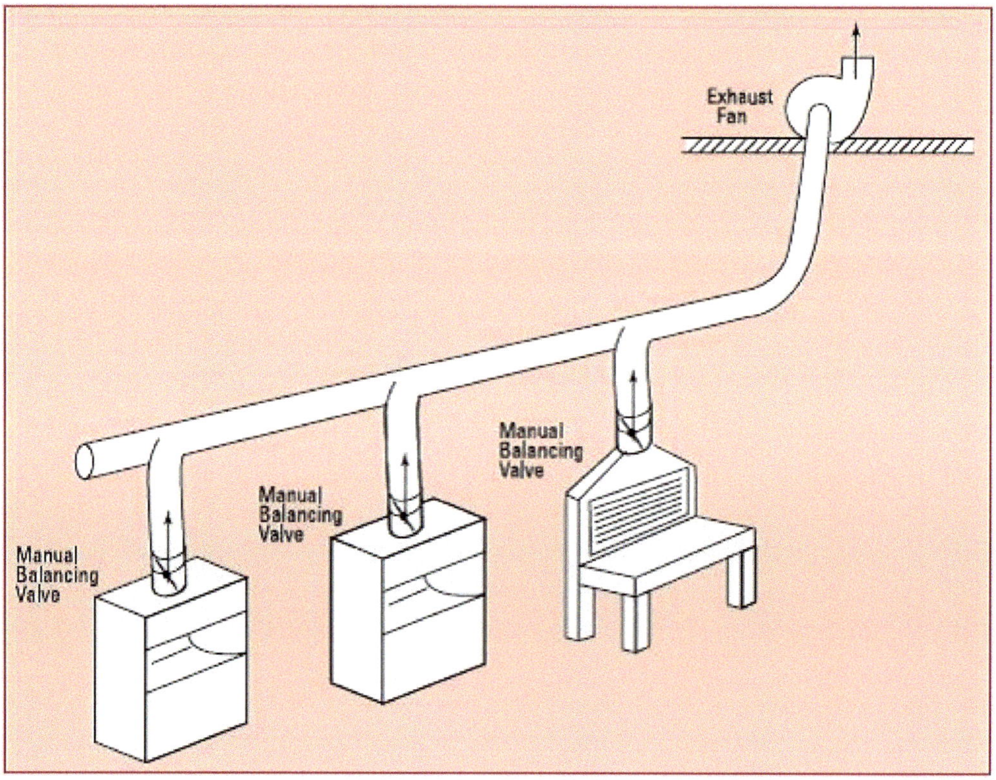

Figure 6-17
Manifolded Exhaust System—Pressure Independent

(Reprinted with permission from Price Industries [Price 2006])

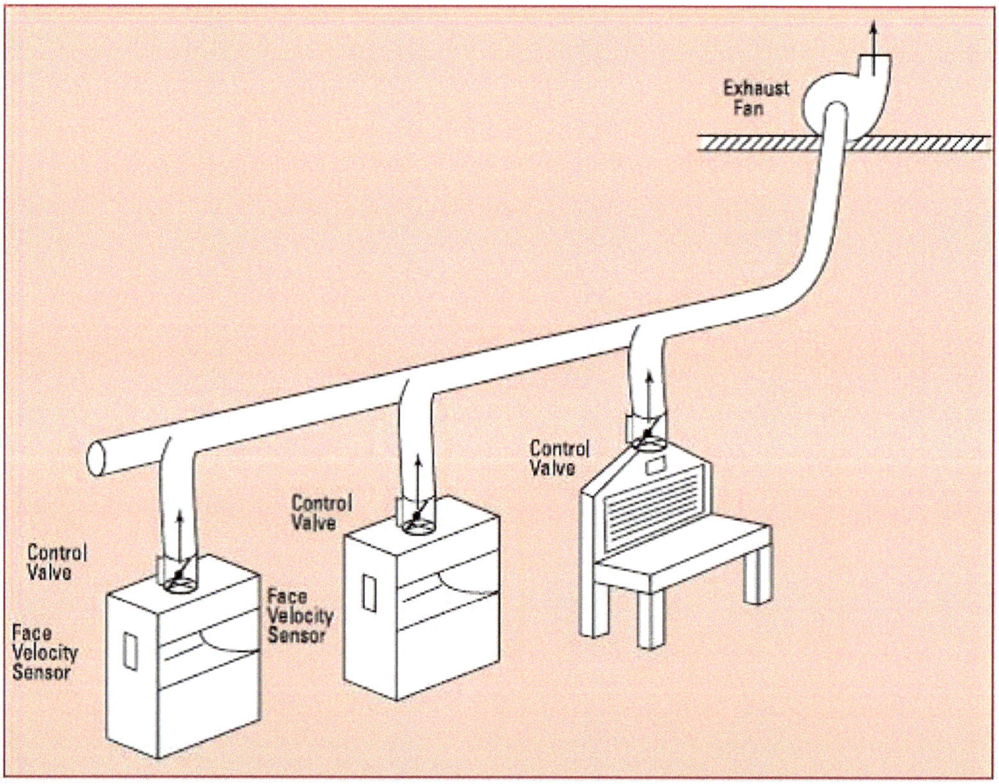

Figure 6-18
Manifolded Exhaust System—Single-Story Building

(Labs21 2007)

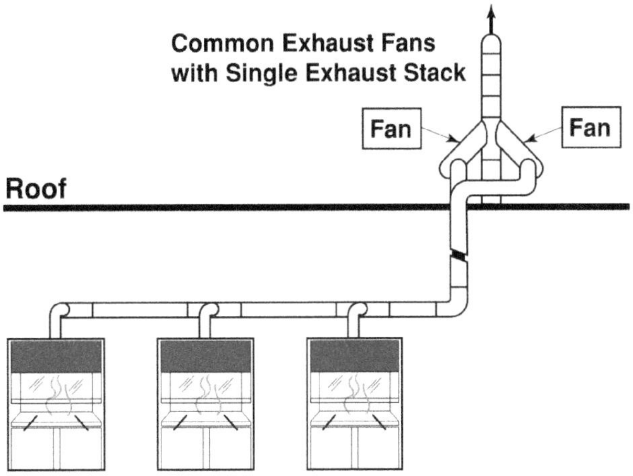

Figure 6-19
Manifolded Exhaust System—Multistory Building

(Labs21 2007)

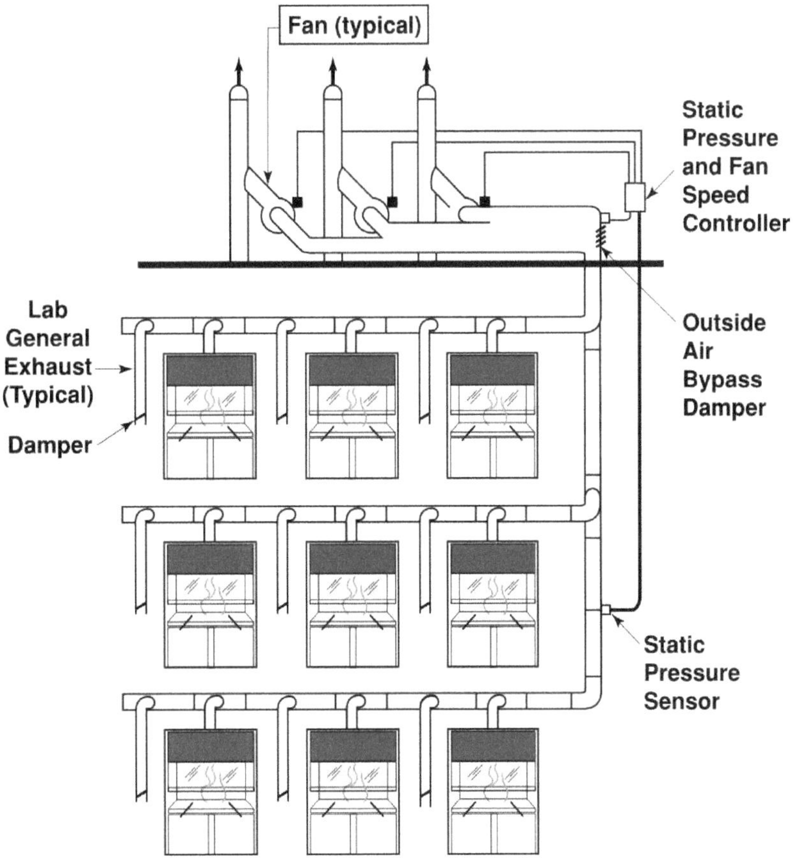

Duct Design

Duct Sizing

Laboratory exhaust duct airflow velocities and static pressures are dependent on numerous aspects of the exhaust materials they are designed to handle. Higher levels of pressure should be applied to laboratories that contain hazardous materials, such as BSL-2 and higher laboratories, to ensure containment. Lower levels of pressurization thus can be used in such instances as teaching laboratories and low-level chemical laboratories. In terms of transport velocities within ducting, these are typically in the range of 1000 to 2000 fpm (5 to 10 m/s) depending on the need to ensure that deposits do not form. This is particularly important in such applications as perchloric acid hood exhaust.

Duct Location

Duct mains should be located above the corridor outside the laboratory spaces, and airflow control devices should be located as close to the corridor wall as practical within the laboratory space being served when using volumetric offset control, in which the flow measurement and control of the system takes place at the perimeter of the space. This minimizes the exposure of the ducts upstream of the supply air terminals and downstream of the exhaust air terminals for which leakage flow is not measured or controlled. When both the supply and exhaust ducts cannot be located outside the laboratory space, locate the supply duct outside the space and the exhaust duct above the space. This arrangement allows exhaust duct leakage to help maintain the space under negative pressure. Other ducts crossing through the ceiling space above the laboratory should be sealed to minimize duct leakage, as this can affect space pressure relationships.

ACCOMPLISH FUME EXHAUST STACK DESIGN

Safe discharge of exhaust air that may contain harmful contaminants is a requirement for all laboratory buildings. For adequate dilution of the exhaust, it must be discharged either from a significant height or at a high vertical momentum. Detailed guidance on the design of exhaust stacks is provided in Chapter 9.

The general characteristics of good fume exhaust stack design are as follows:

- Location away from outdoor air intakes, building entrances, and operable windows
- Consideration of the prevailing and critical wind directions and wind speeds (those likely to result in the highest concentrations at nearby intake locations)
- Sufficient plume rise

SELECT FUME EXHAUST FANS

With the fume exhaust ductwork and stack designed, the exhaust airflow and static pressure losses can be calculated, and with these calculated, the exhaust fans can be selected. However, the proper selection of fans for a specific application goes far beyond just airflow and static pressure. Other items that must be considered include the following:

- Location
- Construction
- Controls
- Reliability/Redundancy

Like the entire design process, the selection and application of an exhaust fan is an iterative process, with each of the above criteria affecting one another.

Location

Although hazard levels vary within different types of laboratories, it is generally good practice to locate the exhaust fan on the roof of the building or within a roof-mounted penthouse enclosure. As is shown in Figure 6-20, the ductwork before the fume exhaust fan inlet is negatively pressurized and after the exhaust fan discharge is positively pressurized. Therefore, if the exhaust fan were to be located inside the building (below the roof level), the primary concern would be the potential of exhaust air leaking from the positively pressurized discharge ductwork downstream of the exhaust fan.

Figure 6-20
Exhaust Fan Pressure Relationships

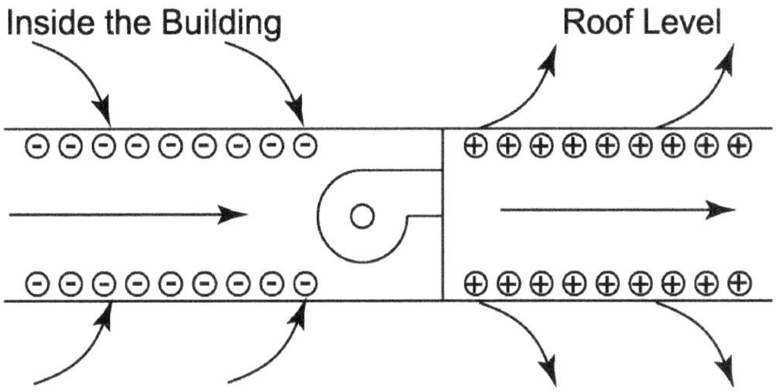

Figure 6-21
Centrifugal Fan Isometric

(ASHRAE [2012a], Chapter 21, Figure 1)

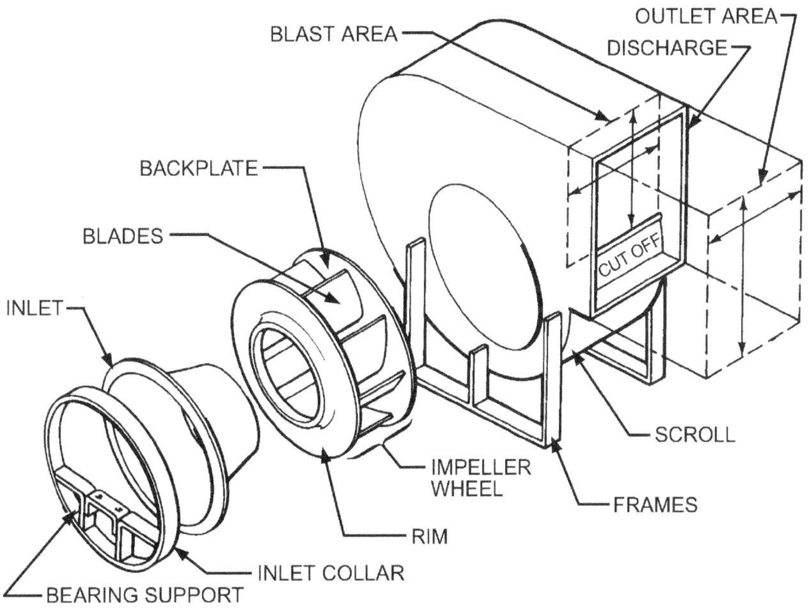

Locating exhaust fans on the roof, however, has a direct impact on the type of fan that can be used, because they must be able to withstand adverse weather conditions if located outdoors. Noise (acoustic) problems should be considered if the fan is to be located inside or outside the building.

Construction

The construction of a laboratory exhaust fan is dependent on several key criteria, including the following:

- Fan type
- Pressure class
- Material composition
- Protective coating
- Special considerations

Fan Type

The three generally preferred types of laboratory fume exhaust fans used are centrifugal, mixed-flow, and special-application fans. Centrifugal fans (see Figure 6-21) operate by a centrifugal force that rotates the inlet air and moves it from inside to outside of an impeller wheel. There is also kinetic energy imparted to the air by the virtue of the air's velocity leaving the impeller.

Direct-drive, rather than belt-driven, fans are generally preferred for laboratory exhaust systems because of their increased reliability, lower maintenance, and improved energy efficiency. Belt failure is a major cause of failure of exhaust systems with belt-drive fans. Major motor manufacturers' rated fatigue life for bearings for belt-drive applications is a third of the life of a typical direct-drive fan (L-10 of 50,000 hours versus 150,000 hours), thereby decreasing reliability and increasing maintenance. ANSI/AIHA/ASSE Z9.5 (AIHA 2012) requires stopping belt-drive fans on a monthly basis to check belt tension, sheave wear, and belt wear. Belt-drive losses can be 3% to 8% for fans over 10 hp (7.46 kW) per AMCA Publication 203 (AMCA 2011). Because these fans typically operate 24/7/365, the energy savings can be substantial for direct-drive fans.

For laboratory applications, centrifugal fans are very reliable, are easy to maintain, can be applied to a wide range of requirements, and have good pressure-volume characteristics. The three centrifugal fan types applied to laboratory systems are airfoil, backward curved, and radial. Table 6-2 lists the application of each of these fans, and Figure 6-22 shows the general performance characteristics.

Mixed-flow fans can be provided for outdoor applications. Mixed-flow fans are often used because the motor can be an integral part of the fan. Figure 6-23 shows a bifurcated mixed-flow fan. In-line mixed-flow fans are more compact than side-inlet centrifugal fans. A primary benefit of the mixed-flow fan is that it is located in line with the airstream and is a more compact design. Additionally, this avoids the pressure losses of the 90° turn required for centrifugal fans. The performance characteristics for a mixed-flow fan are shown in Figure 6-24.

Special-application fans include laboratory induced draft fans and dilution fans. Induced draft fans are used in perchloric acid fume hoods and other highly corrosive or high-temperature applications to avoid damage to the fan components and facilitate maintenance. A perchloric acid exhaust fan system (see Figure 6-25) should be separate from the laboratory's fume hood exhaust system.

Dilution fans, which induce outdoor air and mix it with fume exhaust to dilute the contaminants being exhausted, are designed as either centrifugal (Figure 6-26) or mixed-flow (Figure 6-27). The fans have vertical discharge, are provided with a windband, and may have a mixing intake plenum. FanFlow.avi, a video showing how the air flows from the building and mixes with outdoor air as it flows through an in-line mixed-flow induced draft dilution fan, is available with this Guide at www.ashrae.org/LabDG courtesy of Strobic Air Corporation. Because this type of fan is designed for outdoor application, the ratio of outdoor air to indoor air is adjustable using a bypass air damper. These fans can be designed in multiple sets and provided with sound attenuation. Fan exit velocities can be as high as 7000 to 8000 fpm (35.6 to 40.6 m/s), creating greater plume rise than a conventional fan operating at 3000 fpm (15.2 m/s), but this is done at the cost of higher energy consumption.

Table 6-2 Centrifugal Fan Type Application

(ASHRAE [2012a], Chapter 21, Table 1)

Fan Type	Performance Characteristics	Applications
Airfoil	Peak efficiency at 50% to 60% maximum volume, with maximum volume at peak efficiency	Low-, medium-, and high-pressure applications with relatively clean exhaust air
Backward Curved	Similar to but with lower efficiency than airfoil design	Good for corrosive environment where airfoil could corrode
Radial	Higher pressure characteristic but lower efficiencies	Material handling with high-pressure requirements—rugged wheel is sometimes coated and is simple to repair

Figure 6-22
Centrifugal Fan Performance Characteristics (P_{tf} = fan total pressure, η_t = mechanical efficiency of fan, W_o = power output of fan)

(ASHRAE [2012a], Chapter 21, Table 1)

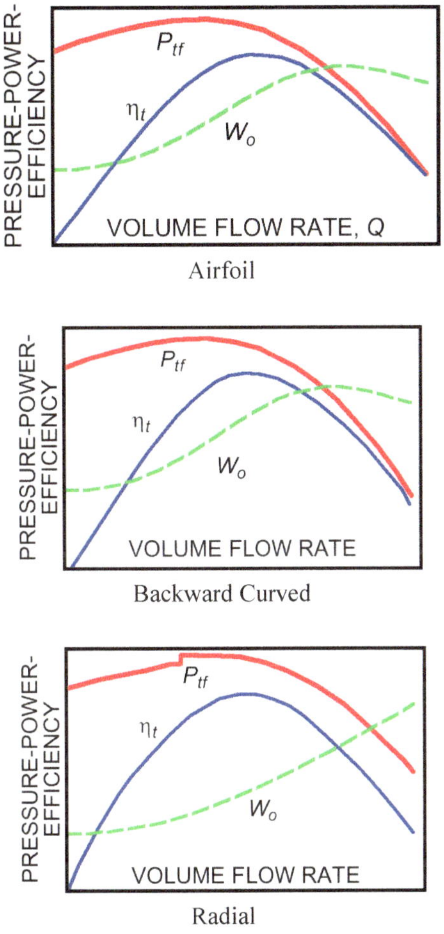

Figure 6-23
Direct-Drive In-Line Bifurcated Mixed-Flow Fan

(Courtesy of Greenheck)

Figure 6-24 Mixed-Flow Fan Performance

(ASHRAE [2012a], Chapter 21, Table 1)

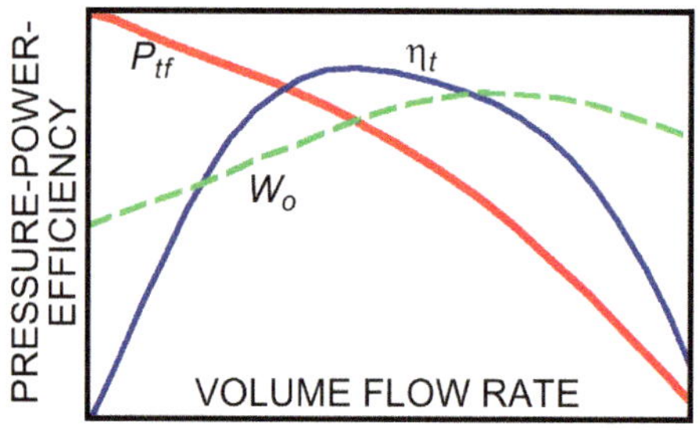

Figure 6-25 Fan-Powered Venturi for Perchloric Acid Fume Hoods

(Reprinted with permission from MK Plastics [2009])

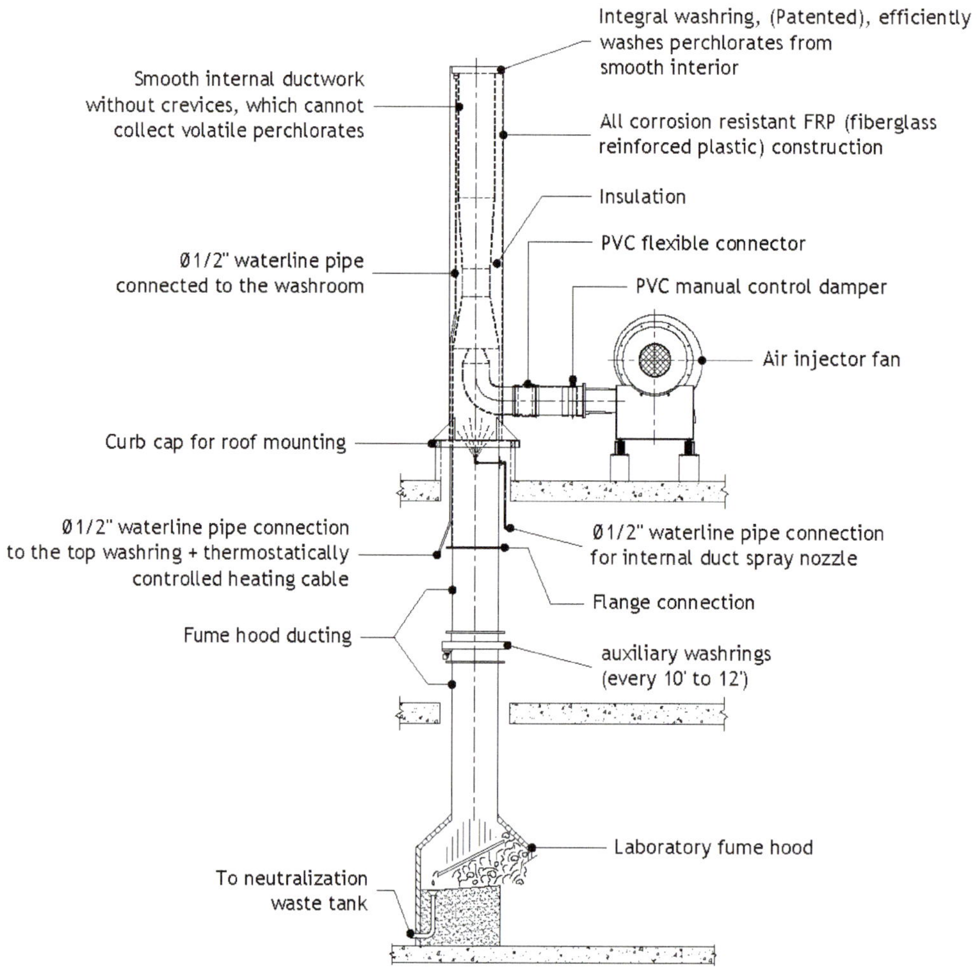

Regardless of the fan type used, the fan must meet the requirements of the Air Movement and Control Association International (AMCA) through ANSI/AMCA Standard 210|ANSI/ASHRAE Standard 51 (AMCA 2007) and those specified in *NFPA 70: National Electrical Code®* (NFPA 2011). AMCA offers independent certified ratings for fan performance (ANSI/AMCA Standard 210 [2007] and AMCA Publication 211 [2013a] for centrifugal and mixed-flow fans, ANSI/AMCA Standard 260 for dilution fans [2013b], and ANSI/AMCA Standard 300 [2014] and AMCA Publication 311 [2010a] for all fan type acoustics). AMCA certified ratings ensure independent third-party verification of fan airflow and acoustic performance as well as fan energy consumption. Induced Flow Fan Testing.mp4, a video that shows the operation of dilution fans as well as the AMCA test methodology used to test and certify fan inlet airflows and total discharge airflows, is available with this Guide at www.ashrae.org/LabDG.

Figure 6-26
Centrifugal
Induced Draft
Dilution Fan

(Reprinted with permission from Greenheck [2012a])

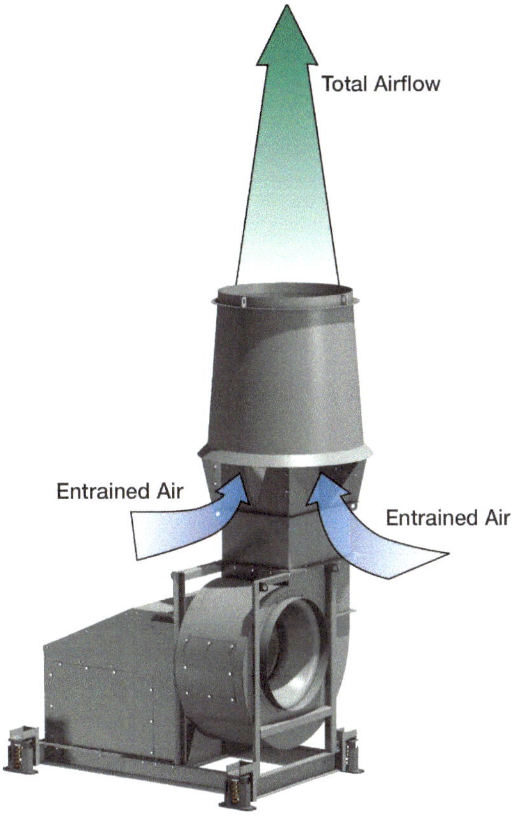

Figure 6-27
In-Line
Mixed-Flow
Induced Draft
Dilution Fan

(Reprinted with permission from Greenheck [2006])

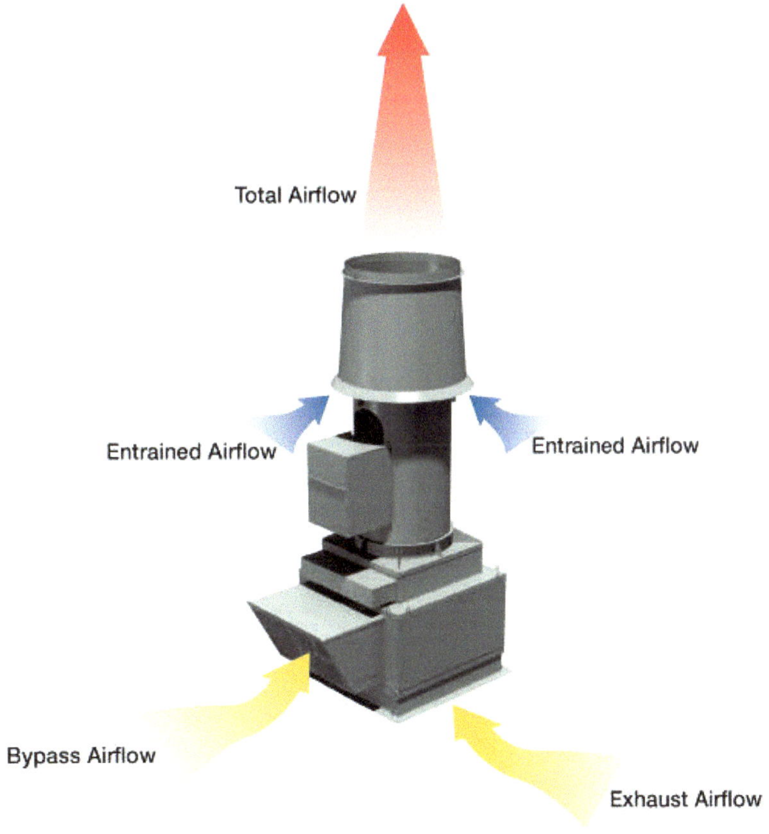

Pressure Class

The fan types previously discussed can easily be applied to low-, medium-, and high-pressure exhaust systems. It is fairly common to have negative pressures in a laboratory system ranging from –0.5 to –12 in. w.g. (0.13 to 3.0 kPa). As the negative pressure increases, sturdier fans and ductwork are required. Higher system exhaust pressures and flows require fans to be adequately constructed. This construction is categorized by fan classes, which range from class I through class IV and are defined by AMCA standards.

Material Composition

The sturdiness of the exhaust fan and the corrosion resistance requirements determine the material composition of the fan. To avoid corrosion problems from the contaminated exhaust air or subsequent condensation, laboratory fans are often coated or constructed using special materials, such as stainless steel, fiber reinforced plastic (FRP), polyvinyl chloride (PVC), or chlorinated polyvinyl chloride (CPVC).

Protective Coating

In addition to the fan material, protective coatings are often added to increase the fan's service life and to reduce required maintenance and system downtime. In many instances, the coatings are used on lower-cost fan materials (such as carbon steel) to avoid the use of higher-cost materials (such as stainless steel).

If the exhaust has the potential to contain hazardous emissions, dispersion modeling should be performed per ANSI/AIHA/ASSE Z9.5 (AIHA 2012) to avoid reentrainment of pollutants into the building and to ensure the exhaust airstream is properly diluted prior to reaching the ground, roof level, or other potentially occupied spaces. The dispersion modeling can be used to identify the optimum stack placement, minimum stack height, minimum exit velocity, and/or volume flow rate.

Special Considerations

There are additional construction considerations that are classified as "special" because they are dependent upon each unique application. First is the use of a drainage connection for the fan. Since the use of caps on exhaust stacks is not recommended due to the adverse impact of the caps on dispersion of the exhaust air, the fan must have either an integral drain installed in it or one installed in the discharge/suction of the fan (refer to Chapter 45 of *ASHRAE Handbook—HVAC Applications* [2015a]).

Second, in those cases where the lower explosive limit (LEL) of the exhaust airstream is greater than 25%, nonsparking wheels and explosion-proof motors are required. The AMCA spark-resistant standards, which depend upon the level of explosion hazard, should be followed (AMCA 2010b). The fan construction requirements defined in these standards are summarized as follows:

- *Type A construction*: Use of nonferrous material for all fan parts that are in contact with the exhaust air. This is to ensure that any mechanical failure or movement of the wheel will not cause a spark.
- *Type B construction*: Use of nonferrous material for the wheel and the ring that the shaft of the fan passes through. This type of construction is suitable for most applications.
- *Type C construction*: Construct fan in a manner so that any two ferrous parts of the fan are unable to make contact. This type of construction is suggested for all exhaust fan constructions for hoods.

Note that FRP and PVC fans must include special graphite liners so that the fans can be grounded (dissipate static charges) to meet the AMCA spark-resistant construction standard (AMCA 2010b).

Third, the fan, fan housing, and exhaust stack must be designed and constructed to withstand a wind load of at least 30 lb/ft^2 (146 kg/m^2). Good practice requires that the exhaust stack be self-supported to avoid excessive loading on the fan housing (Wunder 2000). In hurricane-prone regions, the wind loads will be much higher and the fan components exposed on the roof may also be required to resist the impact of wind-borne debris. A licensed professional structural engineer should be consulted in such cases.

Serviceability of critical laboratory exhaust fans must be included in the design criteria. The elimination of drive components such as couplings, belts, pillow block bearings, and sheaves greatly reduces the maintenance service requirements of the equipment. In addition, balancing the fan better than standard AMCA requirements will extend the life of the motor bearings.

Side-inlet centrifugal fans used should be single width, single inlet (SWSI) AMCA arrangement 1, 9, or 10 for belt drive and arrangement 8 or 4 for direct drive. In-line mixed-flow fans should use bifurcated fan housings that isolate the fan drive components from contaminated exhaust (AMCA arrangement 9 for belt drive and arrangement 2 or 4 for direct drive). These drive arrangements can aid in serviceability of drive components without fan disassembly and access into a contaminated exhaust. AMCA arrangement 4 for centrifugal and mixed-flow fans direct-couples the fan wheel to the motor shaft, eliminating all other drive components, which adds to the reliability and life of the equipment, greatly reduces maintenance cost, and reduces energy consumption (AMCA 2010c).

Controls

Although the hoods, ductwork, fans, and exhaust stacks are designed for peak operating conditions, it is the controls that maintain the safety and consistency of operation at nonpeak conditions. For constant-volume systems, the controls turn equipment on and off. There is no need for modulation of devices since the exhaust air volume is constant.

However, the need for proper controls is critical in VAV systems where the quantity and velocity of exhaust air changes are based on fume hood sash position. For these systems, there are four key control loops that must be evaluated and integrated:

- Hood exhaust and other contaminated air
- General exhaust
- Fan control
- Exhaust stack velocity

Figure 6-28 shows a general control system layout for a VAV laboratory exhaust system. The hood exhaust is controlled based on sash position to maintain a constant hood face velocity. In the past, *NFPA 45: Standard on Fire Protection for Laboratories Using Chemicals* (NFPA 2015) governed minimum hood flow requirements when the sash is lowered and or closed. Since the 2011 edition, *NFPA 45* refers to ANSI/AIHA/ASSE Z9.5 as the sole guideline on hood minimum flow. In 2012, ANSI/AIHA/ASSE Z9.5 (AIHA 2012) was revised to change from a minimum airflow per unit length of hood width or airflow per unit of bench area to a minimum hood flow rate related to the hood's volume and measured in terms of minimum hood air changes per hour (ACH). Although no definitive rates are mentioned or required, a range of 150 to 375 ach (hood) is indicated. For a typical 6 ft (1829 mm) bench type fume hood, these rates correspond to a range of about 100 to 250 cfm (47 to 118 L/s).

Figure 6-28
Typical VAV Exhaust Control Components

(Reprinted with permission from Greenheck [Rossi 2006])

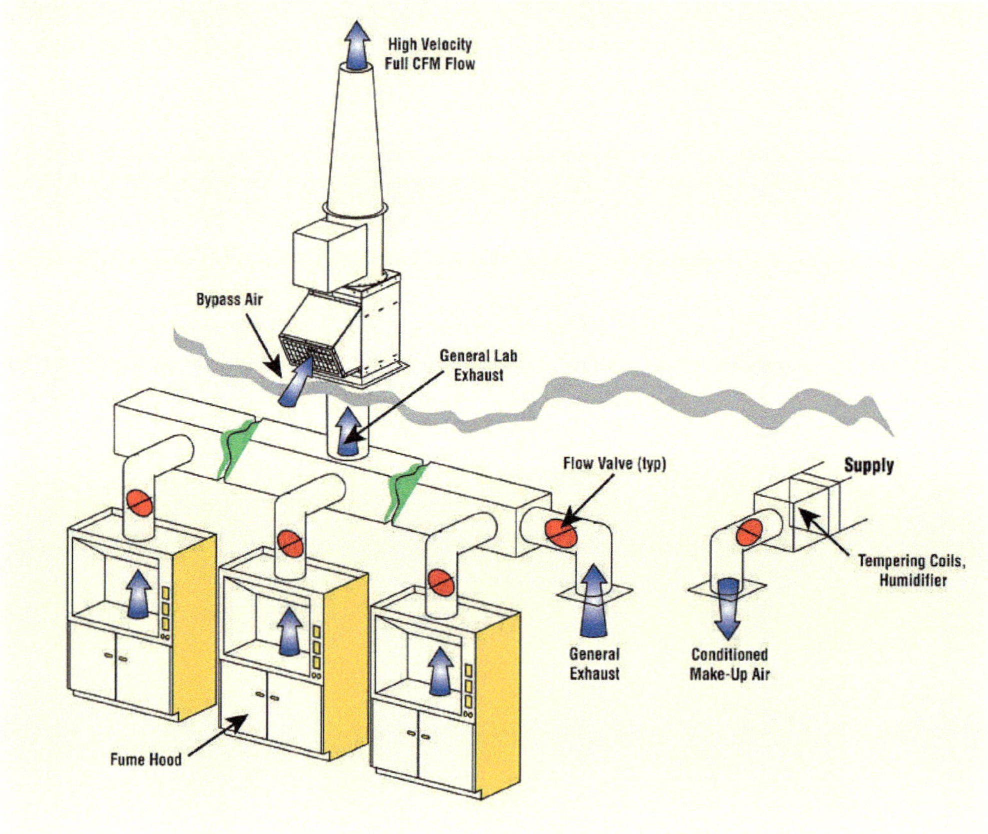

Figure 6-29
Laboratory Exhaust Fan Bypass Air Plenum

(Reprinted with permission from Greenheck [Rossi 2006])

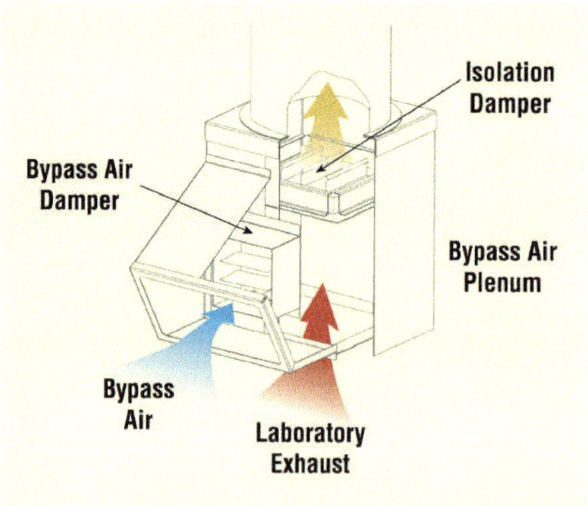

The general exhaust is controlled either to maintain a constant exhaust airflow quantity or to track the hood exhaust flow. Typically, a constant volume is desired to maintain the indoor air quality. The exhaust fan is controlled to maintain a constant negative pressure within the exhaust air duct. This is required for the VAV controllers to function properly.

The final control component for many laboratory exhaust systems is the bypass air damper on the exhaust fan system. With a constant-volume exhaust fan, the proper amount of air feeding into the exhaust fan (see Figure 6-29) is maintained along with the designed static pressure in the exhaust riser by opening and closing the bypass air damper

Figure 6-30 Isolation and Bypass Dampers

(Reprinted with permission from Greenheck [2012b])

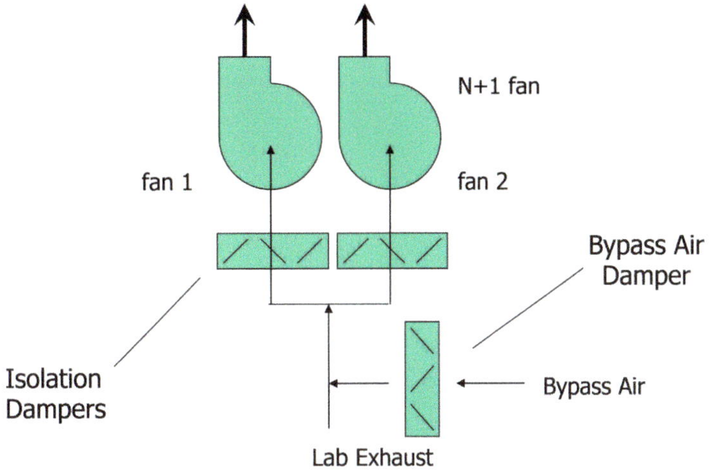

as the laboratory exhaust air volume varies. With a VAV exhaust fan, the variability in the laboratory exhaust air volumes is accounted for by similarly varying the speed of the exhaust fans to maintain the design static pressure in the exhaust riser. When the laboratory exhaust volume flow rate is lower than the minimum required exhaust volume flow rate for the exhaust stack, per code requirements or as determined by dispersion modeling, the bypass air dampers are then controlled to make up the difference. With either exhaust fan system type, the location of the static pressure sensor is critical to preventing the high velocities that are typically present in the exhaust air duct from disrupting the static pressure measurement.

Another critical element is the proper sizing and selection of the bypass air damper to avoid oversizing. In cases of multiple fans mounted on a common plenum (manifolded exhaust), the bypass damper(s) must be sized to bypass the airflow capacity of the operating fan at the required static pressure of the system. The bypass damper is essentially making up for the difference of airflow needed by the building and the requirements of the fan to maintain exhaust outlet velocity and effective stack height. Due to acoustical considerations internal and external to the building, the air velocity through the bypass damper should be sized at a maximum of 3000 fpm (15.2 m/s) velocity at full bypass.

As with any VAV system, the sensors and controllers have a minimum airflow that is required for stable and safe operation. This minimum flow is often described using the turndown ratio defined in Equation 6-2:

$$\text{Turndown ratio} = \frac{\text{minimum airflow}}{\text{design airflow}} 100 \tag{6-2}$$

A minimum turndown ratio of between 10% and 20% is recommended for most systems. For systems where precipitation of particles is a concern, the turndown ratio may be as high as 50% to 80%.

Proper design and selection of the isolation and bypass air dampers is also critical for system control and operation (refer to Figure 6-30). For more detailed information on the control of laboratory HVAC systems see Chapter 11. For example, consider a two-fan laboratory exhaust arrangement. With fan 1 running and fan 2 on standby and exhausting laboratory effluent at 6 in. w.g. (1.5 kPa) static pressure, the differential pressure across the closed isolation damper of fan 2 is 6 in. w.g. (1.5 kPa). Under a moderate to high differential pressure, a commercial quality damper will have blades deflect, deform, and bend, causing the damper to bind and not actuate properly. Additionally, there will be considerable air leakages across the damper and into the exhaust system, reducing containment and causing fan 2 (the standby fan) to spin backwards. With regard to isolation damper control, the standby fan should be started and the damper actuator energized simultaneously. Opening the isolation damper of fan 2 (the standby fan) without energizing fan 2 will reduce containment in the exhaust system, reduce the duct static pressure, and potentially cause fan 2 (the de-energized fan) to spin backwards. If the fan is equipped with a variable-frequency drive (VFD), the fan can be ramped up to a motor frequency that is sufficient to maintain duct static pressure (but not so high as to damage the isolation damper) prior to opening the isolation damper in order to minimize any disruption in the duct static pressure and avoid the possibility of the fan spinning backwards. Isolation dampers should be designed and tested for operation at high differential pressures (up to 15 in. w.g. [3.7 kPa]) and low leakage and corrosion-resistant construction. Isolation dampers are typically parallel blade. Gravity backdraft dampers have high leakage rates at differential pressures and should not be used. An additional note about isolation damper actuator power is that if the fans use soft starters or VFDs, the actuators cannot be powered off the motor power wiring—separate actuator power must be provided.

Bypass dampers must be sized to bypass the required airflow equal to the laboratory exhaust flow turndown (the variable laboratory exhaust volume) at design static and differential pressures. If the bypass damper is too large, there will not be good flow modulation control; if it is too small, the damper will not bypass the required air for turndown. Additionally, since these dampers are sized for bypassing large volumes of air at a high differential pressure, the dampers are typically small. The air velocities of bypass dampers can be as high as 2000 fpm (10.2 m/s). Bypass air damper construction should be equal to the isolation damper construction, but the bypass dampers should have opposed blades for proper flow control.

Reliability/Redundancy

One of the more important issues in laboratory exhaust fan selection and design is the reliability of the exhaust system. For the system to be 100% reliable, the system must be operational 24 hours a day, every day of the year, per *NFPA 45*, which requires the laboratory to be ventilated whenever chemicals are present (NFPA 2015). Therefore, many critical laboratory facilities have redundant (backup) exhaust fans. ANSI/AIHA/ASSE Z9.5 requires at least two exhaust fans on manifolded systems (AIHA 2012). This provides assurance of the safety and operability of the laboratory environment in case one or more of the fans experiences a malfunction or requires maintenance. In such cases, fan downtime can create an intake of outdoor air from the roof through the affected hoods' exhaust air ducts and cause great risks to personnel using those hoods.

Examples of fan system designs that improve reliability are direct-drive fans and multiple VFDs on direct-drive blowers. In terms of cost, redundancy and improved reliability mean higher capital costs, particularly in cases where dedicated exhaust has been chosen for the hoods. In addition, the redundant fan must be cycled into operation on a periodic basis. Exhaust fans should not sit dormant for more than one month so that the fans will be fully operational when required. In a VAV exhaust fan system, all fans, including the redundant fan, can operate at reduced volume flow rates during normal operating conditions as long as the system is designed such that there is sufficient capacity to meet the laboratory airflow demand when one fan is off.

In addition to redundant fans, many laboratory facilities have backup power fuel-fired engine generator sets so that the fans operate when electricity is not available from the main utility grid. In general, as the hazard level increases, the need for a more reliable system increases. The higher reliability is achieved through better-quality materials and equipment and the use of redundant components.

SUPPLY AIR SYSTEM

The primary role of the supply air system is to provide ventilation air to the laboratory space in sufficient quantities for the comfort and safety of the occupants, and in some cases to meet environmental control requirements (pressure, temperature, and humidity) of the particular type of research being conducted in the laboratory. It is the type of exhaust air system (constant volume or VAV), the specific space pressure relationship between the laboratory and adjacent spaces, the required indoor environmental conditions, and the minimum "safe" air exchange rate that determine the type of supply air system (constant volume or VAV) and the volume of supply air required.

With the exhaust system type already determined and the exhaust air duct system designed, the design of the supply air system is simplified. It is common practice to match the type of supply system with the exhaust system. Table 6-3 lists the possible arrangements of exhaust and supply air systems that are commonly used.

Laying out the supply system is often a simple task, because it usually parallels the exhaust system. As was determined during the planning phase (see Chapter 4), the distribution option for ductwork layout is integral to the layout of both spaces and the building. During the layout of ductwork, for both supply and exhaust, the primary concern is accessibility for cleaning/decontamination and for maintenance of other systems (valves, piping, etc.).

With the type and layout determined, the design of the supply system focuses on the central system and controls. The specific steps to follow are as follows:

1. Verify supply air quantity.
2. Select specific system type.
3. Evaluate need for auxiliary air supply.
4. Select air-handling unit (AHU).
5. Determine control strategy.

VERIFY SUPPLY AIR QUANTITY

Because laboratory hood exhaust air quantity is based on capture velocity and the general exhaust air quantity is based on maintaining sufficient air exchange rates for safety, the determination of the supply air quantity simply focuses on maintaining the proper pressure (airflow direction) relationship. With a typical control error of ±5%, the supply air quantity is often sized to be 10% greater than the exhaust air when the room must be positively pressurized and 10% less than the exhaust air when the room must be

Table 6-3 Exhaust and Supply Air Systems

Exhaust System	Supply System	Comments
Constant volume	Constant volume	Good combination—easily maintains pressure requirements but is energy intensive
	VAV	Must be controlled to maintain pressure relationships; good for systems where exhaust sources, such as fume hoods, are turned on and off
VAV	VAV	Excellent combination—supply controlled to maintain pressure relationship
	Constant volume	Do not use—will result in loss of pressure control and serious problems

negatively pressurized. In addition to the amount of air needed for controllability of offset airflow, the engineer should evaluate the minimum amount of offset air required to maintain differential pressure across closed doors and other leakage paths into the space. In situations with poor construction and high levels of infiltration/exfiltration, this differential may need to be increased to 15% or 20%. Thus, for pressure-controlled spaces, better air conservation is achieved as the sealing of the room envelope is improved.

Once the supply air quantity for a laboratory space has been determined, its value must be compared to the minimum ventilation rates. Minimum airflow rates for laboratories are typically in the range of 6 to 10 ach; however, minimum airflow rates for some spaces may be dictated by specific health and safety standards or internal facility policies. This is especially critical for negatively pressurized spaces where the supply air volume is less than the exhaust air volume. In these instances, if the air being drawn into the space is not clean (from an adjacent laboratory), the supply and exhaust air volumes for the space may need to be increased so that the proper ventilation air exchange rate is maintained. The maximum airflow rate for the laboratory should be reviewed to ensure that the appropriate air delivery methods are chosen. A sample template that can be used to document the proper exhaust and supply air rates is available in Chapter 3 as Table 3.1.

SELECT SPECIFIC SYSTEM TYPE

Depending on the distribution layout being used, the supply air system can either be centrally located (central system) or can be located within the room (unitary system). Either of these systems can be constant volume or VAV.

Central System

The specific type of central system selected is either constant volume or VAV. However, there are many options available in meeting the space requirements, including the following:

- Constant-volume single-zone system
- Constant-volume multizone system
- Multispeed system
- VAV single-duct system
- VAV dual-duct system

The constant-volume system is the simplest system. A constant-volume single-zone system serves a single space and produces hot or cold air from unit-mounted heating and cooling coils as required to control the space temperature. A central multizone AHU serving multiple rooms or spaces supplies the same-temperature cool or neutral air to each zone. Space temperature is then controlled using some form of reheat. Figure 6-31 details a constant-volume terminal reheat system. This system produces chilled air at the AHU and uses duct-mounted hot-water or electric reheat coils for zone control.

In an effort to reduce operating costs while still maintaining the capital-cost benefits of a central system, multizone constant-volume systems are often used. These systems divide the space into two or more zones (often called *dual-zone systems* or *multizone systems*, respectively). In a multizone system, the central AHU has two or more coil "decks" where heating and cooling can be individually supplied to each zone. Figure 6-32 is a schematic of a central multizone AHU with hot, cold, and neutral decks, and Figure 6-33 is a schematic of a system layout showing the zone air distribution.

Typically, a multizone system has a cooling and heating coil providing cool or heat. Since it is very inefficient to mix mechanically cooled or heated air, the cooling coil is turned off during the heating season and unconditioned air is bypassed through the cool-

Figure 6-31
Constant-Volume Terminal Reheat System

(McDowall 2007)

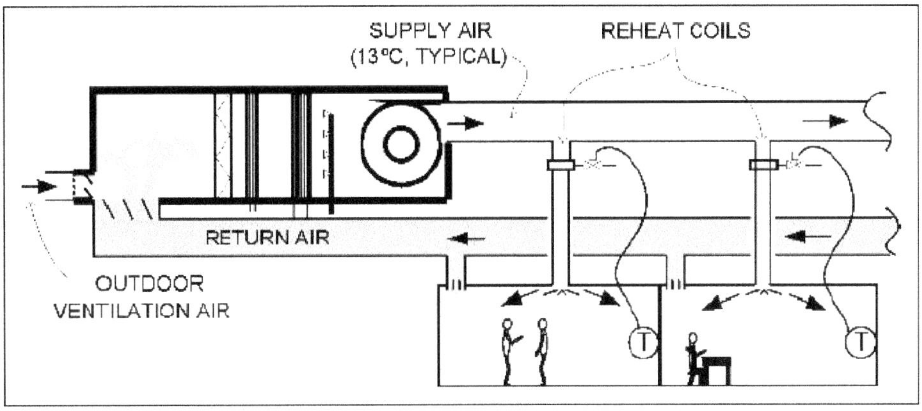

Figure 6-32
Central Triple-Deck Multizone AHU

(Reprinted with permission from Texas A&M University Energy Systems Laboratory [Joo et al. 2008])

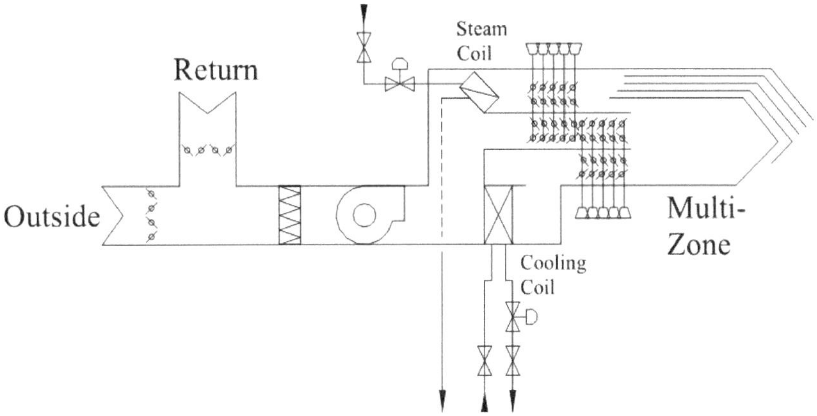

Figure 6-33
Multizone Constant-Volume System

(Courtesy of Johnson Controls)

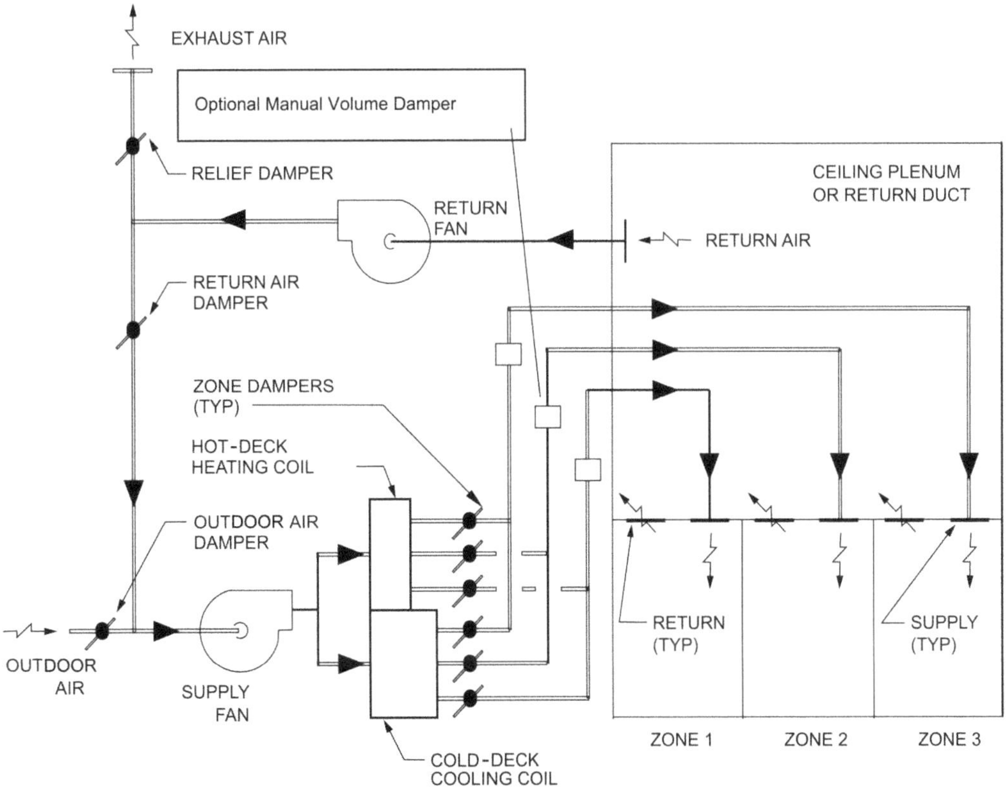

ing coil. The heating coil is then turned off during the cooling season. A way to be able to provide either cooling or heating to a space is to install a bypass section where either cooled air and bypass air or heated air and bypass air is supplied to the space. However, the main problem with multizone systems in laboratories is that any bypass air is unconditioned outdoor air. This results in poor comfort control in the occupied spaces due to lack of humidity control. Therefore, care must be taken in the design and control of multizone systems in laboratories. Additional information on the design and application of single-zone and multizone systems can be found in Chapter 4 of *ASHRAE Handbook—HVAC Systems and Equipment* (ASHRAE 2012a).

A hybrid between a constant-volume and a VAV system is a multispeed system. In this system, the ductwork is the same as for a single-zone constant-volume system, with the central fan having the ability to operate at multiple speeds/airflows. The supply fan is typically controlled based on the status of the exhaust systems, decreasing its speed as the exhaust systems shut down. With the cost of VSDs and systems decreasing, the use of multispeed systems is infrequent.

Because of the superior energy performance and greater control flexibility, most laboratory systems are designed as VAV systems. In VAV systems, the central system provides and conditions air to all spaces. At each space is a terminal unit (air valve) that varies the volume of air to the space to meet thermal comfort, pressurization, hood makeup, and minimum ventilation requirements. When excess air must be provided to the space for pressurization or ventilation requirements, the air must be reheated to maintain comfort. Figure 6-34 shows a schematic of a typical laboratory VAV system.

The two primary types of VAV systems are single-duct and dual-duct. The single-duct VAV system is the traditional system in which conditioned air is supplied to each space in varying quantities through an air terminal unit. The terminal unit (typically referred to as an *air valve* or *VAV box*) is simply a flow sensor, damper, and reheat coil mounted in a sheet metal box with controls. Figure 6-35 illustrates a single-duct VAV system. The single-duct VAV system is simple to design and construct, and it probably has the lowest first cost of the supply systems available. Accurate and stable room temperature control is possible, and the system has the design advantage of being easier to modify in a remodel

Figure 6-34 Typical Laboratory VAV System

(Labs21 n.d.)

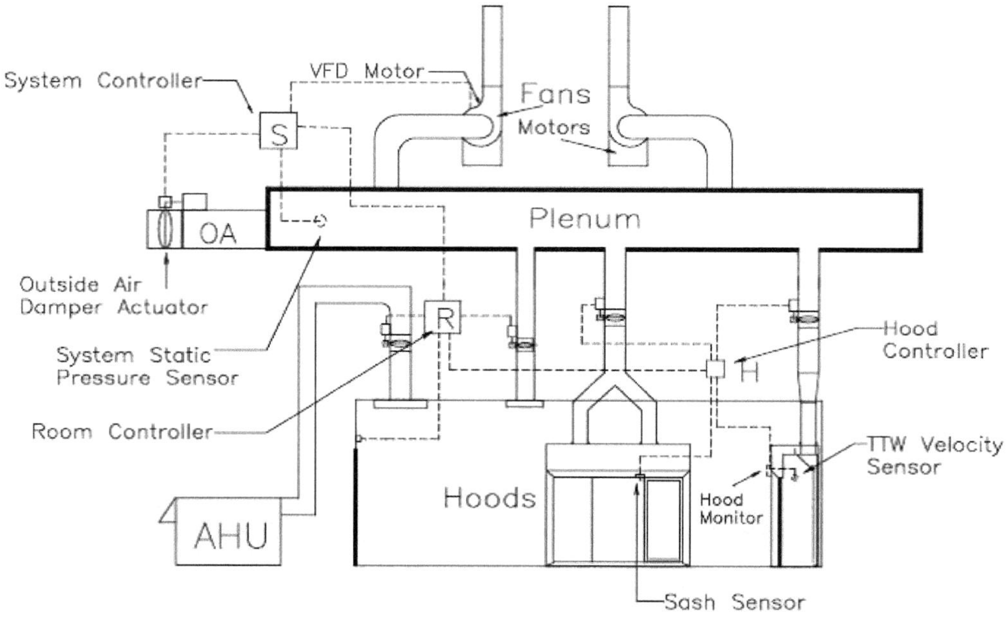

than some other designs. However, this kind of system uses a great amount of energy. In using the flow sensors provided with the typical VAV terminal unit, the engineer must consider the minimum-flow-reading accuracy of the flow sensor and the pressure transducer when using volumetric offset control. Most manufacturers have a minimum velocity that must be maintained across the sensor to maintain the velocity pressure amplification factor (typically two or three times the actual velocity pressure) on the sensor. Thermal dispersion sensors can improve the accuracy and turndown capability of the system.

A dual-duct VAV system uses two supply air ducts—a hot duct and a cold duct. Instead of having a reheat coil as in a single-duct system, which is energy intensive, the dual-duct terminal units mix hot and cold air to deliver the correct temperature air to each space as required to meet thermal comfort conditions. Dual-duct VAV systems can be very energy efficient. However, special attention must be given in selecting and maintaining the two box dampers to minimize leakage of hot or cold air through a closed damper. Further, there must be sufficient space to lay out two parallel duct runs for the hot and cold ducts. The cost of a dual-duct system is typically significantly higher than that of a single-duct system, primarily due to the additional ductwork required. Figure 6-36 illustrates a typical dual-duct VAV system.

Unitary System

In contrast to central systems, unitary systems use individual AHUs to provide conditioning for each laboratory space. Unitary systems are used for several reasons, such as the following:

- Large laboratories
- Isolation of laboratories
- Limited duct distribution space
- To reduce fan energy

As the sizes of individual laboratories increase, there is a point when it is more economical and simpler to install separate air systems for each laboratory. A simple example is that of cleanrooms in industrial and pharmaceutical industries. There is also a need for

Figure 6-35
Single-Duct
VAV System

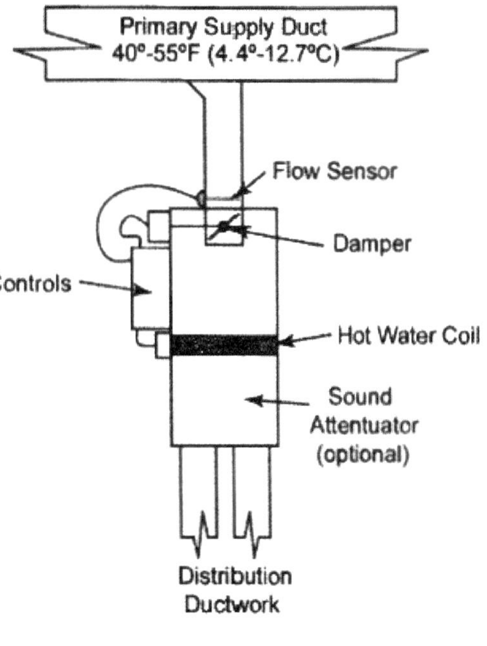

Figure 6-36
Typical
Dual-Duct
VAV System

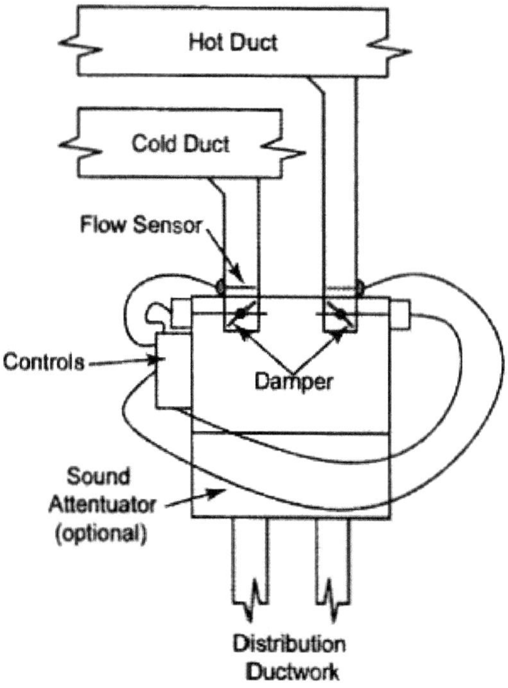

Figure 6-37
Typical
Unitary
System
Layout

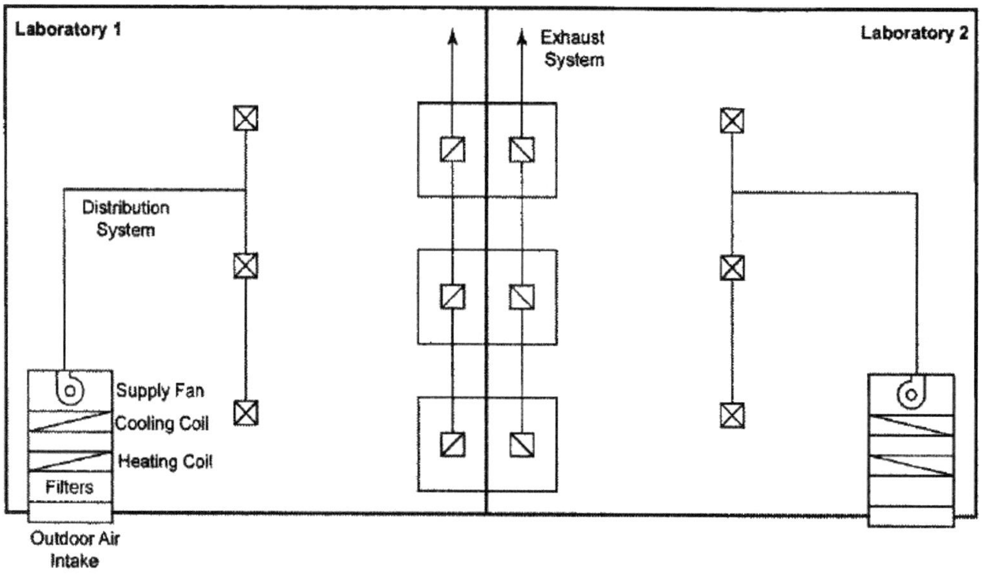

unitary systems when spaces must be physically isolated from one another. This is the case when dealing with highly toxic and dangerous substances in biological laboratories. Another typical reason unitary systems are installed is a lack of duct distribution space. Since the exhaust air ductwork takes precedence, when there is limited space, the supply and outdoor air ductwork often must be contained within the laboratory space, as shown in Figure 6-37. Because unitary systems provide heating and cooling directly to laboratory spaces, the need for inefficient reheating systems is greatly reduced. Unitary systems allow the ventilation system to supply only the tempered air required for minimum ventilation and makeup air for exhaust devices, which reduces the fan energy required.

Similar to central systems in nearly all aspects, unitary systems are sized to meet the specific requirements of each laboratory and can offer immense adaptability. They can also be operated using constant-volume or VAV control methods to match the exhaust control. However, unitary systems do have a number of drawbacks. The primary drawback is that each unitary system must have access to outdoor air. This often requires a separate outdoor air intake louver (envelope penetration) for each system. This often eliminates the potential for central, economical heat reclaim and air treatment. Another drawback is that chilled- and hot-water piping connections must be made to each system. This can be a substantial cost increase.

Finally, in some situations, the unitary system cannot be physically located within the laboratory space but must be in a separate mechanical room. Because of the need for accessibility for maintenance, a separate mechanical room for each unitary system will significantly increase the size of the building or significantly reduce the size of the laboratories. Biological, clinical, and animal research laboratories typically do not allow placement of the unitary system within the laboratory due to concerns for possible system contamination from the experiments (system and ductwork upstream of supply fan under a negative pressure). Also, if the unitary system is located either above the space or adjacent to the space, significant noise may be generated and must, therefore, be controlled.

With the growing concern over reducing laboratory energy use, there has been a resurgence in the application of unitary systems coupled with a dedicated outdoor air system (DOAS) for laboratories in which minimum air change rates are not the primary determinant of the supply airflow. If the required laboratory outdoor air is significantly less than the required supply air, then coupling a DOAS unit with local recirculation type unitary units can potentially save a significant amount of fan energy. This type of system has seen growing popularity in the application of fan-coil units, fan-powered induction units, and active and passive beams. Capital costs for these types of systems can sometimes be significantly higher than those for traditional VAV systems because of the increased amount of piping and the coils at each zone.

Active and passive beams, often referred to as *chilled beams*, are a technology developed more than 30 years ago in Scandinavia that have become popular in Europe and are beginning to make inroads in the United States. This technology allows the majority of the sensible cooling load to be removed at the room level, which decentralizes the cooling. In other words, the sensible cooling load is transferred from the chilled-water coil in the central AHU to the chilled beam suspended from the ceiling in the laboratory space. Additionally, since the amount of air supplied by the AHU is reduced, the ductwork main and branch sizes can be reduced.

The beams come in several forms, including active beams, passive beams, two-pipe beams, four-pipe beams, and custom units integral with the room lights. Active beams (see Figure 6-38) cool the room air through induction by running primary air (from air handlers) through air nozzles across the coils. Most of the air (secondary or induced air) comes from the room itself and therefore uses very little primary air, making the ductwork smaller. This system is beneficial when there is a limited floor-to-ceiling height.

Passive beams (see Figure 6-39) are similar to active beams; however, they do not use any primary air. This type of beam functions solely on induction moving the air around the room with natural-convection loops. As hot air rises and passes through the cooling coil of the chilled beam, it cools and drops back down, creating a natural convection loop.

A two-pipe beam is a cooling-only unit. A four-pipe beam is a cooling and heating unit. Both two- and four-pipe beams can be either active or passive. Chilled beams provide the ability to have individual thermal comfort control for each room. Each space can have its own thermostat that controls a chilled- and hot-water modulating control valve

Figure 6-38
Active Beam

(Labs21 2009)

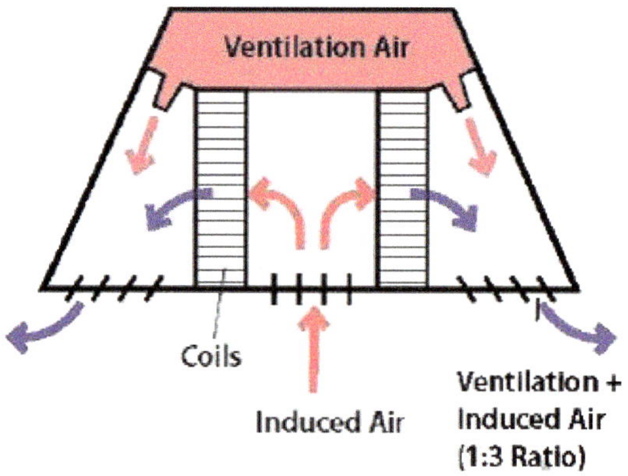

Figure 6-39
Passive Beam

(Labs21 2009)

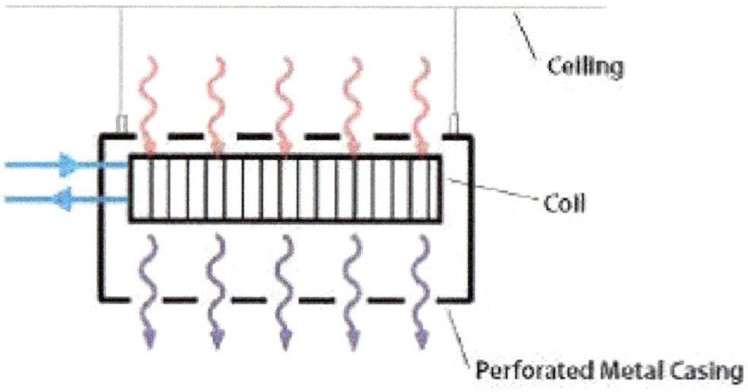

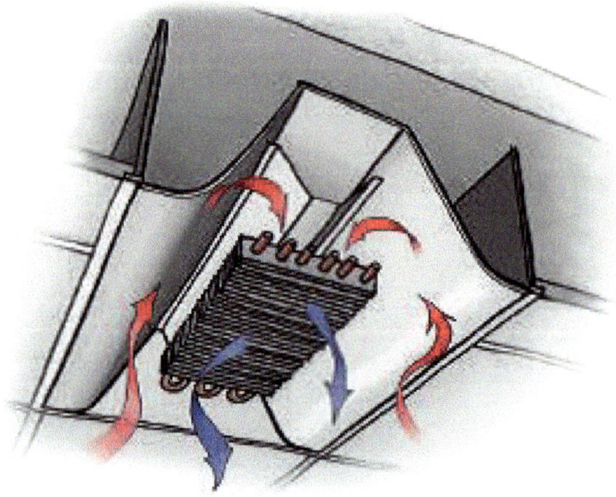

serving the beam's air induction unit. To avoid condensation from the cooling coil in the space, the chilled water must be supplied at a temperature a couple of degrees higher than the space dew-point temperature. Typically, this falls within the range of 57°F to 62°F (13.9°C to 16.7°C).

EVALUATE NEED FOR AUXILIARY AIR SUPPLY

Because of the high volume of air being exhausted through laboratory hoods relative to the volume of supply air required for comfort, methods have been developed to minimize the volume of conditioned air supplied to the laboratory. The basic principle is that unconditioned or less-conditioned (tempered) air is supplied near the face of the hood with the expectation that this tempered air will be directly exhausted and not impact the space conditions. Therefore, the amount of conditioned room air exhausted by the hood is reduced. For laboratory systems, this strategy is called *auxiliary air.*

While each laboratory hood manufacturer designs its auxiliary air systems slightly differently, auxiliary air is typically introduced above the hood opening or along the sides of the hood opening. The required supply volume of conditioned and ventilation air is reduced up to 70% in auxiliary air supply systems, where a constant-speed blower is used to pull in a constant supply of outdoor air.

Tempering Only

Since the auxiliary air is directly exhausted, it is either 1) unconditioned, filtered outdoor air or 2) tempered, filtered outdoor air. Tempering is used to avoid cold drafts on the user of the laboratory hood when the outdoor air temperature falls below about 60°F (15.6°C).

The heating of the outdoor air is typically accomplished by a central system using steam, hot water, direct-fired natural gas, or some combination of the these. Often, multiple direct-fired natural gas makeup air units are used for their simplicity and low capital cost. A summary of auxiliary air heating options is given in Table 6-4.

Effects on Humidity and Latent Loads

The primary concern with auxiliary air systems is the impact that unconditioned or tempered air may have on occupant comfort within the laboratory space. When designed properly, the auxiliary air should not mix with the fully conditioned air in the occupied space. However, in improperly designed or operated systems, introduction of the auxiliary air into the occupied space may increase the space humidity through an increase of the space latent load. The result is an increase in occupant discomfort due to a clammy, sticky perception of the skin. Therefore, to avoid adverse comfort effects, the auxiliary air must be supplied directly to the laboratory hood and not have a chance to mix with the occupied space air. Despite the possibility of occupant discomfort, about a 50% reduction in conditioning costs is often worth evaluating the application of auxiliary air systems.

The effect of introducing the auxiliary air into the space can be determined by a psychrometric analysis of the system. This is accomplished by using a psychrometric chart

Table 6-4
Auxiliary Air Heating Options

Heat Sources	Benefits	Drawbacks
Steam	Simple and economical	Must properly size coil and steam trap to avoid freezing at part loads
Hot water	Good part-load control characteristics	Typically is not used when outdoor air temperature is below 32°F (0°C)
Direct-fired natural gas	Can be easily added to systems	

Figure 6-40
Auxiliary Air Impact on Space Conditions

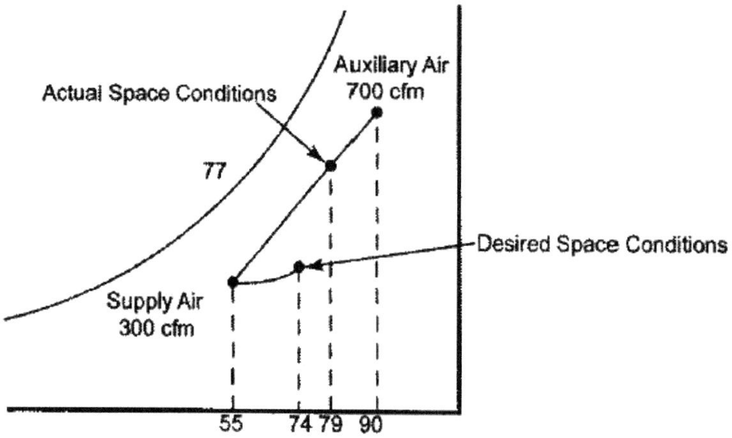

and comparing the supply airstream/auxiliary airstream mixed-air condition with the room air. Figure 6-40 shows an example of such an analysis.

In addition to the problems related to the thermal condition of auxiliary air, the system itself presents some engineering challenges. The auxiliary air must be of reasonably uniform velocity across the discharge area. Balancing the two separate supply air systems for the laboratory can be another engineering challenge. Auxiliary air fume hoods reduce the amount of energy and operating costs used to condition makeup air. However, they do not reduce fan energy consumption because they do not change the amount of exhaust air. Additional considerations of some of the drawbacks of implementing such a system, such as adverse impact on experiments, are discussed in Chapter 5. VAV or reduced-airflow systems are better designs used now in lieu of the older auxiliary air design.

ANSI/AIHA/ASSE Z9.5 (AIHA 2012) discourages the use of auxiliary air fume hoods. In summary, the following difficulties and installation criteria are associated with auxiliary air fume hoods:

- The auxiliary air supply must be introduced outside the fume hood to maintain appropriate velocities past the researcher.
- The flow pattern of the auxiliary air must not degrade the containment performance of the fume hood.
- Auxiliary air must be conditioned to avoid blowing cold air on the researcher; often the air must be cooled to maintain the required temperature and humidity within the hood. Auxiliary air can introduce additional heating and cooling loads in the laboratory.
- Only a vertical sash should be used in the hood.
- Controls for the exhaust, auxiliary, and supply airstreams must be coordinated.
- Additional coordination of utilities during installation is required to avoid spatial conflicts caused by the additional duct system.
- Humidity control can be difficult. Unless auxiliary air is cooled to the dew point of the specified internal conditions, there is some degradation of humidity control; however, if such cooling is done, the rationale for using auxiliary air has been eliminated.

SELECT AIR-HANDLING UNIT

The selection of an AHU, whether central, unitary, or auxiliary, is the same as for a typical nonlaboratory application. The uniqueness with the selection for laboratories is that 100% outdoor air is required. Therefore, special consideration must be given to the

Figure 6-41
Typical AHU

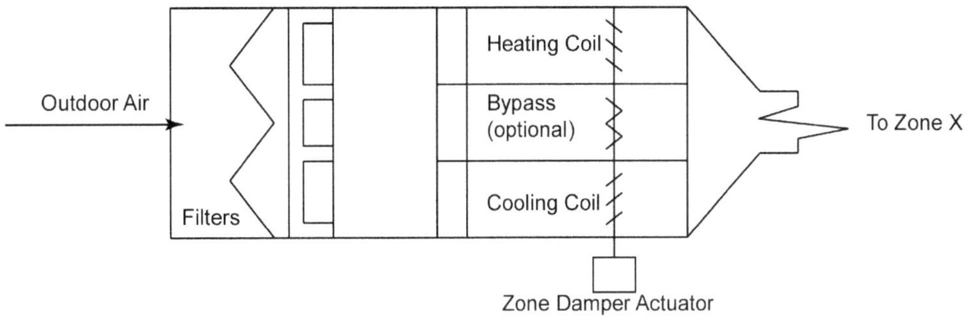

selection of the heating and cooling coils, filtration, and fans. A typical layout for an AHU is shown in Figure 6-41.

Heat recovery systems are discussed in detail in Chapter 10. Once through the heat recovery system, if present, the air is filtered, heated or cooled, humidified or reheated, then supplied to the individual spaces. The role of the AHU is to take the supply air at varying conditions and condition it to a consistent setpoint. Inlet air can range from very dry and cold to very hot and humid. In this section, selection and sizing of each of the main components for laboratory AHUs are discussed.

Filtration

The treatment and filtration of outdoor air is discussed in detail in Chapter 8. In general, unless there is potential reentrainment of exhaust air or the outdoor air is highly polluted, treatment of the outdoor air is limited to filtration of particles.

Filtration of the outdoor air is meant to protect the air-conditioning systems (clean coils, ductwork, etc.) and to protect the processes/experiments located in the laboratory space. Therefore, the level of filtration required for a laboratory is typically determined by the use of the laboratory. Table 6-5 summarizes the possible filtration levels typically used in laboratories.

While higher filtration is often desirable, there is a trade-off between better filtration and higher filter and operating costs. The operating costs increase with filter efficiency due to higher filter pressure drops and subsequent fan energy increase. In addition, when selecting and designing the filtration system, several criteria must be evaluated:

- *Eliminate moisture carryover*. Because filters collect and hold dust, dirt, and pollen, the introduction of moisture into filters can result in the growth of fungi and mold. To avoid this, any visible moisture from the outdoor air must be eliminated. The primary concern is carryover of rainwater through the outdoor air intake.
- *Loading pressure effect*. As a filter is used, it becomes loaded with materials. This results in an increase in the pressure drop across the filters. The effect of the increase in pressure drop from clean to fully loaded on the system airflow must be addressed. This is of primary concern in constant-volume systems but can also be deleterious in VAV systems if not addressed.
- *Filter bypass*. The typical construction of the rack holding the filters allows for some bypass. Consideration of the rack design must be done to avoid any bypassing of the filters.
- *Maintenance*. To ensure laboratory operations and system integrity are not compromised, adequate access and safety requirements for servicing and testing of the filters must be addressed during the design.

Table 6-5
Laboratory Filtration Options

MERV Rating	Composite Average Particle Size Efficiency*, % in Size Range, µm			Average Arrestance, %	Application	Notes
	Range 1 0.30–1.0	Range 2 1.0–3.0	Range 3 3.0–10.0			
7	n/a	n/a	50 ≤ E3 < 70	n/a	Prefilter	Used to extend life of higher-efficiency filters
11	n/a	65 ≤ E2 < 80	85 ≤ E3	n/a	Office, nonlaboratory spaces	Prefilter typically not required
13	E1 < 75	90 ≤ E2	90 ≤ E3	n/a	General laboratory	Minimum efficiency
15	85 ≤ E1 < 95	90 ≤ E2	90 ≤ E3	n/a	Biological and clinical laboratories	Sometimes used as final filters
>17	n/a	n/a	n/a	n/a	Animal and biomedical research and cleanrooms	HEPA filters used as final filters
—	n/a	n/a	n/a	n/a	Gas-phase filtration	Used when outdoor air is polluted

MERV = minimum efficiency reporting value
HEPA = high-efficiency particulate air
**ANSI/ASHRAE Standard 52.2 (ASHRAE 2012b) defines twelve particle size ranges that are placed in three larger groups according to the following schedule: ranges 1–4 (or E1, which is 0.3 to 1.0 µm), ranges 5–8 (or E2, which is 1.0 to 3.0 µm), and ranges 9–12 (or E3, which is 3.0 to 10.0 µm). Averaging the composite minimum efficiency (CME) for each of these groups will calculate the average particle size efficiency (PSE), and the resulting three percentages (E1, E2, E3) are then used to determine the MERV.*

Filtration for the air supply depends on the requirements of the laboratory. Conventional chemistry and physics laboratories commonly use minimum efficiency reporting value (MERV) 11-13 filters (ASHRAE 2012b). Biological and biomedical laboratories usually require MERV 13-15 filtration. HEPA filters should be provided for spaces where research materials or animals are particularly susceptible to contamination from external sources. HEPA filtration of the supply air is necessary for such applications as environmental studies, studies involving specific pathogen-free research animals or nude mice, dust-sensitive work, and electronic assemblies. In many instances, BSCs or laminar-flow clean benches (which are HEPA filtered) may be used rather than HEPA filtration for the entire laboratory.

Heating and Cooling Coils and Humidification

The purpose of the heating and cooling coils and humidification is to provide a consistent temperature/humidity supply air to the space. The actual sizing of the components is no different than any other application. However, because of the higher than normal capacities (temperature/humidity differentials) in laboratory applications, there are a few key selection guidelines that should be followed:

- *Design conditions*. Design the heating coil and system for the absolute lowest temperature expected to ensure that the coil does not freeze. If hot water is used as the heating medium, adding glycol to the desired concentration to avoid freezing may be desirable in some climates. Since laboratory air handlers typically condition 100% outdoor air, the engineer should investigate both the peak sensible cooling and dehumidification loads to see which is higher. In warm and humid climates, the peak dehumidification condition may be more critical (Harriman et al. 1997).

Figure 6-42
Plenum Fan Array

(Courtesy of Greenheck)

- *Cooling coil cleaning*. Due to the high latent loads, the cooling coil depth in a laboratory is typically at least eight rows. To simplify maintenance and condensate management, it is recommended that the cooling coil be split into two or more sections. This can be two six-row coils or a similar combination to make cleaning the coils easier. An access section, of at least 18 in. (457 mm) and preferably 24–36 in. (610–914 mm), must be left between all coils to facilitate cleaning.
- *Cooling coil face velocity*. To avoid carryover of condensate from the cooling coil to the supply air, the cooling coil face velocity should not exceed 500 fpm (2.5 m/s). In situations with high condensate levels, a maximum face velocity of 400 fpm (2.0 m/s) is recommended.
- *Humidification*. Sufficient space upstream and downstream of humidifiers is required to obtain even flow distribution and avoid condensation of moisture on ductwork or supply fans. In some applications (cleanrooms, hospitals) the humidifier must be located upstream of the final filters.

Supply Air Fan Types

Types of fans used for supply are not very different from fans for exhaust. There is no need for special protective coating of these fans, as the supplied air is not as contaminated as the exhaust air. However, in certain laboratory settings such as cleanrooms, where the supply air is HEPA filtered, there will be a need for more powerful fans to force the air through these filters. There also may be a requirement to provide multiple fans for redundancy. Increasingly fan arrays, also called *fan walls*, using multiple direct-driven plenum fans (also called *plug fans*) with VFDs are used to meet this goal (see Figure 6-42).

DETERMINE CONTROL STRATEGY

Details on the selection and application of control strategies for air-handling systems are contained in Chapter 11.

Air Quality

The supply air system plays a key role in determining the air quality and comfort of the conditioned laboratory space. The key aspects in which the AHU directly affects the indoor air quality (IAQ) are as follows:

- *Filters*. The filters primarily determine the cleanliness of the supply air from a particulate contamination viewpoint. However, the filters can also be a site of fungal and mold growth if moisture is present. Consideration should then be given to application of air purification technologies such as ultraviolet germicidal irradiation (UVGI). If removal of outdoor pollutants is a concern, gas-phase filtration (e.g., using activated carbon or potassium permanganate) may also be required.
- *Cooling coil*. If dirt is present on the coils or the condensate pan is not sloped to drain properly, mold and fungi can grow in these locations, compromising the IAQ. Again, UVGI or another form of treatment may be in order.
- *Humidifier*. Improper placement or maintenance of the humidifier can result in standing water in the ductwork or downstream wetting and subsequent fungal and mold growth.
- *System integrity*. Poor integrity of the AHU (due to air leakage) between the filters and the supply fan can result in unfiltered or conditioned air moving into the system. Poor integrity downstream of the AHU due to excessive duct leakage or inadequate duct insulation can result in inadequate capacity to condition the laboratory space or condensation and subsequent fungal and mold growth on ductwork or the building components.

DUCT CONSTRUCTION

Duct construction is an important aspect of the HVAC system in any building, but it is especially important in laboratories. This is due to the hazardous nature of the materials contained in the exhaust airstreams and the high energy use of laboratory HVAC systems to condition and move large volumes of nonrecirculated air. Duct construction for laboratories includes general parameters that are common to all types of buildings and duct material selection and application.

GENERAL PARAMETERS

In the construction of ductwork for laboratories, several parameters are universal. These primarily deal with specifications established by the Sheet Metal and Air Conditioning Contractors' National Association (SMACNA) that pertain to almost all ductwork of any building type. They include such things as the following:

- Duct fabrication
- Duct system components
- Performance testing
- Duct material selection and application

Duct Fabrication

As do all types of buildings, laboratories require dimensional stability in the ductwork to prevent shape deformation and maintain the strength of the ductwork. Dimensional stability for laboratories is especially important, as laboratories may require higher than normal velocities in the exhaust ductwork to keep contaminants suspended in the airstream. A ductwork shape that deforms can create additional pressure losses, cause the air velocity to decrease, and increase the fan energy required for air transport. Also, frequent and repeated shape deformation can cause fatigue and reduce the strength of the ductwork.

Containment of the airstream (by minimizing ductwork leakage) is needed in laboratories, as in other types of buildings, to minimize energy losses from conditioning and moving of the air that leaks and to provide the desired room conditions. Additionally, laboratories must minimize ductwork leakage to control the spread of hazardous contaminants that are contained in the exhaust airstream, as this could result in the reintroduction of hazardous materials into the occupied and mechanical equipment spaces in the laboratory building. Also, ductwork leakage should be minimized for the supply ductwork, as considerable energy (due to the larger fans needed to overcome the pressure drop of high-efficiency filters) can be lost when conditioned 100% outdoor air or highly filtered supply air is not delivered to the desired spaces.

Ductwork should be tested to confirm that the specified leakage rates have been attained. Leaks from positive-pressure exhaust ductwork can contaminate the building, so they must be kept to a minimum. Designs that minimize the amount of positive-pressure ductwork are desirable. All positive-pressure ductwork should be of the highest possible integrity. The exhaust fan discharge should connect directly to the vertical discharge stack. Careful selection and proper installation of airtight flexible connectors at the exhaust fans are essential. Some engineers prefer that flexible connectors be used on the exhaust fan inlet only. If flexible connectors are used on the discharge side of the exhaust fan, they must be of high quality and included on a preventative maintenance schedule, because a connector failure could result in the leakage of hazardous fumes into the equipment room. Another viewpoint contends that the discharge side of the exhaust fan should be hard-connected to the ductwork without the use of flexible connectors. The engineer should evaluate these details carefully. The potential for vibration and noise transmission must also be considered. Machine rooms that house exhaust fans should be ventilated to minimize exposure to exhaust effluent (e.g., leakage from the shaft openings of exhaust fans).

As with all ductwork, laboratory ductwork exposure to damage needs to be minimized. Types of damage include exposure to weather, wind forces and windborne debris, seismic forces, temperature extremes, corrosive atmospheres, and biological contamination. As laboratory exhaust systems may frequently be exposed to corrosive chemicals, biological materials, and temperature extremes, the materials used to construct the ductwork should be selected to withstand exposure to the conditions and contaminants that will be present in the laboratory.

Ductwork support includes both hangers for horizontal sections and supplemental structural framing for vertical riser sections. There are various methods available for hanging ductwork in buildings, including laboratory buildings. Hangers consist of three components: the upper attachment to the building structure, the hanger itself, and the lower attachment to the ductwork. Hangers with a maximum spacing of 8 to 10 ft (2.4 to 3 m) typically can support ductwork, although the weight limits for the individual hangers must be considered. Wide ducts typically require closer spacing of hangers to reduce the load on the individual hangers. Ductwork risers should be supported by steel angles or channels fastened to the sides of the duct with welds, bolts, sheet metal screws, or rivets. When fastening supports to the ductwork, caution must be used for ducts greater than 30 in. (762 mm) wide, as the internal pressure of the duct may cause expansion, which could tear fasteners out of the duct material. Riser supports should be provided at one- or two-story intervals (e.g., every 12 to 24 ft [3.7 to 7.3 m]). As with other building components, ductwork construction, hangers, and risers should follow appropriate seismic and wind restraint requirements as applicable for the location of the laboratory. Typically, seismic and wind restraint for ductwork involves using some form of vibration isolation or increased strength materials for the hanger and riser supports of the ductwork. For additional information on seismic and wind restraint design, see Chapter 55 of *ASHRAE Handbook—HVAC Applications* (2015a).

The thermal conductivity of ductwork should always be considered, both for supply and exhaust air systems. During cooling, heat gain needs to be minimized for supply air ductwork so that sufficient cooling capacity can be delivered to the desired conditioned laboratory spaces. Similarly, heat loss from supply air ductwork must be minimized during heating to provide sufficient capacity to maintain the desired space temperature. During cooling, condensation can form on supply air ductwork that has a high thermal conductivity (e.g., uninsulated steel ductwork). For exhaust systems, protecting against heat loss and condensation on the interior of the ductwork are important considerations for laboratories. Heat loss through exhaust ductwork, which primarily is used to remove heat from equipment and heat-producing chemical reactions, can increase the temperature of surrounding spaces and increase the cooling load. Condensation can form on the inside of exhaust ductwork if the exhaust air has a high amount of water/chemical vapors in it and the exhaust air is cooled significantly by the surrounding air.

SMACNA (2005) defines seven static pressure classes for ductwork: 0.5, 1, 2, 3, 4, 6, and 10 in. w.g. (0.12, 0.25, 0.50, 0.75, 1.0, 1.5, and 2.5 kPa). Each class of ductwork is for static pressures above the previous class and up to the number designating the class. For example, 2 in. w.g. (0.50 kPa) ductwork is for ductwork with a static pressure greater than 1 in. w.g. (0.25 kPa) up to 2 in. w.g. (0.50 kPa). For static pressures up to 0.5 in. w.g. (0.12 kPa), 0.5 in. w.g. (0.12 kPa) ductwork would be used. For each size of ductwork within a given pressure class, a minimum duct wall thickness, joint specification, and reinforcement specification are provided (SMACNA 2005). HVAC design drawings should identify each class of ductwork being used.

Duct systems used for perchloric acid fume hoods are seam-welded Type 316L stainless steel, FRP, PVC, or CPVC and generally vertical. Duct systems used in high-containment biological laboratories (BSL-3 and BSL-4 laboratories) should have completely sealed joint construction to facilitate gas decontamination (Wunder 2000).

Duct System Components

For new construction, all air-handling ducts and plenums installed as part of a laboratory HVAC air distribution system should be thermally insulated in accordance with ANSI/ASHRAE/IES Standard 90.1 (ASHRAE 2013a), while existing buildings should meet the requirements of ANSI/ASHRAE/IES Standard 100 (ASHRAE 2015b). Due to the large amount of nonrecirculated air used in laboratories, it may be economically feasible to use more insulation for laboratory ductwork than the minimum values required by the applicable standards. Additional insulation and vapor retarders may be needed to limit vapor transmission and condensation. The selection of insulation determines the heat gain or loss from the ductwork, which impacts the supply air quantities, supply air temperatures, and coil loads. Therefore, to ensure that the required conditions for the laboratory are met, it is important to verify that the insulation requirements selected during the design phase are correctly implemented during construction.

Various sound-control devices are available to minimize the noise created by the air distribution system. First, a vibration isolator can be used to support the supply or exhaust fan that serves the section of ductwork. Fan flexible fabric connectors should be used when the fan is connected to the ductwork. When fan vibration rates are held to a minimum level through dynamic and static balancing, vibration isolators and flexible connections may be eliminated. Flexible connections are a maintenance item that must be replaced, as the fabric deteriorates and leaks form. Also on exhaust systems, the fabric duct is pulled into the airstream, reducing the effective area of the duct. This can create system effect issues with the exhaust fan. Vibration isolators are also used to support air terminal units, to connect the outlets of AHUs and terminal units to ductwork, and sometimes to support the ductwork itself, especially for ductwork located near a fan. Other sound-control methods are also available, such as sound attenuators, duct linings, and sound enclosures. Care should be taken when using duct linings so they are not exposed to moist airstreams, and they should typically not be used in exhaust airstreams because of their potential to trap harmful contaminants.

One very important difference between laboratory and typical commercial building ductwork systems is how fire, smoke, and volume control are achieved. Laboratory exhaust systems cannot use automatic fire and/or smoke dampers because a malfunction or false alarm would cause the hazardous exhaust air to back up into the laboratory and compromise personnel safety (see section 6-10.3 of *NFPA 45* [2015]). Similarly, fire detection and alarm systems should not automatically shut down laboratory exhaust fans. For these reasons, multiple-story laboratory buildings usually require a separate fire-rated mechanical shaft for the ductwork and other utilities servicing each floor/laboratory control unit. However, specific fire/smoke separation and duct dampering requirements should always be verified with the authority having jurisdiction.

In addition to national codes and standards from ASHRAE, SMACNA, National Fire Protection Association (NFPA), American National Standards Institute (ANSI), and others, many laboratories may also have to conform to state and local codes that pertain to the construction of ductwork systems. Common items included in state and local codes are insulation requirements for smoke and flame spread, smoke evacuation procedures, and fire safety controls. High-containment biological laboratories may have requirements for gastight dampers and bag-in/bag-out filter housings to control the release of gas sterilants and pathogens.

Performance Testing

SMACNA defines three different classes for duct sealing requirements (SMACNA 2005), which are listed in Table 6-6. Sealant types include liquids, mastics, gaskets, and tapes. Welding can be used to seal some types of laboratory system ductwork, such as the exhaust ductwork for perchloric acid fume hoods.

The permissible leakage rate and static pressure of the ductwork according to Equation 6-3 define the SMACNA duct leakage classification:

$$C_L = \frac{Q}{\Delta P^{0.65}} \qquad \text{(I-P)} \quad (6\text{-}3)$$

$$C_L = \frac{710 \cdot Q}{\Delta P^{0.65}} \qquad \text{(SI)} \quad (6\text{-}3)$$

where

C_L = leakage class number

Q = leakage rate, cfm per 100 ft^2 (L/s per m^2) of duct surface

P = static pressure, in. w.g. (Pa)

Table 6-7 lists applicable leakage rates based on pressure and sealant classifications. It is up to the laboratory designer, in consultation with laboratory staff and based on the materials being handled in the ductwork, to determine the appropriate leakage values for each ductwork system in a laboratory.

The values contained in Table 6-7 are suggested minimum values, which may need to be exceeded for exhaust system ductwork that handles certain types of hazardous materials. These leakage classifications are shown as a corresponding leakage based on percentage of airflow in Table 6-8.

Table 6-6
SMACNA Duct Sealing Classes

(SMACNA 2005)

Seal Class	Sealing Requirements	Applicable Static Pressure Construction Class
A	All transverse joints, longitudinal seams, and duct wall penetrations	4 in. w.g. (1.0 kPa) and up
B	All transverse joints and longitudinal seams only	3 in. w.g. (0.75 kPa)
C	Transverse joints only	2 in. w.g. (0.49 kPa)
In addition to the above, any VAV system duct of 1 in. w.g. (0.25 kPa) and 1/2 in. w.g. (0.13 kPa) construction class that is upstream of the VAV boxes is to meet Seal Class C.		

Table 6-7
Applicable Leakage Rates

Duct Class	0.5, 1, 2 in. w.g. (0.12, 0.25, 0.50 kPa)	3 in. w.g. (0.75 kPa)	4, 6, 10 in. w.g. (1.0, 1.49, 2.49 kPa)
Seal Class	C	B	A
		Leakage Class	
Rectangular Metal Duct	24	12	6
Round Metal Duct	12	6	3

Duct leakage pressure tests are highly recommended for laboratory HVAC systems in order to verify that the leakage requirements of the ductwork are met and safety and energy use will not be negatively impacted. The procedure for testing laboratory ductwork leakage is the same as for typical buildings. Ductwork leakage testing can be performed on either a whole ductwork system or multiple portions of the ductwork system. Other equipment, such as AHUs, coils, fans, and VAV boxes, should be isolated from the section of ductwork being tested. Leakage testing of ductwork is performed by using a flowmeter, a differential pressure sensor, and a variable-flow fan to positively pressurize the ductwork as shown in Figure 6-43.

Before performing the leakage test, calculate the volume of air that is permissible and confirm that the fan has enough capacity to supply that amount of air. Take care to not overpressurize the ductwork, as the sealant and connections may develop leaks if exposed to excessive pressures. Starting the variable-flow fan at a low speed or with the inlet damper closed and using an appropriately sized fan will help to avoid this problem. The speed of the fan should be increased until the pressure in the ductwork being tested reaches the desired value for the test. When this occurs, record the flow rate from the flowmeter at regular intervals (every 2 to 5 min) for approximately 15 to 30 min to verify that the flow rate and pressure can be sustained for an extended period of time. Calculate the leakage rate of the ductwork and compare it to the leakage requirement. If the leakage exceeds the requirement, inspect the ductwork for audible leaks and use a smoke test or soap solution to locate the leaks in the ductwork. Once the leaks have been located, depressurize the ductwork, repair the leaks, allow the sealant to cure, and retest the ductwork.

Duct Material Selection and Application

Numerous types of materials are available for the construction of ductwork for both the supply and exhaust systems. The choice of duct material depends on which system (supply or exhaust) it will be used for, the types of substances to which the ductwork will be exposed, and applicable codes and standards. For instance, supply air ductwork for laboratories can generally be fabricated from galvanized steel, as for typical commercial buildings. However, the selection of exhaust ductwork for laboratories requires a specific set of criteria for material selection and application that differs significantly from that of typical commercial buildings, as laboratory exhaust may contain a wide variety of hazard-

Table 6-8 Leakage as a Percentage of Airflow[a,b]

(ASHRAE [2009], Chapter 21, Table 7)

Leakage Class	System cfm per ft^2 Duct Surface[c]	Static Pressure, in. of water					
		0.5	1	2	3	4	6
	2	15	24	38	49	59	77
	2.5	12	19	30	39	47	62
48	3	10	16	25	33	39	51
	4	7.7	12	19	25	30	38
	5	6.1	9.6	15	20	24	31
	2	7.7	12	19	25	30	38
	2.5	6.1	9.6	15	20	24	31
24	3	5.1	8.0	13	16	20	26
	4	3.8	6.0	9.4	12	15	19
	5	3.1	4.8	7.5	9.8	12	15
	2	3.8	6	9.4	12	15	19
	2.5	3.1	4.8	7.5	9.8	12	15
12	3	2.6	4.0	6.3	8.2	9.8	13
	4	1.9	3.0	4.7	6.1	7.4	9.6
	5	1.5	2.4	3.8	4.9	5.9	7.7
	2	1.9	3	4.7	6.1	7.4	9.6
	2.5	1.5	2.4	3.8	4.9	5.9	7.7
6	3	1.3	2.0	3.1	4.1	4.9	6.4
	4	1.0	1.5	2.4	3.1	3.7	4.8
	5	0.8	1.2	1.9	2.4	3.0	3.8
	2	1.0	1.5	2.4	3.1	3.7	4.8
	2.5	0.8	1.2	1.9	2.4	3.0	3.8
3	3	0.6	1.0	1.6	2.0	2.5	3.2
	4	0.5	0.8	1.3	1.6	2.0	2.6
	5	0.4	0.6	0.9	1.2	1.5	1.9

Leakage Class	System L/s per m^2 Duct Surface[c]	Static Pressure, Pa					
		125	250	500	750	1000	1500
	10	15	24	38	49	59	77
	12.7	12	19	30	39	47	62
68	15	10	16	25	33	39	51
	20	7.7	12	19	25	30	38
	25	6.1	9.6	15	20	24	31
	10	7.7	12	19	25	30	38
	12.7	6.1	9.6	15	20	24	31
34	15	5.1	8.0	13	16	20	26
	20	3.8	6.0	9.4	12	15	19
	25	3.1	4.8	7.5	9.8	12	15
	10	3.8	6	9.4	12	15	19
	12.7	3.1	4.8	7.5	9.8	12	15
17	15	2.6	4.0	6.3	8.2	9.8	13
	20	1.9	3.0	4.7	6.1	7.4	9.6
	25	1.5	2.4	3.8	4.9	5.9	7.7
	10	1.9	3	4.7	6.1	7.4	9.6
	12.7	1.5	2.4	3.8	4.9	5.9	7.7
8	15	1.3	2.0	3.1	4.1	4.9	6.4
	20	1.0	1.5	2.4	3.1	3.7	4.8
	25	0.8	1.2	1.9	2.4	3.0	3.8
	10	1.0	1.5	2.4	3.1	3.7	4.8
	12.7	0.8	1.2	1.9	2.4	3.0	3.8
4	15	0.6	1.0	1.6	2.0	2.5	3.2
	20	0.5	0.8	1.3	1.6	2.0	2.6
	25	0.4	0.6	0.9	1.2	1.5	1.9

[a] *Adapted from* HVAC Air Duct Leakage Test Manual *(SMACNA 1985).*

[b] *Percentage applies to airflow entering a section of duct operating at an assumed pressure equal to average of upstream and downstream pressures.*

[c] *Ratios in this column are typical of fan volumetric flow rate divided by total system surface. Portions of systems may vary from these averages.*

Figure 6-43
Duct Leakage Testing Apparatus

(Reprinted with permission from SMACNA [2012])

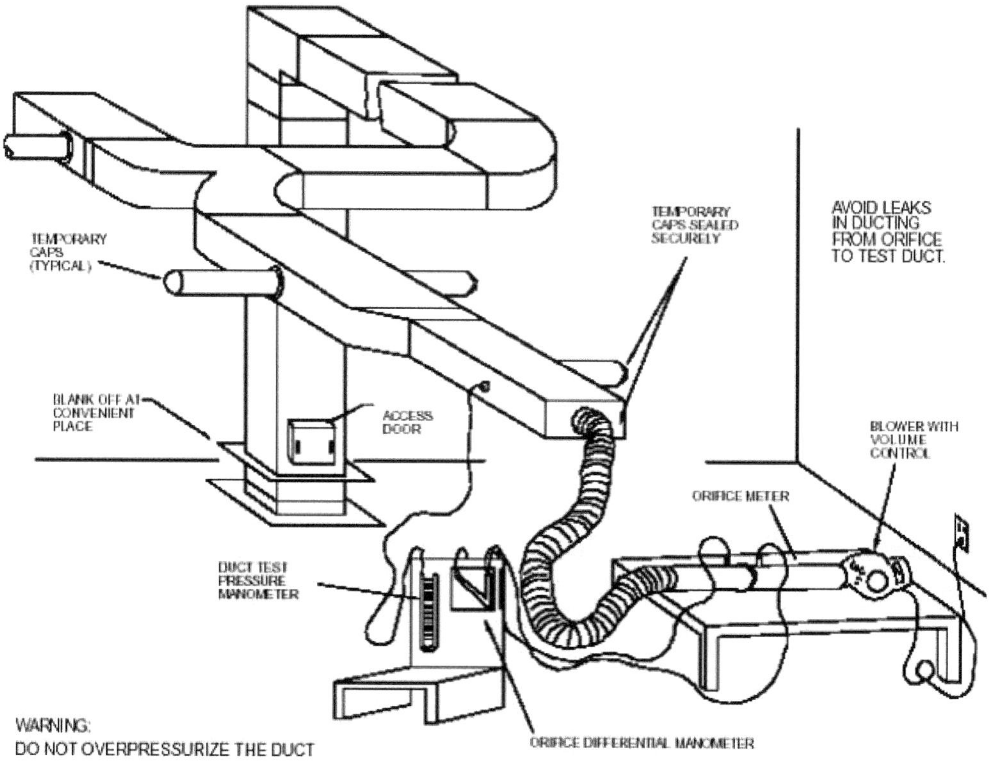

ous materials. Considerations in selecting materials include resistance to chemical attack and corrosion, reaction to condensation, flame and smoke ratings, ease of installation, ease of repair or replacement, and maintenance costs. Selection criteria for both supply and exhaust ductwork are described in detail in the following sections.

Supply Ducts

Typically, general practices for all buildings can be used in the selection of supply duct materials for laboratories. The reasoning for this is that most supply ductwork will not come in contact with sufficient concentrations of hazardous laboratory materials to warrant special criteria such as those used for exhaust systems. Ducts are less subject to corrosion when runs are short and direct, the flow is maintained at reasonable velocities, and condensation is avoided.

Therefore, the following three items are the general considerations to make when designing supply ducts:

- The *ambient temperature* of the space surrounding the supply duct should be considered when selecting supply duct materials. Higher ambient temperatures relative to the supply air temperature can create condensation on the outside of the ductwork. Condensation on ductwork can eventually cause corrosion of the ductwork, leading to small leaks developing or premature failure of the ductwork. Also, condensation can lead to increased formation of mold and fungal organisms, which can cause unacceptable odors or health impacts on occupants. To counter the effects of condensation, sufficient insulation, proper insulation sealing, and nonferrous duct materials can be used. Horizontal ductwork may be more susceptible to corrosion if condensate accumulates in the bottom of the duct. Applications with moist airstreams (cage washers, sterilizers, etc.) may require condensate drains that are connected to chemical sewers.

The design should include provisions to minimize joint or seam corrosion problems. If flow through the ductwork is intermittent, condensate may remain for longer periods because it will not be able to reevaporate into the airstream. Moisture can also condense on the outside of ductwork exhausting cold environmental rooms.

- *Duct velocities and pressures* also need to be considered, as they affect the choice of materials and the overall energy use of the supply air system. For instance, low-pressure ductwork is often made of thinner material than higher-pressure ductwork, but it requires a larger duct to move the same volume of air as a high-pressure system. High-pressure systems require engineering analysis for sound control. The selection of duct velocities and pressures will also affect the type, size, and energy use of the supply air fan.
- The *length and arrangement of supply ductwork* need to be determined. Lengthy runs of ductwork result in larger supply air heat gains or losses and require additional cooling and heating capacity or additional insulation. Simplifying the arrangement of the supply ductwork by minimizing offsets and fittings contributes to limiting the static pressure losses in the system, thereby improving energy efficiency through a reduction in required fan horsepower.

Exhaust Ducts

Because of the wide variation in types of hazardous materials that may be present in the exhaust airstream, exhaust ducts have more design considerations related to the selection of materials and applications than supply ducts.

Exhaust ductwork requires materials that will not react or degrade when exposed to the hazardous materials in the exhaust air. In determining the appropriate duct material and construction, the HVAC engineer should ascertain and evaluate a variety of factors:

- Determine the types of effluents (and possibly combinations) handled by the exhaust system.
- Classify effluents as either organic or inorganic, and determine whether they occur in the gaseous, vapor, or liquid state.
- Classify decontamination materials.
- Determine the concentration of the reagents used and the temperature of the effluents at the hood exhaust port (this may be impossible in research laboratories).
- Estimate the highest possible dew-point temperature of the effluent.
- Determine the ambient temperature of the space housing the exhaust system.
- Estimate the degree to which condensation may occur.
- Determine whether flow will be constant or intermittent (intermittent flow conditions may be improved by adding time delays to run the exhaust system long enough to dry the duct interior prior to shutdown).
- Determine whether insulation, watertight construction, or sloped and drained ductwork are required.
- Determine if fire sprinkler protection within the ductwork is required.
- Select the materials and construction most suited for the application.

Appropriate materials can be selected from standard references and by consulting with manufacturers of specific materials. For example, exhaust ductwork for perchloric acid fume hoods is typically made from stainless steel, and fume exhaust that contains large amounts of corrosive acids may require a protective interior coating (e.g., asphaltum,

Table 6-9
Chemical-Resistant Properties and Flame Ratings

Material	Acids[1]		Alkalies[1]		Organic Solvents[1]	Flammability[2]
	Weak	Strong	Weak	Strong		
Aluminum[3]	N	N	N	N	N	G
Asphalt-coated steel[4]	Y	Y	Y	Y	N	G
Epoxy-coated steel	Y	Y	Y	Y	Y	G
Galvanized steel[5]	N	N	N	N	Y	G
Epoxy glass fiber reinforced[6]	Y	Y	Y	N	Y	SL
Polyester glass fiber reinforced[7]	Y	Y	Y	N	Y	SL
Polethylene fluorocarbon[8]	Y	Y	Y	Y	Y	SE
Polyvinyl chloride (PVC)[9]	Y	Y	Y	Y	N	SE
Polypropylene[10]	Y	N	Y	N	N	SE
316 Stainless steel[11]	Y	Y	Y	Y	Y	G
304 Stainless steel[11]	Y	N	Y	N	Y	G

[1] *N = attacked severely; Y = no attack or insignificant*
[2] *G = good fire resistance; SL = slow burning; SE = self-extinguishing*
[3] *Aluminum is not generally used due to its subjectivity to attacks by acids and alkalies.*
[4] *Asphalt-coated steel is resistant to acids, subject to solvent and oil attacks.*
[5] *Galvanized steel is subject to acid and alkaline attacks under wet conditions.*
[6] *Epoxy glass fiber reinforced is resistant to weak acids and weak alkalies and is slow burning.*
[7] *Polyester glass fiber reinforced can be used for all acids and weak alkalies but is attacked severely by strong alkalies and is slow burning.*
[8] *Polyethylene fluorocarbon is an excellent material for all chemicals.*
[9] *PVC is an excellent material for most chemicals and is self-extinguishing but is attacked by some organic solvents.*
[10] *Polypropylene is resistant to most chemicals and is self-extinguishing but is subject to attack by strong acids, alkalies, gases, anhydrides, and ketones.*
[11] *Types 3016 and 304 stainless steel are subject to acid and chloride attacks, varying with the chromium and nickel content.*

polyvinyl, or polytetrafluoroethylene (PTFE). Commonly used materials for chemical fume exhaust systems and their characteristics are included in Table 6-9.

Each of these materials should be evaluated based on its reactivity to chemicals and flammability. Effluents may be classified generically as organic or inorganic chemical gases, vapors, fumes, or smoke and qualitatively as acids, alkalies (bases), solvents, or oils. Exhaust system ducts, fans, dampers, flow sensors, and coatings are subject to corrosion, which destroys metal by chemical or electrochemical action; dissolution, which destroys materials such as coatings and plastics; and melting or yielding, which can cause a catastrophic failure, weakening of the duct, or thinning of the coating and can occur in certain plastics and coatings at elevated temperatures.

Common reagents used in laboratories include acids and bases. Common organic chemicals include acetone, ether, petroleum ether, chloroform, and acetic acid. The HVAC engineer should consult with the safety officer and scientists of the laboratory because the specific research to be conducted determines the chemicals used and therefore the necessary duct material and construction.

The following criteria should be considered when selecting the materials and applications for exhaust ducts:

- The *nature of hood effluents* plays a significant role in determining the type of duct material to use and determines the sealant type and application procedure as well as the possible need for protective coatings. Therefore, knowledge of the current and future effluents and their possible concentrations is necessary to assess the materials required to handle the exhaust. The resistance strengths of various materials to effluent attacks are provided in Table 6-9.

Table 6-10
Exhaust Duct Velocities

(SMACNA 2010)

Contaminant	Examples	Desired Velocity, fpm (m/s)
Vapors, gases, smoke	All vapors, gases, and smokes	1400–2000 (7.1–10.2)
Fumes	Zinc and aluminum oxide fumes	1400–2000 (7.1–10.2)
Very fine light dust	Cotton lint, wood flours, litho power	2000–2500 (10.2–12.7)
Dry dust and powders	Cotton dust, light shavings	2500–3500 (12.7–17.8)
Average industrial dust	Sawdust, grinding dust	3500–4000 (17.8–20.3)
Heavy dust	Metal turnings, lead dust	4000–4500 (20.3–22.9)
Heavy or moist dust	Budding lint (sticky), lead dust with small chips	4500+ (22.9+)

- The *ambient temperature* of the space surrounding the exhaust duct should be considered when selecting exhaust duct materials. As with supply ducts, the ambient air surrounding the exhaust ductwork can create condensation. However, since the exhaust air is generally warmer and more moist than the ambient air, the condensation will form on the inside of the exhaust ductwork. This may create problems for some exhaust systems, as the hazardous materials in the exhaust air may react with the duct after condensation occurs.
- *Effluent temperature* is also a concern for condensation. High-temperature effluents can cool off significantly in a lengthy section of duct. For selection purposes, an estimate of the lowest possible dew-point temperature of the effluent should be determined.
- *Duct velocities and pressures* affect the choice of duct materials and overall energy use of the exhaust air system, as is the case for supply air systems. However, the duct velocities for exhaust air systems are also determined by the type of materials in the exhaust airstream. Table 6-10 lists recommended design duct velocities for different types of exhaust air contaminants. Higher velocities may be required at the exhaust stack than are listed here for the ductwork.
- The *length and arrangement of ducts* is very important in the selection of exhaust duct materials and applications. Some materials that are exhausted, such as perchloric acid, should have minimal or no horizontal runs to prevent condensation from forming. When using manifolded exhaust ductwork systems, the arrangement of the ductwork must be such that the exhausts from all of the sources are compatible with each other and the exhaust duct material.
- *Variance in exhaust air volume* can present a problem in maintaining needed duct velocities. In instances where VAV control is used for exhaust fans, the minimum acceptable duct velocity must be determined and it must be verified that it will be met during operation from minimum to maximum flow of the VAV exhaust fan.
- *Flame spread and smoke developed ratings for duct materials* should also be considered (ASTM 2015). Depending upon the national, local, and state regulations, some materials may not be acceptable for use as duct material in some applications. Therefore, knowledge of the rating for the materials considered for ductwork is required. Table 6-9 presents the chemical-resistant properties and flame ratings for these materials.

ENERGY EFFICIENCY

In the laboratory, energy consumption is typically high due to the use of 100% outdoor air, minimum required air exchange rates, and reheat energy. With the numerous items of equipment and the need to have constant exhaust and operation of hoods, energy-efficient methods and equipment become all that more important. For this reason, several possible methods for lowering or using energy more efficiently are discussed here.

The electrical power requirements of the fan system are represented by the combined supply and exhaust fan power. Fan input power can be estimated by Equation 6-4:

$$\text{Fan input power (bhp)} = \frac{\text{Airflow} \times \text{System air pressure drop}}{6345 \times \text{Fan system efficiency}(\eta_{fan} \times \eta_{motor} \times \eta_{drive})} \quad \text{(I-P)} \quad (6\text{-}4)$$

$$\text{Fan input power (W)} = \frac{\text{Airflow (m}^3\text{/s)} \times \text{System air pressure drop (Pa)}}{\text{Fan system efficiency}(\eta_{fan} \times \eta_{motor} \times \eta_{drive})} \quad \text{(SI)} \quad (6\text{-}4)$$

Reducing the energy consumed by a laboratory fan system requires changing one or more of the three variables in the equation: fan system efficiency, airflow, or system pressure drop. Table 6-11 indicates areas of opportunity for reducing the power requirement of a laboratory fan system.

FAN SYSTEM EFFICIENCY

Standard design practice usually results in an efficiency of around 62%. Careful selection of a direct-drive fan, where appropriate, and use of high-efficiency motors can increase that efficiency to around 72%, resulting in a power reduction of about 15% at best. While this is an important aspect of design, opportunities are minimal; if energy efficiency is emphasized as an important design criterion, however, conventional design methods can optimize the fan efficiency.

AIRFLOW

The airflow through the system is typically set by the requirements of the facility. Two key design decisions separate typical laboratory facilities from energy-efficient ones. The design decision with the greatest impact on airflow energy use is to use a variable-flow exhaust system rather than a constant-flow or constant-volume system. Varying supply and exhaust flows based on actual usage immediately captures the significant savings possible from reducing the flow—a 25% reduction in airflow results in about a 58% reduction in the fan power required. Note that ANSI/ASHRAE/IES Standard 90.1 requires VAV systems, with some exceptions (ASHRAE 2013a). The second design decision typically impacting laboratory airflow levels is the determination of the minimum laboratory airflow rate. A minimum airflow rate exceeding 1 cfm/ft^2 (0.47 L/s·m^2) should be carefully investigated to verify whether it is necessary.

Table 6-11 Potential for Fan Energy Savings in Traditional Laboratory Designs

(Labs21 2005)

Parameter	Savings Potential	Comment
Fan system efficiency	5% to 15%	Minor potential, traditional design is often OK
Airflow	0% to 60%	VAV supply and exhaust systems provide big savings in fan and conditioning energy when compared with constant-flow systems; actual savings depend on facility usage
System pressure drop	30% to 65%	Traditional design results in energy-intensive laboratory systems; large reductions are possible in numerous areas

SYSTEM PRESSURE DROP

System pressure drop typically offers the greatest potential for energy savings in the ventilation system. Despite the huge impact of the ventilation system on yearly energy consumption, it is not uncommon to find laboratory buildings with a supply and exhaust system combined total of 8 to 12 in. w.g. (2 to 3 kPa) pressure drop. As shown in the fan power equation, this very high pressure drop directly results in a ventilation system with high fan power consumption. To reduce the ventilation system's energy consumption, all that is required is to reduce the system's pressure drop.

The pressure drop in a laboratory fan system should be addressed throughout both the supply and exhaust sides of the system. Although VAV systems inherently reduce pressure drop during nonpeak conditions by reducing the volume of airflow, designers using VAV systems should still consider opportunities for further reduction in pressure drop. The extended hours of operation for laboratory buildings (typically continuous, 24/7 operation) results in higher fan energy use compared to similar nonlaboratory buildings that are scheduled OFF or in setback mode at night. Therefore, the economically optimum pressure drop for a laboratory VAV system is typically lower than that for a nonlaboratory building. Oversizing the ducts also provides significant flexibility to add load to the system in the future. A variable-flow supply and exhaust system reduces the airflow and the power required to run it by approximately the cube of the reduction in flow. The greatest challenge in applying VAV systems in laboratories is ensuring that the balance between supply and exhaust is maintained properly. Typically, these systems make use of one of two general methods: direct pressure-independent measurement of airflows or through-the-wall airflow or pressure measurement.

While these methods are radically different, the results are comparable. The primary difference is that the pressure drop associated with pressure-independent flow measurement valves is about 0.60 to 0.30 in. w.g. (0.15 to 0.75 kPa), in comparison to about 0.05 in. w.g. (0.13 kPa) pressure drop across a typical butterfly control damper. The energy savings associated with a 0.25 in. w.g. (0.62 kPa) pressure difference on the supply and exhaust sides adds up quickly when the entire laboratory facility's airflow is considered.

The fan energy required to supply the required high exhaust discharge velocity, which is part of the total pressure drop as velocity pressure, and the pressure drop in the exhaust stack itself are the most significant components of the exhaust system pressure drop. In a constant-volume system, the stack pressure drop can be minimized using conventional duct design techniques. Minimizing the pressure drop in a VAV exhaust system is made more difficult by the varying exhaust flow. Sometimes laboratory owners may allow for a lower minimum exhaust velocity based on wind tunnel modeling results but still require that the system be designed for a higher minimum exit velocity. In these cases, VFDs may be used to modulate exhaust flow between the minimum and the design exit velocities, allowing for lower discharge velocities during periods of reduced exhaust flows. More complex control methodologies include varying the exit velocity based on outdoor wind speed and measured contamination levels in the exhaust airstream.

AIR RECIRCULATION

Air recirculation is one method available to laboratory designers that can lower the operational costs of central systems. The idea behind this method is to recirculate or reintroduce exhaust from clean spaces into the supply air and laboratory spaces, thus requiring less conditioned air volumes. However, a few standards must be followed within a laboratory with respect to the amounts and types of exhaust air that can be reentrained into supply air.

Chemical and Fume Exhaust

ANSI/ASHRAE Standard 62.1 (ASHRAE 2013b) classifies the exhaust from laboratory hoods as Class 4 air, which is not allowed to be recirculated or transferred to any space or recirculated within the space of origin. *NFPA 45* also specifically states that air exhausted from chemical fume hoods and other special local exhaust systems is not to be recirculated (NFPA 2015). However, some exhaust hoods do allow recirculation of exhaust air within the hood and laboratory depending upon filtration and hazard levels; they are described in Chapter 5 of this Guide.

From Administrative Spaces

One method of air recirculation that is commonly allowed and practiced in laboratories is the use of exhaust air from administrative spaces and nonlaboratory zones. Except for laboratories that deal with highly toxic or hazardous materials, a constant leakage of air from administrative zones is allowed to enter laboratory areas. This is due to the negative pressurization of most laboratories and helps to maintain containment of pollutants. It is possible to actively transfer or recirculate return air from administrative spaces into the laboratory, but caution must be taken to minimize cross-contamination and maintain appropriate pressure relationships. Also, if exhausts from the laboratory or hoods are combined into the general building exhaust, recirculation in this manner should not be used.

Within the Laboratory

Recirculation of air exhausted from the general laboratory space (as distinguished from laboratory chemical hoods) is restricted by ANSI/AIHA/ASSE Z9.5. The reasoning for this is that not all contaminants may be captured by the hoods in the laboratory. Thus, introducing exhaust from the laboratory space would serve only to increase the concentrations of trace chemicals and contaminants to unsafe levels. For this reason it is typical that 100% of the air supplied to the laboratory is exhausted unless the specific criteria outlined in Section 5.4.7.1 of ANSI/AIHA/ASSE Z9.5 are met (AIHA 2012).

However, one instance where laboratory exhaust recirculation is acceptable and often used is in cleanroom laboratories, since high degrees of air change rates (600 to 900 ach) are required for filtration and it requires substantial costs for conditioning of these air volumes. Another case is where ventilation and temperature control are decoupled, such as with a fan-coil or active or passive beam that recirculates room air locally within the same space, for temperature control, with minimum ventilation and exhaust handled separately.

HEAT RECOVERY

Within the laboratory, a large portion of heat that is generated by equipment in the laboratory is exhausted into the atmosphere. This is a significant energy loss for laboratories and a major component of the high operating costs in most laboratory buildings. Therefore, heat recovery is an area of interest for the design engineer to lower annual energy costs for certain climates.

Four commonly used energy recovery systems are often considered for laboratories: energy recovery wheels, flat-plate air-to-air heat exchangers, heat pipes, and runaround coils. All of these systems have unique pros and cons. *NFPA 45* (NFPA 2015) prohibits the use of energy conservation devices that could result in recirculation of exhaust air or exhausted contaminants unless designed in accordance with Section 4:10.1, "Nonlaboratory Air," and Section 5.4.7.1, "General Room Exhaust," of ANSI/AIHA/ASSE Z9.5 (AIHA 2012). To use this technology, careful planning and understanding of the operation and limitations of the equipment as it pertains to the laboratory are necessary. For this reason, a thorough discussion of possible energy recovery options for laboratories is presented in Chapter 10 of this Guide.

REDUCED ROOM VENTILATION RATES

Most laboratories (with the exception of cleanrooms) can be handled with a minimum airflow rate in the range of 6 to 12 ach. However, recent research (Klein et al. 2009) indicates that minimum ventilation rates at the lower end of the 6 to 12 ach range may not be appropriate for all laboratories. Chapter 16 of *ASHRAE Handbook—HVAC Applications* states that "minimum ventilation rates should be established on a room-by-room basis considering the hazard level of materials expected to be used in the room and the operation and procedures to be performed" (ASHRAE 2015a, p. 16.8). The chapter further states that as the operation, materials, and hazard level of a room change, an increase or decrease in the minimum ventilation rate should be evaluated.

Active sensing of air quality in individual laboratories (Sharp 2010) is an alternative approach for dealing with the variability of appropriate ventilation rates, particularly when energy efficiency is important or when less may be known about the hazard level. With this approach, the minimum airflow rate is varied based on sensing the laboratory's actual air quality level or *air cleanliness*. Sensors used to determine air quality should be evaluated for their ability to detect the chemicals being used in the space. When air contaminants are sensed in the laboratory above a given threshold, the minimum air change rate is increased proportionally to an appropriate level to purge the room. When the air is "clean" and contaminants are below the threshold, lower minimum airflow rates may be appropriate.

Extensive studies of laboratory room environmental conditions (Sharp 2010) have shown that the air in laboratories is typically "clean" over 98% of the time. As such, a laboratory can normally operate safely at airflow rates as low as 2 ach with some adjustments to thermal requirements and changes in hood sash heights. In the event that a spill occurs, such as someone dropping a hazardous chemical in the laboratory, a sensor would detect this and increase the airflow into a range of from 8 ach to as high as 16 ach to increase the air dilution until the emergency event is under control, the contaminants are eliminated, and the airflow is reduced.

This demand-based control approach can use either a set of individual sensors in each laboratory room or a single set of sensors integrated into the building's management system for monitoring and control of up to 20 laboratory spaces. However, monitoring using many sets of individual dedicated sensors for each laboratory can drive the first cost, maintenance, and recalibration costs up. Dedicated sensors may be warranted for high-hazard or special laboratories.

There is no way to detect every possible substance or compound in a laboratory with this approach, nor is there a need to do so based on the fact that dilution ventilation at 6 or 8 ach provides less personal protection as compared to the increased dilution ventilation protection and clearance times provided when demand-based control detects laboratory air contaminants. In fact, to provide increased safety over traditional fixed ventilation levels, demand-based control need only sense a large majority of the compounds used in laboratories, not all of them.

For example, the use of a photoionization detector (PID), a type of total volatile organic compound (TVOC) sensor, in combination with a metal oxide chemical sensor can accurately detect hundreds of commonly used laboratory chemicals. Even if a compound is not detectable by itself, many compounds are mixed with solvents that allow the mixture to be detected. These sensors, when coupled with a laser-based particle-counting sensor to identify aerosol vapors and smoke, can detect acid spills or other out-of-control reactions or other events needing increased ventilation. The appropriateness and cost-effectiveness of a laboratory demand-based ventilation control strategy should be carefully reviewed based on the expected chemical use, monitoring frequency, and cost to operate and maintain the sensing equipment.

REFERENCES

AIHA. 2012. ANSI/AIHA/ASSE Z9.5-2012, *Laboratory ventilation*. Falls Church, VA: American Industrial Hygiene Association.

AMCA. 2007. ANSI/AMCA Standard 210-07|ANSI/ASHRAE Standard 51-07, *Laboratory methods of testing fans for certified aerodynamic performance rating*. Arlington Heights, IL: Air Movement and Control Association International.

AMCA. 2010a. AMCA Publication 311-05 (Rev 10/10): *Certified ratings program—Product rating manual for fan sound performance*. Arlington Heights, IL: Air Movement and Control Association International.

AMCA. 2010b. AMCA Standard 99-0401: *Classification for spark resistant construction*. Published in ANSI/AMCA Standard 99-10: *Standards handbook*. Arlington Heights, IL: Air Movement and Control Association International.

AMCA. 2010c. AMCA Standard 99-2404: *Drive arrangements for centrifugal fans*. Published in ANSI/AMCA Standard 99-10: *Standards handbook*. Arlington Heights, IL: Air Movement and Control Association International.

AMCA. 2011. AMCA Publication 203-90 (R2011): *Field performance measurement of fan systems*. Arlington Heights, IL: Air Movement and Control Association International.

AMCA. 2013a. AMCA Publication 211-13: *Certified ratings program—Product rating manual for fan air performance*. Arlington Heights, IL: Air Movement and Control Association International.

AMCA. 2013b. ANSI/AMCA Standard 260-13: *Laboratory methods of testing induced flow fans for rating*. Arlington Heights, IL: Air Movement and Control Association International.

AMCA. 2014. ANSI/AMCA Standard 300-14: *Reverberant room method for sound testing of fans*. Arlington Heights, IL: Air Movement and Control Association International.

Amon, D., T. Smith, P. Mathew, and O. VanGeet. 2007. *System static pressure optimization*. Washington, DC: Laboratories for the 21st Century.

ASHRAE. 2009. Chapter 21, Duct design. In *ASHRAE handbook—Fundamentals*. Atlanta: ASHRAE.

ASHRAE. 2012a. *ASHRAE handbook—HVAC systems and equipment*. Atlanta: ASHRAE.

ASHRAE. 2012b. ANSI/ASHRAE Standard 52.2, *Method of testing general ventilation air-cleaning devices for removal efficiency by particle size*. Atlanta: ASHRAE.

ASHRAE. 2013a. ANSI/ASHRAE/IES Standard 90.1-2013, *Energy standard for buildings except low-rise residential buildings*. Atlanta: ASHRAE.

ASHRAE. 2013b. ANSI/ASHRAE Standard 62.1-2013, *Ventilation for acceptable indoor air quality*. Atlanta: ASHRAE.

ASHRAE. 2015a. *ASHRAE handbook—HVAC applications*. Atlanta: ASHRAE.

ASHRAE. 2015b. ANSI/ASHRAE/IES Standard 100-2015, *Energy efficiency in existing buildings*. Atlanta: ASHRAE.

ASTM. 2015. ASTM E84-15, *Standard test method for surface burning characteristics of building materials*. West Conshohocken, PA: ASTM International.

Caplan, K.J., and G.W. Knutson. 1977. The effect of room air challenge on the efficiency of laboratory fume hoods (RP-70). *ASHRAE Transactions* 83(1).

Caplan, K.J., and G.W. Knutson. 1978a. Laboratory fume hoods: A performance test. *ASHRAE Transactions* 84(1).

Caplan, K.J., and G.W. Knutson. 1978b. Laboratory fume hoods: Influence of room air supply. *ASHRAE Transactions* 84(2).

DiBerardinis, L.J., J.S. Baum, M.W. First, G.T. Gatwood, E. Groden, and A.K. Seth. 2013. *Guidelines for laboratory design: Health, safety, and environmental considerations*, 4th Ed. New York: John Wiley & Sons.

Greenheck. 2006. Laboratory exhaust systems—Vektor™-MD performance. Schofield, WI: Greenheck Fan Corporation. http://old.greenheck.com/pdf/lab_exhaust / VektorMDPerformanceOctober2006.pdf.

Greenheck. 2012a. Greenheck product application guide: Fan application FA/128-12, An analysis of induced flow laboratory exhaust fan systems and the benefit of AMCA 260 certified performance. Schofield, WI: Greenheck Fan Corporation. www.greenheck.com/media/articles/Product_guide/FA128-12_AnalysisLabExhaust.pdf.

Greenheck. 2012b. Damper application considerations for laboratory exhaust systems. *Greenheck industry white papers: From the inside out*. Schofield, WI: Greenheck Fan Corporation. www.greenheck.com/media/pdf/industryarticles/Inside%20Out_Damper%20Applications.pdf.

Harriman III, L.G., D. Plager, and D. Kosar. 1997. Dehumidification and cooling loads from ventilation air. *ASHRAE Journal* 39(11):37–45.

Joo, I.-S., L. Song, M. Liu, and B. Douglas. 2008. Optimal control in three-deck multi-zone air-handling units: A case study. Presented at the Sixteenth Symposium on Improving Building Systems in Hot and Humid Climates, Plano, TX, December 15–17. http://digitalcommons.unl.edu/cgi/viewcontent.cgi?article=1033&context=archengfacpub.

Klein, R., C. King, and A. Kosior. 2009. Laboratory air quality and room ventilation rates. *Journal of Chemical Health and Safety* 9/10.

Labs21. n.d. Labs21 advanced course series—High-performance laboratory exhaust devices. Washington, DC: Laboratories for the 21st Century. http://labs21.lbl.gov/workshop/AdvCourse-HPLabExhDev-5.pdf.

Labs21. 2005. *Laboratories for the 21st Century: Best practices—Low-pressure-drop HVAC design for laboratories*. Washington, DC: Laboratories for the 21st Century. www.i2sl.org/documents/toolkit/bp_lowpressure_508.pdf.

Labs21. 2007. *Laboratories for the 21st Century: Best practice guide—Manifolding laboratory exhaust systems*. Washington, DC: Laboratories for the 21st Century. www.i2sl.org/documents/toolkit/bp_manifold_508.pdf.

Labs21. 2009. *Laboratories for the 21st Century: Best practice guide—Chilled beams in laboratories: Key strategies to ensure effective design, construction, and operation.* Washington, DC: Laboratories for the 21st Century. www.i2sl.org/documents/toolkit/bp_chilled-beam_508.pdf.

McDowall, R. 2007. *Fundamentals of HVAC systems*, SI edition. Atlanta: ASHRAE.

MK Plastics. 2009. Venturi perchloric acid exhaust system. Montreal, Quebec, Canada: MK Plastics Corporation. www.mkplastics.com/documents/literature/Venturi _Catalog_50-01_FEBRUARY_2009.pdf.

NFPA. 2011. *NFPA 70: National electrical code®*. Quincy, MA: National Fire Protection Association.

NFPA. 2015. *NFPA 45: Standard on fire protection for laboratories using chemicals.* Quincy, MA: National Fire Protection Association.

NIH. 2013. Lab module design considerations. *Technical news bulletin*, December. Bethesda, MD: National Institutes of Health, Office of Research Facilities, Division of Technical Resources. http://orf.od.nih.gov/PoliciesAndGuidelines/Documents/Technical%20Bulletins/Lab%20Module%20Design%20Considerations%20December %202013%20Bulletin_508.pdf.

Price. 2006. *Engineering guide—Critical environments*. Suwanee, GA: Price Industries.

Rossi, T. 2006. Exhausting considerations: Designing exhaust systems can be tricky. *Pollution Engineering*, March.

Sharp, G.P. 2010. Demand-based control of laboratory air change rates. *ASHRAE Journal* 52(2):30–41.

SMACNA 1985. Appendix A. In *HVAC air duct leakage test manual*. Chantilly, VA: Sheet Metal and Air Conditioning Contractors' National Association.

SMACNA. 2005. *HVAC duct construction standards—Metal and flexible*, 3d ed. Chantilly, VA: Sheet Metal and Air Conditioning Contractors' National Association.

SMACNA. 2010. *HVAC systems applications*, 2d ed. Chantilly, VA: Sheet Metal and Air Conditioning Contractors' National Association.

SMACNA. 2012. *HVAC air duct leakage test manual*, 2d ed. Chantilly, VA: Sheet Metal and Air Conditioning Contractors' National Association.

Wunder, J.S. 2000. Personal communication from operating experiences with laboratory equipment. University of Wisconsin–Madison.

BIBLIOGRAPHY

ASHRAE. 1995. ASHRAE Standard 110-1995, *Method of testing performance of laboratory fume hoods*. Atlanta: ASHRAE.

ASHRAE. 2013. *ASHRAE handbook—Fundamentals*. Atlanta: ASHRAE.

Labs21. 2003. *Laboratories for the 21st Century: Best practices—Energy recovery for ventilation air in laboratories*. Washington, DC: Laboratories for the 21st Century. www.nrel.gov/docs/fy04osti/34349.pdf.

Labs21. 2005. *Laboratories for the 21st Century: Best practice guide—Minimizing reheat energy use in laboratories*. Washington, DC: Laboratories for the 21st Century. www.i2sl.org/documents/toolkit/bp_reheat_508.pdf.

Labs21. 2008a. *Laboratories for the 21st Century: Best practice guide—Optimizing laboratory ventilation rates*. Washington, DC: Laboratories for the 21st Century. www.i2sl.org/documents/toolkit/bp_opt_vent_508.pdf.

Labs21. 2008b. *Laboratories for the 21st Century technical bulletin: Aerosol ductwork sealing in laboratory facilities*. Washington, DC: Laboratories for the 21st Century. www.i2sl.org/documents/toolkit/bulletin_lab_duct_seal_508.pdf.

Labs21. 2008c. *Laboratory modeling guideline using ASHRAE 90.1-2007 appendix* G. Washington, DC: Laboratories for the 21st Century. www.i2sl.org/documents/toolkit/ashrae_appg_2007_508.pdf.

Labs21. 2011. *Laboratories for the 21st Century: Best practices—Modeling exhaust dispersion for specifying acceptable exhaust/intake designs*. Washington, DC: Laboratories for the 21st Century. www.nrel.gov/docs/fy11osti/52017.pdf.

Murphy, J. 2010. Selecting efficient fans. *ASHRAE Journal* 52(4):64–65.

PG&E. 2011. *High performance laboratories*. A Design Guidelines Sourcebook. Sacramento, CA: Pacific Gas and Electric Company. www.pge.com/includes/docs/pdfs/mybusiness/energysavingsrebates/incentivesbyindustry/Labs_BestPractices.pdf.

Process Cooling

7

OVERVIEW

Within a laboratory, particular pieces of equipment may require cooling that cannot be efficiently or sufficiently supplied by air cooling using mechanisms discussed in Chapter 6. In these cases, using process cooling becomes necessary to ensure the safety of personnel, equipment, and experiments.

Process cooling is the supply of water or other fluid to various types of equipment for cooling purposes. The fluid is pumped through a series of pipes attached to equipment, and heat is transferred from the equipment to the fluid.

For process cooling it is imperative to understand the following:

- Types of water-cooled loads
- Water treatment and quality requirements
- Temperature and pressure requirements
- System pumping configurations

TYPES OF WATER-COOLED LOADS

Several types of equipment require process cooling. Usually, this equipment generates high amounts of heat in very short periods that cannot be adequately handled solely by the primary air system. Typical types of equipment in a laboratory that require process cooling includes lasers, centrifuges, vacuum and diffusion pumps, and various others.

LASERS

Typically used in physics and analytical laboratories, lasers can rapidly create large amounts of heat and thus need to be cooled. Because excessive heat within laser equipment can cause equipment failure and potential fire hazards, process cooling is used for personnel, experiment, and equipment safety. To further maintain safety, lasers are often equipped with interlocks that disable the heat production in the event of insufficient cooling. Smaller lasers, 1 to 2 in. (25 to 50 mm) and with capacities in the milliwatt range, are mostly used for telecommunication. These lasers are typically cooled with thermoelectric coolers. Larger lasers, 2 to 6 ft (0.6 to 1.8 m) in length with capacities ranging from 2 to 20 W, need to be water cooled with their own dedicated cooling system.

The source of the heat produced by lasers may originate from its amplifying medium (solid, liquid, or gas) or from a particular target on which the laser is focused for cutting or boring. In the case of the latter, cooling may be needed to prevent the target from annealing or hardening.

CENTRIFUGES

A centrifuge is a piece of equipment used for liquid/solid separation purposes. It does this by rotating samples at varying speeds using a rotor and an attached motor. Although the centrifuges generally found in clinical and biological laboratories are typically air cooled by an integral fan system, those found in more industrial laboratories often have the option of using process cooling to maintain acceptable internal temperatures during their operation.

VACUUM AND DIFFUSION PUMPS

Vacuum and diffusion pumps are used in processes where distillation or concentration of substances is required. Smaller versions of this equipment typically do not require any supplementary conditioning. However, larger operations, and thus larger pumps, usually do require the introduction of water cooling to ensure the safe operation and longevity of the pump.

OTHER

Other types of equipment, such as large blast ovens, tube furnaces, incubators, and autoclaves, may require the use of process cooling. Generally, the deciding factor for this type of cooling depends on manufacturer requirements and power consumption, since larger consumption can be translated into greater heat gains from the equipment.

WATER TREATMENT AND QUALITY REQUIREMENTS

Where water is used for process cooling, it must be properly treated to obtain and maintain quality. Treatment is used to combat various water problems, such as the following:

- *Corrosion.* This develops when metal is destroyed by chemical or electrochemical reactions within the process water. Corrosion inhibitors and pH control chemicals may be added to the water to minimize damage by reducing the amount of oxygen present.
- *Biological growth.* This is more prevalent in open systems and can cause blockages in the water distribution system. Excessive slime buildup can be expelled with the use of microbicidal materials.
- *Scale formation.* This typically results from the precipitation of calcium carbonate and can be controlled by regulating the pH of the process water.
- *Solid solutes.* When present in systems, these reduce heat transfer and interfere with process water flow. Thoroughly flushing the system after new piping is installed and before system operation can eliminate these suspended materials. In addition, the use of strainers removes larger solids.

In treating the process cooling water to reduce these problems, care must be taken when administering chemicals. Too high or too low a concentration can cause adverse effects. For example, at too-high concentrations of chlorine, corrosion can actually speed up. However, at too-low concentrations, pitting can develop.

Addressing the following quality requirements is helpful to maintaining effective water treatment:

- Power capacity should be available and adequate for chemical injection pumps.
- Domestic water supply should be available for mixing chemicals.
- A drain is necessary for bleed-off waste.
- Turbulent flow through pipes is needed for water treatment chemicals to contact the metal surfaces to be protected.
- Sufficient space must be provided for chemical equipment and storage.

Consult Chapter 49 of *ASHRAE Handbook—HVAC Applications* (ASHRAE 2015) for further details on water treatment.

TEMPERATURE AND PRESSURE REQUIREMENTS

For process cooling, the temperature and pressure requirements of the water used are dependent on the cooling needs of the equipment or the desired effect the cooling is to accomplish. The typical temperature range for most chilled-water systems is 40°F to 45°F (4.4°C to 7.2°C) at a pressure of 120 psi (827 kPa), which is suitable for most applications. However, some equipment may have greater cooling needs. Designers of process cooling systems should consult with the original equipment manufacturer (OEM) or distributor regarding the needs and requirements of the specific equipment.

Because additional cooling of the water for individual equipment can be costly, higher flow rates are often used. Higher flow rates are synonymous with higher process cooling costs because of this increased pumping power, but these costs are usually much lower than the costs associated with an increase in compressor power for the refrigerant subsystem used to cool the water.

There are some instances where the cooling needs of a piece of equipment warrant the use of a dedicated system or the use of domestic water. However, this may require additional piping and can also be costly. Any proposed design should include the manufacturer's specifications for the pressure and temperature thresholds of the equipment.

SYSTEM PUMPING CONFIGURATIONS

This section presents the basic architecture of a chilled-water system used for process cooling. Various pumping subsystems with their alternative pumping configurations are emphasized, and discussions include general practices for good design. More detailed information on overall system design is available in Chapter 13 of *ASHRAE Handbook—HVAC Systems and Equipment* (ASHRAE 2012).

SYSTEM BASICS

Most chilled-water systems, including the ones used for process cooling, are closed instead of open. The main difference between closed and open water systems is the number of interfaces that exist between the water and a compressible gas (e.g., air). Whereas an open system has two or more such interfaces, closed systems have only one. The basic closed chilled-water system comprises the following components/subsystems:

- *Source subsystem:* the point where heat is removed from the process cooling system, e.g., chillers, heat pump evaporators, and heat exchangers.
- *Load:* the point where heat flows into the cooling system from the process, rendering it cooler (e.g., lasers and associated heat-producing equipment).

- *Pump subsystem:* the mechanical devices used and configured to provide sufficient energy to circulate the water in the process cooling system.
- *Distribution subsystem:* the piping system that connects the various components of the entire cooling system and the main conduit through which all the water flows.
- *Expansion chamber:* also known as an *expansion tank* or *compression tank*; provides a space into which the noncompressible water can expand or from which it can contract as changes in temperature influence a volumetric change in the water.

PUMPING SUBSYSTEMS

The pumping system uses circulating pumps that may vary in size. These include small in-line pumps that can deliver 5 gpm (1.1 kg/h) at 6 or 7 ft (1.8 or 2.1 m) head as well as base-mounted or vertical pipe-mounted pumps that can handle much larger flows with head pressures restricted only by the overall system characteristics.

To effectively size a pump for a given process cooling system, the system operating requirements must be prudently matched to the pump operating characteristics. When selecting pumps, the following steps are generally good practice:

- Select for design flow rates and use pressure-drop charts that illustrate the actual closed-loop system piping pressure drops.
- Select slightly to the left of the maximum efficiency point of the pump curve to avoid undesirable pump operation or overloading.
- Select a pump with a flat curve to make allowance for unbalanced circuitry and to provide a minimum differential pressure increase across two-way control valves.

Pumping systems are arranged using the following common configurations:

- Parallel pumping
- Series pumping
- Compound pumping
- Two-speed/variable-speed pumping

Parallel Pumping

Pumps configured in parallel each operate at the same head pressure while providing their own share of the system flow at that pressure. Typically, the pumps used should be of the same size, and thus the characteristic curves for parallel pumps are attained by doubling the flow of the single pump curve. Overloading can be prevented during single-pump operation provided that the pumps in a parallel configuration are constantly powered. A single pump can be used for standby service of up to 80% of design flow, depending on a given system curve and pump curve.

Series Pumping

Pumps configured in series each operate at the same flow rate while providing their own share of the total head pressure at that flow. Like the parallel pump configuration, a single pump in series configuration can provide up to approximately 80% flow for standby service, which is often adequate.

Compound Pumping

Compound pumping, also known as *primary-secondary pumping*, is typically used for larger systems. The main advantages of using this type of configuration are as follows:

- A designer is able to achieve differing water temperatures and temperature ranges in different elements of the process cooling system.
- The control, operation, and analysis of large systems are less complex because individual circuits are hydraulically isolated (i.e., they function dynamically independent of each other).
- The primary and secondary circuits can be designed for different flow characteristics. For instance, the secondary circuit can incorporate two-way valves for improved control and energy efficiency, and the primary circuit can be operated at a constant flow to prevent freezing of chillers.

Two-Speed/Variable-Speed Pumping

Standard two-speed motors used in pumps are quite common. They are available in 1750/1150 rpm, 1750/850 rpm, 1150/850 rpm, and 3500/1750 rpm speeds. At a decrease in system flow, these pump motors will reduce overpressurization—the difference between the pump pressure and the system pressure and a condition that occurs in controlled flow systems.

In contrast to two-speed pumping, variable-speed pumping uses variable-frequency drives (VFDs) that are normally available with an infinite speed range. The pump with the correct controls will be able to follow the system curve and, therefore, like the two-speed pump, avoid any overpressurization effects.

SYSTEMS DESIGN PROCEDURE

Provided the cooling loads, temperature and pressure requirements, and limitations of the equipment to be cooled are understood, engineers and designers may proceed with designing the process cooling system. For a given application, the following basic steps should be followed:

1. Preliminary equipment layout
   a. Determine water flow rates in mains and laterals. Starting from the most remote terminal and progressing toward the pump, sequentially list the cumulative flow in each of the mains and branch circuits for the entire distribution system.
   b. Determine preliminary pipe sizes.
   c. Determine preliminary pressure drop. For each pipe size determined in step 1b, determine the pressure drop in each segment of the distribution system.
   d. Select preliminary pump. The pump should be selected based on its ability to meet the determined capacity requirements.
2. Final pipe sizing and pressure drop determination
   a. Determine final piping layout. Make the necessary adjustments of the preliminary layout to optimize the design based on economics and functionality.
   b. Determine final pressure drop. Repeat step 1c based on the final piping layout.
   c. Select final pump. Based on the final pressure drop calculation in step 2b, plot a pump curve and a system curve and select the pump that operates closest to the design point (the intersection of the two curves).

REFERENCES

ASHRAE. 2015. Chapter 49, Water treatment: Deposition, corrosion, and biological control. In *ASHRAE handbook—HVAC applications*. Atlanta: ASHRAE.

ASHRAE. 2012. Chapter 13, Hydronic heating and cooling. In *ASHRAE handbook—HVAC systems and equipment.* Atlanta: ASHRAE.

Air Treatment

8

OVERVIEW

Treatment of air, either before it is supplied to a space or before it is exhausted to the environment, is often required for the safety and health of occupants or people near the building or for processes that require special conditions. Air treatment involves an understanding of the following:

- Requirements for acceptable and safe levels of pollutants
- Air treatment technologies

This chapter reviews the requirements for allowable concentration limits of typical gases and particulates for state and local ambient concentration limits for short-term exposure, occupational health limits for long-term exposure, and maximum concentration levels for preventing unwanted odors. This is followed by a discussion of the technologies available to achieve acceptable levels.

REQUIREMENTS FOR ACCEPTABLE AND SAFE LEVELS OF POLLUTANTS

The requirements for acceptable and safe levels of pollutants are based on various codes and standards typically developed to protect industrial workers who are exposed to the chemicals. In this section, the allowable concentrations for chemicals are reviewed, followed by a discussion of the consequences of exposure to excessive concentrations. Requirements for industrial processes (electronic manufacturing, etc.) are not discussed, as the laboratory designer will not determine them but rather be given the requirements as a prerequisite.

ALLOWABLE CONCENTRATION LIMITS

The allowable concentration limit of a pollutant depends on the type of pollutant and the length of a person's exposure to it. State and local ambient concentration limits are maximum limits for short-term exposure, whereas occupational health limits are for exposures that are relatively constant for an extended period of time (e.g., eight hours). Therefore, the state and local ambient code limits are typically significantly lower than those for occupational health limits.

State and Local Ambient Concentration Limits for Short-Term Exposure

Most state and local ambient concentration limits are intended to account for instantaneous or short-term exposure. Because these limits account for the exposure of sensitive individuals, they have a safety factor of 40 to 300 (depending on the chemical and the state) compared to the occupational health limits, which must not be exceeded (Petersen and Ratcliff 1991).

To properly account for actual exposures during worst-case releases, usually spills or accidents, it is necessary to calculate the evaporation rate of the liquid chemical. Typically, information on chemical emission rates from existing or proposed local laboratory sites are insufficient.

Once armed with preliminary information of the key chemicals to be used at a local site, Equation 8-1, developed by Kawamura and MacKay (1987), can be used to calculate the evaporation (mass emission) rate, which is primarily dependent on the saturation vapor pressure of the liquid:

$$E = k \cdot M \cdot P(T_s)/(RT) \tag{8-1}$$

where

E = evaporation rate per unit area, g/m^2h
k = mass transfer coefficient, m/h
M = molecular weight
$P(T_s)$ = vapor pressure of the chemical evaluated at the surface of the pool, Pa
R = gas constant (8.314 $Pa \cdot m^3/mol \cdot K$)
T = absolute temperature, K

Using the calculated emission rates of various chemicals along with their corresponding health threshold concentrations, a ratio of emission rate to thresholds can be used to rank chemicals in terms of their relative potential to be harmful. That is, a chemical with a high emission rate and low odor or health threshold would cause the most adverse effects. Table 8-1 presents a list of estimated emission rates and health thresholds for a few selected chemicals.

Occupational Health Limits for Long-Term Exposure

Occupational health limits are typically expressed as an 8-hour time-weighted average (TWA) concentration to which healthy individuals can be exposed during a 40-hour work week with no significant health effects (Petersen and Ratcliff 1991). The Occupational Safety and Health Administration (OSHA) and the American Conference of Governmental Industrial Hygienists (ACGIH) publish lists of, respectively, permissible exposure limits (PELs) and threshold limit values (TLVs) for various chemicals. The National Institute for Occupational Safety and Health (NIOSH) publishes recommended exposure limits (RELs), which are limits based on human and animal studies and recommended to OSHA. Whereas RELs and TLVs are recommendations and guidelines, PELs are legally binding and OSHA has the authority to warn, cite, or fine violators if workers' exposure exceeds PELs.

Refer to the latest OSHA-published PELs and ACGIH-published TLV-TWA thresholds for chemicals (OSHA n.d.; ACGIH 2001a). The thresholds are listed as parts by volume of gas or vapor per million parts by volume of contaminated air.

Detailed rationale for the determination of the TLVs is available from ACGIH (2001a). Basically, the exposure is the average exposure per hour divided by the length of the workday (eight hours). The calculated exposure must not exceed the given limit.

Table 8-1
Estimated Emission Rates and Health Thresholds for Selected Chemicals
(Adapted from Petersen and Ratcliff [1991])

Chemical	Emission Rate (Q), lb_m/s (g/s)	Health Limit (HL), lb_m/ft^3 ($\mu g/m^3$)	Limit Type	$HL/Q \cdot 1$ g/s,[5] lb_m/ft^3 ($\mu g/m^3$)
Isopropyl mercaptan[4]	3.68×10^{-6} (0.00167)	N/A	N/A	N/A
Bromine—100 g spill[2]	8.82×10^{-4} (0.4)	1.25×10^{-7} (2000)	ACGIH STEL	3.12×10^{-7} (5000)
Hydrogen fluoride[4]	8.82×10^{-4} (0.2)	1.62×10^{-7} (2600)	ACGIH CEIL	8.11×10^{-7} (13 000)
Ethylene oxide[4]	6.67×10^{-3} (3.026)	5.61×10^{-7} (9000)	NIOSH—10 min	1.86×10^{-7} (2974)
Chloroform[1]	9.44×10^{-3} (4.28)	6.26×10^{-7} (10 000)	NIOSH—60 min	1.46×10^{-7} (2336)
Ammonia—2 lb spill[3]	2.38×10^{-3} (1.08)	1.50×10^{-6} (24 000)	ACGIH STEL	1.39×10^{-6} (22 222)
Dichloromethane[1]	9.39×10^{-3} (4.26)	1.09×10^{-5} (174 000)	ACGIH TWA	2.55×10^{-6} (40 845)
Acetonitrile[1]	2.07×10^{-3} (0.94)	6.31×10^{-6} (101 000)	ACGIH STEL	6.70×10^{-6} (107 447)
Tetrahydrofuran[1]	5.64×10^{-3} (2.56)	4.61×10^{-5} (738 000)	ACGIH STEL	1.80×10^{-5} (288 281)
Acetone[1]	5.05×10^{-3} (2.29)	1.49×10^{-4} (2 380 000)	ACGIH STEL	6.49×10^{-5} (1 039 301)
Hexane[1]	6.62×10^{-3} (3)	1.12×10^{-4} (1 800 000)	ACGIH—15 min	3.74×10^{-5} (600 000)
Ethyl acetate[1]	4.48×10^{-3} (2.03)	8.75×10^{-5} (1 400 000)	ACGIH TWA	4.30×10^{-5} (689 655)

STEL = short-term exposure limit
CEIL = ceiling limit
TWA = time-weighted average

[1] *Emission rate estimated assuming a 1 gal (4 L) spill and 6 × 6 ft (2 x 2 m) pool; thermal effects included.*
[2] *Emission rate estimated assuming a 0.08 gal (310 mL) spill and 1.5 × 1.5 ft (0.5 x 0.5 m) pool; thermal effects included.*
[3] *Emission rate estimated assuming a 0.3 gal (1 L) spill and 3 × 3 ft (1 x 1 m) pool; no thermal effects included (little variation of vapor pressure with temperature).*
[4] *Emission rate estimated with supplied mass and release durations.*
[5] *These numbers are used for comparison with wind tunnel results when Q = 0.22 lb/s (1 g/s).*

Some substances are also given a ceiling value that cannot be exceeded during any 15-minute period. Other chemicals have an eight-hour weighted maximum value, an acceptable ceiling value (ACV), and a maximum peak value (MPV) for a given duration. The concentration is allowed to exceed the ACV but not the MPV.

The general guidance for acceptable pollutant concentrations in commercial office buildings is one-tenth the TLV (ASHRAE 2013a).

Preventing Unwanted Odors

Odor perception is a very complicated process with significant individual variation. Chemicals with molecular masses greater than 300 are generally odorless (ASHRAE 2013b). Humans can perceive chemicals with molecular weights less than 300 if the concentration is above an odor threshold limit, which can be as low as 0.1 parts per billion (ppb). This is often too low to be detected by a direct reading instrument. Refer to the latest ACGIH-published TLV odor level thresholds (ACGIH 2001b). For more discussion of odors, refer to Chapter 12 of *ASHRAE Handbook—Fundamentals* (ASHRAE 2013b). A safety factor of greater than 1 is required to ensure that occupants can detect the odor before the hazardous levels of the gases are present. Ruth (1986) suggests that both upper and lower limits for odors be documented to account for differences in individuals and experimental techniques.

It is important from an occupant satisfaction basis to recognize that only chemicals with safety factors greater than 1 and with odor threshold limits below the occupational limit are important. The reason for this is that occupants will smell the odor before it is harmful, and their satisfaction with the environment decreases with increasing odors. Conversely, from a health basis, it is those chemicals with odor threshold limits above occupational limits that are of concern. The reason is that once the occupants smell the chemical, they are already at risk for health problems.

In sensitive areas, such as universities and mixed-use spaces, it is recommended that lower limits (i.e., concentrations where only a few individuals can detect the odor) be used. An average between the lower and upper limits can be used in less sensitive areas such as chemical manufacturing plants (Petersen and Ratcliff 1991).

CONSEQUENCES OF EXPOSURE TO EXCESSIVE CONCENTRATIONS

Inadequate performance of systems that results in exposure of users to contaminants above limits given in codes and standards can cause both reversible and irreversible health effects. While skin irritation and headaches are milder reactions that usually are reversible, cancer, chronic respiratory disease, and heart diseases are reactions to exposure to hazardous substances that are often irreversible. Unfortunately, the causes of symptoms are hard to distinguish from other causes, such as viruses and stress reactions. Therefore, identifying the substance causing the symptoms is difficult, as odors can cause milder symptoms such as headaches, nausea, and loss of appetite without exposures that exceed safe limits.

AIR TREATMENT TECHNOLOGIES

Air treatment technologies are applied in the exhaust or the supply based on what contaminants need to be reduced to an acceptable level. Technologies available to achieve acceptable levels of pollutants include fan-powered dilution, filtration, scrubbing, condensing, and oxidation.

FAN-POWERED DILUTION

Using fan-powered dilution systems is the most common method for handling laboratory exhaust effluent, and some laboratory experts suggest that the systems should be thoroughly evaluated before selecting air treatment processes (Wunder 2000). These dilution systems induce clean (outdoor) air into the contaminated airstream and thereby dilute the contaminant exit concentration. The dilution increases the air volume out of the stack and thereby increases the effective stack height, which also ensures good mixing. This strategy can be used with variable-air-volume (VAV) systems to ensure adequate stack exit velocity under all operating conditions. The use of dilution fans is briefly discussed in the Select Fume Exhaust Fans section of Chapter 6.

FILTRATION

Filtration is a treatment process that traps particles or gas in a filter that, when it fills up, can be removed or cleaned. Often, different filtration media are used in series to extend the life of the more expensive, higher-efficiency filter. For example, pleated filters are usually used as prefilters to bag, electrostatic, or activated carbon filters.

Terms commonly used to describe the particles that get trapped by the filters include the following (ASHRAE 2013c):

- Dust—Particles less than 100 μm.
- Fume—Solid particles formed by condensation of vapors on solid materials.
- Smoke—Small solid and liquid particles (and gas) produced by incomplete combustion.

- Bioaerosol—Virus, bacteria, fungi, and pollen.
- Mist—Airborne small droplets.
- Fog—Fine airborne droplets (smaller than mist) normally produced by condensation.
- Smog—Air pollution of particles (solid, liquid, and gas) that impairs visibility and is irritating or harmful.

Particulate Filters

Particulate filters are mechanical devices that use five mechanisms to collect particles:

- *Straining.* The membrane openings of the filter are smaller than the larger particles, which are blocked from passing through the filter.
- *Direct interception.* The particle follows a flow line that results in direct contact with a filter fiber and thus becomes attached to the fiber.
- *Inertial deposition.* The particle initially follows a flow line around a fiber but, because of sufficient size or inertia, deviates from the flow line and comes in contact with a fiber in the filter and attaches to the fiber. A higher percentage of the particles will collide with the fiber at high velocities due to increased inertia, but because of increased high drag and bouncing forces at high velocities, the fibers should have an adhesive coating to ensure that the particles stay in the filter.
- *Diffusion.* Very small particles have a random movement about their streamline (Brownian motion) that can cause the particle to contact a fiber and thereby become attached. This effect is increased with decreasing velocity and decreasing particle size.
- *Electrostatic effect.* The particles are charged and pass by an oppositely charged surface to which they are attracted and attached.

Common types of particulate filters include the following:

- *Fibrous media unit filters.* With this type of filter, particles accumulate on the filter, causing the filtration efficiency and the filter differential pressure to increase as the filter becomes loaded. However, at high particulate loading, the efficiency of some filters decreases due to offloading. These types of filters should be replaced when there is reduced filtration efficiency or the filters reach the upper limit for pressure drop. The efficiency of high-efficiency particulate air (HEPA) filters, however, increases with loading, but less air is moved due to an increase in pressure drop, thus leading to the need for their replacement. Fibrous media filters can range from coarse low-efficiency filters to ultrafine high-efficiency filters (with greater than 99.999% efficiency for 0.12 μm particles). Particulate-matter filters having a minimum efficiency reporting value (MERV) of 6 should be used (ASHRAE 2013a). Examples of typical fibrous media unit filters are shown in Figure 8-1.
- *Renewable media filters.* These filters remain at constant pressure and efficiency by periodically introducing new fibrous filter media into the filtration stream.
- *Electronic air cleaners.* This technology charges particles and collects them on an oppositely charged plate. Electronic air cleaners have relatively constant pressure loss and high filtration efficiency.

Any of the above can be combined to create an efficient filtration system.

Gas-Phase Filters

Gas-phase filters adsorb gaseous contaminants by Van der Waals forces, which are created by physically broken or heated surfaces. Activated carbon is a commonly used adsorbent, as it has an enormous internal surface area compared to its mass (1000 m^2 to 1 g [10,800 ft^2 to 0.035 oz] of activated carbon). Other common adsorbents include acti-

Figure 8-1 Typical Fibrous Media Unit Filters

(Modified with permission from Jasun Envirocare, PLC)

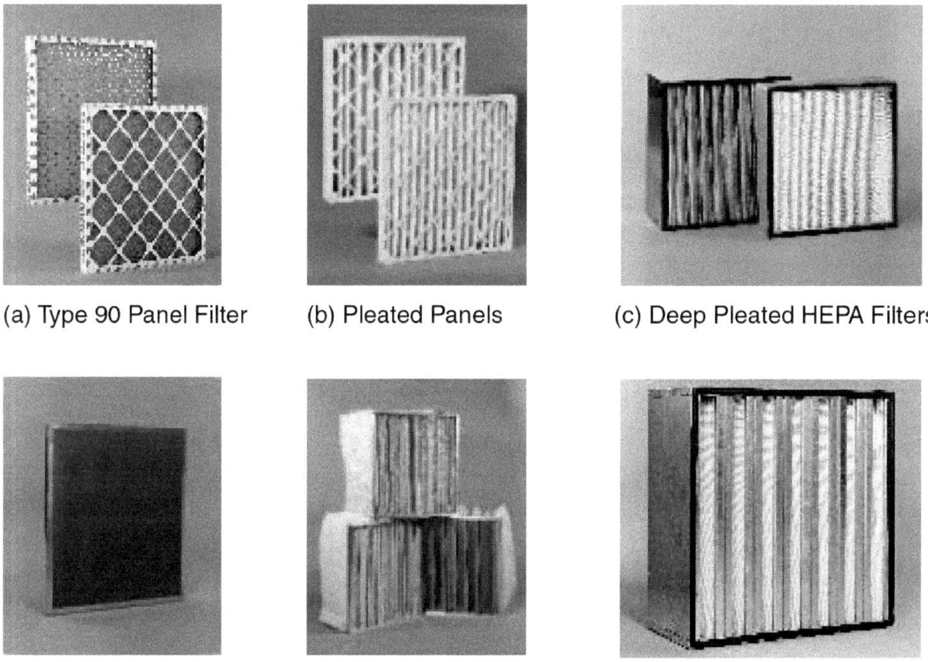

(a) Type 90 Panel Filter (b) Pleated Panels (c) Deep Pleated HEPA Filters

(d) Activated Carbon Panel (e) Bag Filters (f) HEPA Filter

vated alumina silica gel and molecular sieves. Alumina silica gel will absorb water to the exclusion of other chemicals and is for that reason often used to dry airstreams. Molecular sieves are used for specialized pollution control where a specific contaminant needs to be removed.

The adsorbent in gas filters can be impregnated with a chemically reactive substance to contain contaminants that are too loosely bound to the adsorbent alone to improve filtration efficiency. The temperature, concentration, and molecular mass of the contaminant determine the capacity of the adsorbent. The adsorbent process is reversible, and high temperatures or solutions of contaminants in a liquid can reactivate the adsorbent.

There are three types of equipment for adsorption:

- Fixed beds—The absorbent is periodically regenerated or replaced.
- Moving beds—Moving the adsorbent as granular in cycles allows regeneration of the adsorbent and thereby better pollution control.
- Fluidized beds—Suspends fine particles of adsorbent in the gas stream.

Filter Retaining System

Particulate filters and gas-phase filters used in laboratory applications may have to be secured in nuclear-grade bag-in/bag-out housings to protect maintenance personnel from exposure to hazardous materials during change-out procedures. Also, without an engineered filter clamping mechanism, filter bypass can take place, contaminating downstream ductwork and associated equipment. These systems are custom-designed systems supplied by the manufacturer.

Figure 8-2 shows a filter housing with an isolation valve on each end, pressure differential readings across each filter section, and bag-in/bag-out housing.

SCRUBBING

Scrubbers are air treatment systems that use a liquid (typically water) to dissolve or react with the gases to be removed. In the process of removing the gases, scrubbers may

Figure 8-2
Typical
Filter Housing

(Courtesy of Flanders Corporation)

also remove particles that the liquid comes into contact with. The liquid and air are in direct contact by spraying, by dripping the liquid onto a packing through which the airstream is passed, or by bubbling the exhaust through the liquid. The by-product of scrubbing is a liquid solution or solid particles. The liquid can be disposed of or cleaned when the concentration of pollutant in the water is at a level where the efficiency of the scrubbing process is reduced, and the particles can be removed by particulate filters. The three main categories of scrubbers are dry, wet, and particulate.

Dry scrubbing uses an alkaline water solution sprayed into an acid airstream. The acid is absorbed onto the water solution droplet and reacts with the alkaline to form a salt. Solid salt particles then form as the water evaporates, and filters capture the particles. It is important that sufficient time be allotted between the introduction of the alkaline solution and the particulate filter to ensure that the solution and acid reacted and that the water evaporated. Gases typically cleaned with this type of scrubber are hydrochloric acid (from biological waste incinerators), sulfuric acid, sulfur trioxide, sulfur oxides, and hydrogen fluoride.

Wet packed scrubbers remove gaseous contaminates by absorption on the water surface. Particles are removed by impingement of the particles on the liquid droplets. While there is no limit to the particulate removal, the gaseous removal is limited by the partial vapor pressure of the gas. As long as the partial vapor pressure of the gas with respect to the liquid is above the partial pressure of the gas in the exhaust, the gas will be absorbed into the water. Packing material is typically used to distribute the water to provide a large contact area. The packing material should have a high void ratio to ensure a low pressure drop. Common wet packed-scrubber configurations are these:

- Horizontal co-current scrubber
- Vertical co-current scrubber
- Cross-flow scrubber
- Countercurrent scrubber

Particulate scrubbers remove particles by impingement of the particles onto water droplets. Direct interception, inertial deposition, and diffusion are mechanisms in particulate scrubbers that are similar to mechanisms in particulate filters. An additional mechanism in particulate scrubbers is condensation. Condensation occurs when the air or gas is cooled below its dew point. The vapor condenses on the dust particles, which serve as condensation nuclei. The dust particles become larger and the chance of removal is increased. For a given type of particulate scrubber, the efficiency increases as the power applied increases.

The advantages of particulate scrubbers are as follows:

- Constant operating pressure
- No secondary dust source
- Small-parts requirements
- Ability to collect both gases and particles
- Low cost
- Ability to handle both high-temperature and high-humidity gas streams
- Reduced possibility of fire or explosions
- Reasonably small space requirements

The disadvantages of particulate scrubbers are as follows:

- High susceptibility to corrosion
- High humidity in discharge airstream can cause visible exhaust plume
- Large pressure drop, high power requirement
- Possibly difficult or incurs high cost to dispose of wastewater
- Rapidly decreasing efficiency for particles less than 1 μm
- Freeze protection may be required in cold climates

The different types of particulate scrubbers are these:

- *Spray towers and impingement scrubbers.* The gas stream or airstream passes through a single spray or several sprays or a series of irrigated baffles. These scrubbers are low-energy scrubbers (up to 2 kJ/m^3, 0.25 to 1.5 kPa) and generally have a low degree of particle removal (50% to 99% efficiency for particle sizes down to 2 μm).
- *Centrifugal-type collectors.* The gas stream or airstream enters the scrubber tangentially. These scrubbers are medium-energy scrubbers (2 to 6 kJ/m^3, 1.5 to 4.5 kPa).
- *Orifice-type collectors.* The gas stream or airstream passes through narrow openings where the gas/air and the liquid interact. These scrubbers are medium-energy scrubbers (2 to 6 kJ/m^3, 1.5 to 4.5 kPa).
- *Venturi scrubber.* The gas stream or airstream is accelerated up to 60 m/s through a venturi. The liquid is injected right before or at the throat. The rapid acceleration shears the water into a fine mist, which increases the chance of impact. These scrubbers are high-energy scrubbers (>6 kJ/m^3, >4.5 kPa).
- *Electrostatic augmented scrubber.* This is a combination of a particulate scrubber and electrostatic filter.

For a more detailed discussion of scrubbers, refer to Chapter 30 of *ASHRAE Handbook—HVAC Systems and Equipment* (ASHRAE 2012).

CONDENSING

Condensing is a process in which gas or vapor in an airstream condenses on a cold surface or on particles in the airstream. The airstream to be cleaned is cooled to a temperature lower than the dew-point temperature of the gas/vapor in the airstream. The condensed gas/vapor can be removed as a liquid or by removing the larger particles containing the contaminant particle. The cooling capacity of the condenser must be larger than the latent and sensible heat that must be removed, including the latent heat capacity of other gases/vapors that condense.

Gases are removed to a level equal to the partial vapor pressure for the contaminant at the leaving air temperature. When condensing is used to remove particles, the efficiency of the filtration process depends on how much the particle sizes are increased by the condensing liquid and how efficient the filtration equipment is after the condensing has occurred.

OXIDATION

Oxidation of gas is a treatment that breaks down the contaminant into chemicals that are inert or less harmless. This process is, in essence, the same as incineration, but the contaminant is usually at a concentration so low that ignition is impossible. The treatment can be accomplished by using catalysts (catalytic incineration), by increasing the temperature to initialize the wanted reaction (thermal incineration), or by bleeding a strong oxidizer such as ozone into the airstream.

Adsorption and incineration can be combined to reduce the size of the incinerator and the energy consumption of the incineration process as much as 98%. In this combination, the contaminants are accumulated in the adsorption material until it is saturated. The gas stream is then switched to another bed of adsorption material, while the contaminants in the saturated adsorption material are driven off by hot inert gas and incinerated. The volume of this contaminated gas stream is significantly lower, reducing the energy needed for incineration.

REFERENCES

ACGIH. 2001a. *Documentation of the threshold limit values and biological exposure indices*, 7th ed. Cincinnati: American Conference of Governmental Industrial Hygienists.

ACGIH. 2001b. *Threshold limit values for chemical substances and physical agents and biological exposure indices*. Cincinnati: American Conference of Governmental Industrial Hygienists.

ASHRAE. 2013a. ANSI/ASHRAE Standard 62.1-2013, *Ventilation for acceptable indoor air quality*. Atlanta: ASHRAE.

ASHRAE. 2012. Chapter 30, Industrial gas cleaning and air pollution control. In *ASHRAE handbook—HVAC systems and equipment*. Atlanta: ASHRAE.

ASHRAE. 2013b. Chapter 12, Odors. In *ASHRAE handbook—Fundamentals*. Atlanta: ASHRAE.

ASHRAE. 2013c. Chapter 11, Air contaminants. In *ASHRAE handbook—Fundamentals*. Atlanta: ASHRAE.

Kawamura, P.I., and D. MacKay. 1987. The evaporation of volatile liquids. *Journal of Hazardous Materials* 15:343–64.

OSHA. n.d. Table Z-1 Limits for air contaminants. 29 CFR Part 1910, Occupational safety and health standards, toxic and hazardous substances. Washington, DC: U.S. Department of Labor, Occupational Safety and Health Administration. www.osha.gov/pls/oshaweb/owadisp.show_document?p_table=standards&p_id=9992.

Petersen, R.L., and M.A. Ratcliff. 1991. An objective approach to laboratory stack design. *ASHRAE Transactions* 97(2):553–62.

Ruth, J.H. 1986. Odor thresholds and irritation levels of several chemical substances: A review. *Journal of American Industrial Hygienists Association* 47:A-142.

Wunder, J.S. 2000. Personal communication from operating experiences with laboratory equipment, University of Wisconsin–Madison.

BIBLIOGRAPHY

ACS. 1995. *Guide for chemical spill response planning in laboratories*. Washington, DC: CEI/CCS Task Force on Laboratory Waste Management, American Chemical Society. www.acs.org/content/acs/en/about/governance/committees/chemicalsafety/publications/guide-for-chemical-spill-response.html.

ASHRAE. 2012. Chapter 29, Air cleaners for particulate contaminants. In *ASHRAE handbook—HVAC systems and equipment*. Atlanta: ASHRAE.

NIOSH. 2010. *Pocket guide to chemical hazards department of health and human services*. Cincinnati, OH: National Institute for Occupational Safety and Health.

Exhaust Stack Design

9

OVERVIEW

A critical function of a laboratory system is properly exhausting polluted air from a building to avoid pollutant concentrations that are higher than the limits given by codes and standards in sensitive areas, such as air intakes, sidewalks, and building entrances.

This chapter examines the following topics:

- Elements of stack design
- Dispersion modeling

In these sections, the required elements for good stack design and the modeling techniques available to verify a design are presented in detail. Finally, models that can help determine the effects of the design are explained. These models document how the design of the exhaust will perform.

ELEMENTS OF STACK DESIGN

The key elements of stack design are the parameters that influence the dispersion of the gas and the airflow around buildings. The latter is of particular interest, because it has a significant impact on where the pollutants go and is a very complex issue involving the effects of a building's shape and its neighboring buildings on airflow as well as the effects of wind direction and static pressure around the building. The design should avoid reentrainment into building air intakes and contamination of building entrances, exits, and adjacent buildings. These design parameters are put into context in the Design Issues subsection. The issues include aesthetics, energy use, noise and vibration, weather control and drainage, mixing of contaminated airstreams, and maintaining sufficient plume height. Problems with exhaust stacks are release in turbulent airflow, insufficient plume height, stack-tip downwash, and contamination of elevated receptors.

STACK DESIGN PARAMETERS

The key parameters that affect stack design are as follows:

- Stack height
- Exit velocity/exit diameter/volumetric flow rate
- Intake locations
- Surrounding structures

The plume height must be sufficient to ensure that the exhaust plume is sufficiently diluted when it reaches sensitive areas such as building air intakes, entrances, operable windows, and outdoor plazas. The appropriate plume height is a function of the plume rise and the physical stack height for the exhaust system being designed and the subsequent dispersion, or concentration levels, at the aforementioned sensitive locations. While the dispersion modeling process is described in detail in the Dispersion Modeling section, this section focuses on the method for calculating plume height. The information presented in this section can be used to compare the plume heights for different exhaust systems. If two exhaust systems give the same plume height, the dispersion and resulting concentration levels will be identical. The plume height calculation can also be used to determine the most energy-efficient fan system to achieve the same plume rise. For example, the same plume rise may be achieved by adding 5 to 10 ft (1.5 to 3.0 m) to the physical height of the stack and decreasing the exit velocity/volume flow rate, which results in a more energy-efficient system.

The plume height should be calculated at the downwind distance for each air intake of interest based on the stack height and not corrected for buoyancy but reduced if stack-tip downwash occurs:

$$h_e = h_s + h_r - h_d \tag{9-1}$$

where

h_e = plume centerline height, ft (m)
h_s = stack height, ft (m)
h_r = plume rise above stack top at a given downwind distance, ft (m)
h_d = stack-tip downwash, ft (m)

This is a conservative approach that includes safety margins.

For a plume with a low momentum flux (F_m) and a sufficiently high wind speed, the plume rise (h_r) will be close to zero and the lower edge of the plume will spread downward approximately one unit for every five units the plume goes forward. Since it is important that the plume be high enough to avoid contact with building objects and recirculation wake regions, significant plume rise and/or stack height is beneficial (see Figure 9-1). Enhanced plume rise can be accomplished by increasing the momentum flux. If a rain cap is added to the exhaust stack, the plume rise is zero. Therefore, rain caps are

Figure 9-1
Plume Rise

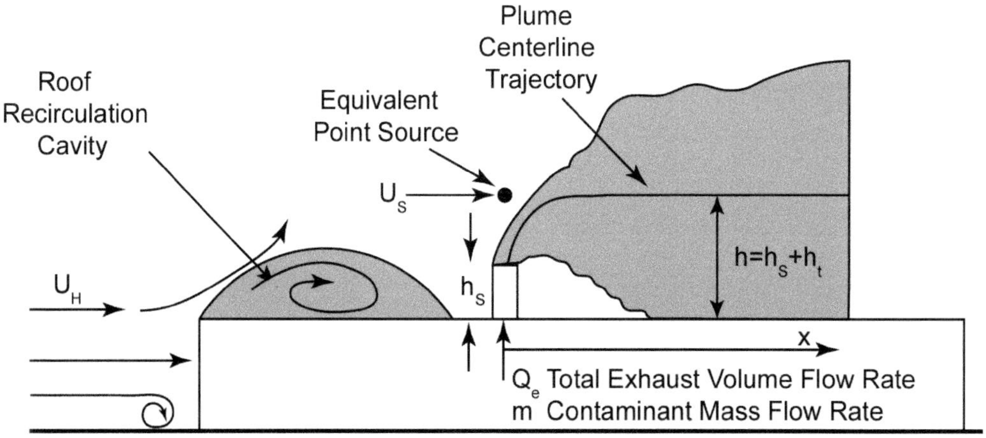

not recommended for laboratory exhaust stacks. The plume rise versus its downwind distance from a stack can be calculated using Equation 9-2:

$$h_r = \min\left\{\left(\frac{3F_m \cdot x}{\beta_j^2 \cdot U_H^2}\right)^{1/3}, h_f\right\} \tag{9-2}$$

where

$$F_m = V_e^2\left(\frac{d^2}{4}\right), \text{ momentum flux, ft}^4/\text{s}^2\ (\text{m}^4/\text{s}^2) \tag{9-3}$$

$$\beta_j = 0.4 + \left(\frac{1.2U_H}{V_e}\right), \text{ jet entrainment coefficient} \tag{9-4}$$

$$h_f = \frac{0.9[F_m \cdot U_H/U^*]^{1/2}}{[U_H \cdot \beta_j]}, \text{ final plume rise, ft (m)} \tag{9-5}$$

$$U_H/U^* = 2.5\ \ln(H/z_o), \text{ logarithmic wind profile equation} \tag{9-6}$$

and

β = design parameter; 1.0 without cap, 0 with cap
x = distance downwind of stack, ft (m)
V_e = stack exit velocity, fpm (m/s)
d = stack diameter, ft (m)
U_H = wind speed at stack top, fpm (m/s)
H = stack height above ground level (e.g., building height plus stack height), ft (m)
U^* = friction velocity, ft (m)
z_o = surface roughness length, ft (m)

Table 9.1 provides examples of z_o values for a range of sites. For example, if z_o equals 0.6 m (suburban) and H = 13 m, substituting into the logarithmic wind profile equation (Equation 9.6) gives $U_H/U^* = 7.7$. The z_o values in Table 9-1, specifically those for the suburban and urban environments, are on the high end for these categories. As such, they tend to result in conservative (i.e., low) plume rise calculations. Site-specific z_o can be obtained through such programs as AERSURFACE (EPA 2013), a free program by the U.S. Environmental Protection Agency (EPA) that can be downloaded at www.epa.gov/scram001/dispersion _related.htm.

Table 9-1
Terrain Factors

Terrain	z_o, ft (m)	a	δ, ft (m)
Flat, water, desert	0.03 (0.01)	0.10	640 (195)
Flat, airport, grassland	0.16 (0.05)	0.14	820 (250)
Suburban	2.0 (0.6)	0.22	1130 (344)
Urban	6.0 (1.8)	0.33	1400 (427)

z_o = surface roughness length, a = exponent that varies with the roughness of the terrain for the local point, and δ = boundary layer thickness of airstream approaching the building.

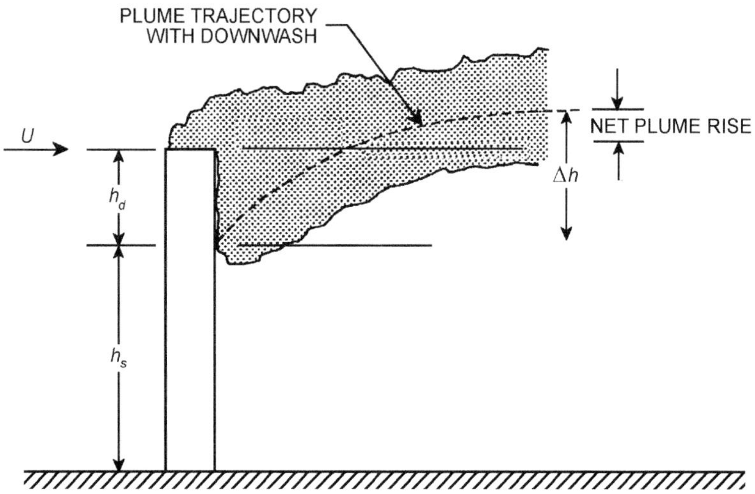

Figure 9-2 Stack Downwash

(ASHRAE [2015b], Figure 3)

A counteractive force to the plume rise is the stack-tip downwash that occurs when the exhaust air is caught within the wake on the leeward side of the stack. This occurs when the wind speed is high compared to the stack exit velocity. ASHRAE (2015b) recommends that the stack velocity be at least 1.5 times higher than the design wind speed (exceeded 1% to 5% of the time) to minimize the effect of wind-induced downwash and provide good initial dilution. For situations where downwash will occur ($V_e / U_H < 1.5$), the additional stack height needed to compensate for the reduced plume rise (see Figure 9-2) can be calculated using Equation 9-7 (ASHRAE 2015b, Equation 9):

$$h_d = d_e(3 - \beta V_e / U_H) \tag{9-7}$$

where h_d is the additional stack height needed to compensate for downwash.

It should be noted that Snyder and Lawson (1991) showed that Equation 9-7 is extremely conservative for most large-diameter stacks (those greater than about 10 in. [254 mm]). Their work showed that stack-tip downwash does not occur for any practical wind condition (i.e., when $V_e / U_H < 0.3$). This work shows that significant energy savings can be achieved by using lower exhaust velocities and confirming that the dispersion is adequate to avoid fume reentry.

Equations 9-2 and 9-7 rely on an accurate estimation of the wind speed. Because the average wind speed is typically not measured at the actual building location but at a meteorological station, the measured wind speed must be corrected for the actual location (see Figure 9-3). This is accomplished using Equation 9-8 (ASHRAE 2013):

$$U_H = U_{met}\left(\frac{\delta_{met}}{H_{met}}\right)^{a_{met}} \cdot \left(\frac{H}{\delta}\right)^{a} \tag{9-8}$$

where

U_H = wind speed at stack top, fpm (m/s)
U_{met} = wind speed at the meteorological station, fpm (m/s)
δ_{met} = boundary layer thickness at the meteorological station, ft (m)
H_{met} = height of the meteorological station, ft (m)
a_{met} = roughness factor for the meteorological station, dimensionless
H = stack height above ground, ft (m)

Figure 9-3
Wind Calculation from Meteorological Station to the Point of Interest

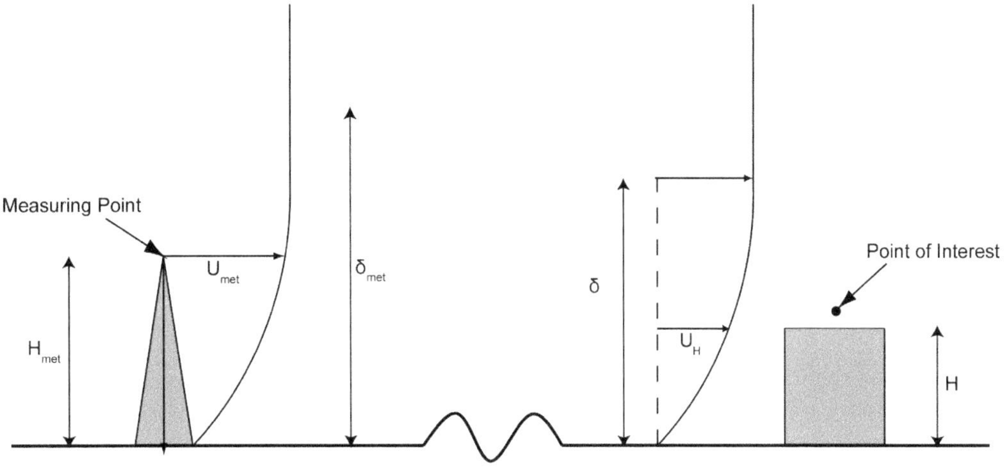

δ = boundary layer thickness of airstream approaching the building

a = exponent that varies with the roughness of the terrain for the local point, dimensionless

The boundary layer is defined as the transition from no velocity (ground) to uniform velocity (gradient height). The height of the boundary layer (δ) and the exponent (a) depend on the roughness factor of the local upwind terrain. Values for these variables have been experimentally defined for the selected terrain types shown in Table 9-1.

For variable-air-volume (VAV) systems, it is important to design the stack for all volumetric flow rates so that adequate dispersion is obtained at the lowest exhaust flow rate. This can be accomplished by carrying out a dispersion modeling analysis as described in the Dispersion Modeling section. The dispersion modeling analysis may show that a higher exit velocity is needed at the low flow setting and that the exhaust diameter will have to be sized accordingly or the stack height increased. Alternatively, outdoor air can be introduced into the exhaust stream before the exhaust fan to maintain the minimum volumetric flow rate and exhaust velocity needed to avoid fume reentry. A variable-geometry exhaust stack that maintains constant velocity regardless of flow can be used but may not be feasible, because the plume rise is a function of both exit velocity and volumetric flow rate, as Equation 9.2 shows. Therefore, caution should be used when implementing a variable-geometry exhaust stack. Maintaining a constant exit velocity does not correspond to maintaining constant plume rise, as can be verified using the previously presented plume rise formulas. When the diameter is decreased, so is the plume rise, even though the exit velocity has remained constant. Therefore, to evaluate the stack performance, dispersion modeling should be conducted using the minimum potential stack diameter to provide assurance that downwind concentrations remain within acceptable limits.

When using a variable-geometry exhaust stack, an iris-type device at the stack exit that opens and closes to increase or decrease the exit velocity should be avoided at all times. This device creates a trip that significantly increases the turbulence intensity of the exhaust flow through the stack, increasing the static pressure drop and ultimately reducing plume rise. Dampers at the end of the stack can create similar issues, because the leading edge of each louver can trip the flow, and the blades are typically too narrow to reestablish fully developed flow. Rather, the variable-geometry device should provide a smooth transition from the larger diameter of the exhaust stack to the smaller diameter of the device while maintaining fully developed flow throughout the transition.

A final design parameter to consider when locating and sizing exhaust stacks is the location of the outdoor air intakes. It is critical that the intake locations be carefully evaluated to avoid contamination from the laboratory exhaust system. In addition, the air intakes should be located so as to avoid other sources of contaminants, such as surrounding building emission sources, dust, fumes from traffic, kitchen exhaust, cooling towers, plumbing vents, loading docks, and leaf-shedding trees.

AIRFLOW AROUND BUILDINGS

The local climate, in particular the wind, is an important parameter for designing stacks for laboratories. Knowledge of the airflow around and over buildings is necessary to avoid contaminating sensitive areas such as building air intakes, building entrances and exits, sidewalks, and adjacent buildings' sensitive areas.

Contaminating the air around building intakes and entrances results in reentrainment of the contaminants exhausted from the laboratories and in contamination of the surrounding buildings. However, determining the airflow around buildings is complicated even for simple geometrical structures. The important parameters for understanding the effect wind has on pollution reentrainment are airflow patterns, relative static pressures around buildings, and adjacent building effects. Wind data can be retrieved from the National Climatic Data Center in Asheville, North Carolina, and local meteorological stations.

Airflow Patterns

The upwind velocity profile, the upwind turbulence, the angle of the approach wind, and the shape of the building—all influence the airflow pattern around a building. The upwind terrain influences the upwind velocity profile. Rougher terrain has higher boundary layers and the free wind speed is reached at higher levels. The wind speed that influences the flow around the building (U_H) will be higher for a building where the terrain is flat compared to that of a building in an urban setting where the terrain is rougher when the free wind speed is the same. The upwind terrain also influences the turbulence of the wind hitting the building. This turbulence created by an upwind building can cause parts of the building to be in the wake or otherwise complicate the airflow pattern, which can be important when considering the stack height. The angle of the approach wind significantly influences the pressure, surface flow pattern, and size and shape of wake regions around the building (see flow patterns in Figure 9-4).

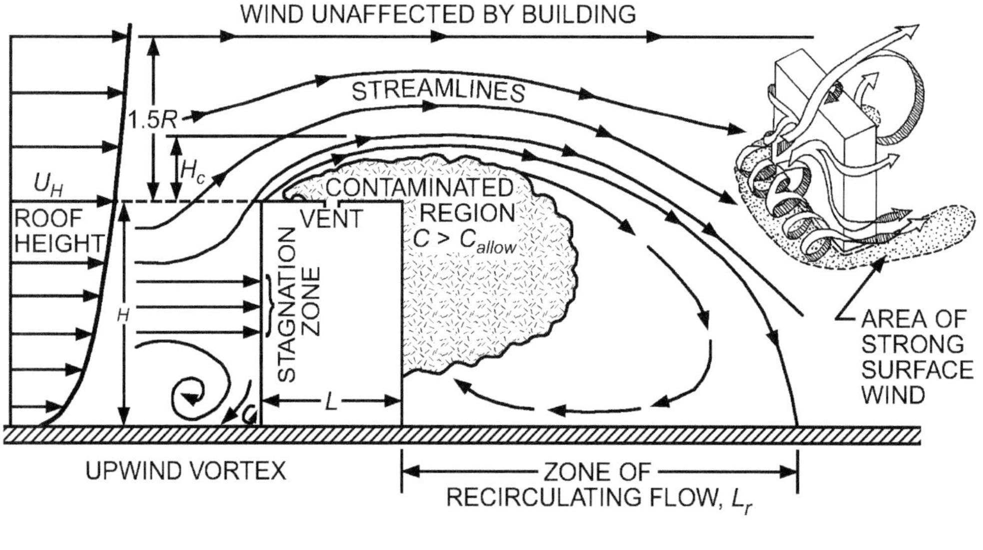

Figure 9-4 Flow Patterns around Rectangular Buildings

(ASHRAE [2015b], Figure 4)

A stagnation point exists on the windward wall and separation occurs at sharp edges. Separation of the airflow creates turbulence and recirculation. The airflow can reattach to the building surface if the building extends far enough downwind. Figure 9-4 illustrates the flow pattern around rectangular buildings. As wind speeds increase, a larger stagnation pressure occurs at the top of the building, resulting in a downwash on the lower one-half or two-thirds of the upwind wall. The top one-quarter or one-third has a flow that is directed upward. However, a horizontal flow can exist in the intermediate zone if the building height is more than three times the width of the building. The size of the recirculating zones created by separation of the flow at sharp edges can be estimated with the help of the scaling length, R (see Equation 9-11), which can be applied to the different locations of separation. While the sizes of these zones will vary over time due to the turbulent flow in the approaching wind and the wake region, the important fact to understand is where air is recirculating and where exhaust flows should not be located to avoid recirculation into the building.

Relative Static Pressures around Buildings

The difference between the static pressures outside buildings and inside buildings influences where exfiltration and infiltration occur. The static pressure distribution on the outside of the building, according to ASHRAE (2013), varies with the angle of attack of the wind. In general, the static pressure caused by wind will be positive on the upward wind surface until the angle of attack exceeds 45°, the static pressure will be negative on the upward wind surface when the angle of attack exceeds 75°, and both positive and negative pressures exist when the angle of attack is between 45° and 75°. On the leeward side, the static pressure will always be negative. The pressure on the roof caused by wind will, on average, be negative, but roof angles steeper than 20° will have parts of or the whole upwind side at a positive pressure. The static pressure can be calculated from Equation 9-9:

$$P_s = C_P \cdot P_v \tag{9-9}$$

where

P_s = static pressure, in. w.g. (Pa)

C_P = static pressure coefficient, dimensionless

P_v = velocity pressure, in. w.g. (Pa)

C_P for different wind directions, building shapes, and locations on the building can be found from figures and tables in *ASHRAE Handbook—Fundamentals* (ASHRAE 2013). The spatially average static pressure coefficient summarized around the building (all four walls) is approximately C_P , sum = –0.2.

The velocity pressure can be calculated from Equation 9-10:

$$P_v = \frac{\rho \cdot U_H^2}{2C_2} \tag{9-10}$$

where

P_v = velocity pressure, in. w.g. (Pa)

ρ = density of air, lb/ft^3 (kg/m^3)

U_H = approach wind speed at upwind wall height H, ft/s (m/s)

C_2 = conversion coefficient, which in I-P units = g_c (= 3.44×10^{-8} ft^4/[lb$_f$·s]) and in SI units = 1

Wind effects can also disturb fume hood operation by creating volume surges resulting in inadequate exhaust. This is unacceptable if highly toxic materials are handled. The exhaust hoods should be tested under both low and high wind conditions. If wind effects are suspected to cause inadequate exhaust, fume hoods should be provided with a flow-sensing monitor with both visual and audio alarms. The static pressure on the inside of the building is influenced by the stack effect and the pressurization caused by the air-handling system. The stack effect is caused by the movement of air in vertical enclosures (stairways, elevator shafts, etc.) and is induced by the density difference between the air in the enclosure and the ambient air (Dorgan and Dorgan 1996). Tall buildings in cold climates are significantly influenced by the stack effect in the heating season; this causes negative pressure in the bottom half of the building and positive pressure in the top half of the building. The air-handling system can be used to pressurize the whole building, or parts of the building, to ensure exfiltration or infiltration or to overcome or reduce the pressure difference created by the stack effect or the wind pressure. Supplying more air than is exhausted and enough to obtain an interior pressure that is higher than the exterior pressure ensures exfiltration, while supplying less air than is exhausted and enough to obtain a pressure that is lower than the exterior pressure ensures infiltration. The difference in the amounts of air supplied and exhausted depends on the pressure that is to be achieved and the leakage of the building structure. However, too high or too low a pressure inside the building can cause difficulties in opening or closing doors and windows and can create drafts, and most HVAC systems are designed to create a neutral pressure.

Adjacent Building Effects

Adjacent upstream buildings can change the upstream flow pattern around the building by creating wakes around parts of the building or the whole building, by creating turbulent flow, or by changing the wind profile. Exhaust gases can contaminate downstream buildings if the exhaust is not diluted enough or does not pass over the downstream buildings. Figure 9-5 shows some examples of how an adjacent building can influence the plume from an exhaust stack. Figure 9-6 shows plume behavior from a shorter laboratory building impacting the side of a taller neighboring building when the wind is directing the plume toward the taller building. However, the impact is not limited to when the wind is blowing from this direction. As shown in Figure 9-7, the plume will also impact the taller building when the wind is blowing from the exact opposite direction. In this case, the plume is caught within the downwind wake region of the taller building and is literally sucked into the leeward side of the taller building.

Flow visualization videos of the plume behaviors shown in Figures 9-6 and 9-7 are available with this Guide at www.ashrae.org/LabDG. As illustrated in these videos, the effect by or on adjacent buildings is often extremely complicated and may require wind tunneling tests or computer simulations. However, some general guidelines are suggested by Wilson et al. (1998):

- Designers should avoid locating stacks near the edge of a roof, where the high wind speed can deflect the plume into the roof edge recirculation and reduce dilution by factors of 2 to 10.
- With the emitting building upwind, a lower roof on an adjacent building will always have higher dilution (lower concentration) than would occur on a flat roof at the emitting building height. Ignoring the step down produces conservative designs.
- Designers should consider increasing the stack height, volume flow rate, and/or exit velocity to increase the plume rise so that the exhaust is not trapped in the recirculation cavity from a taller upwind adjacent building.

Figure 9-5
Adjacent Building Effect on Stack Plume

(Wilson et al. 1998)

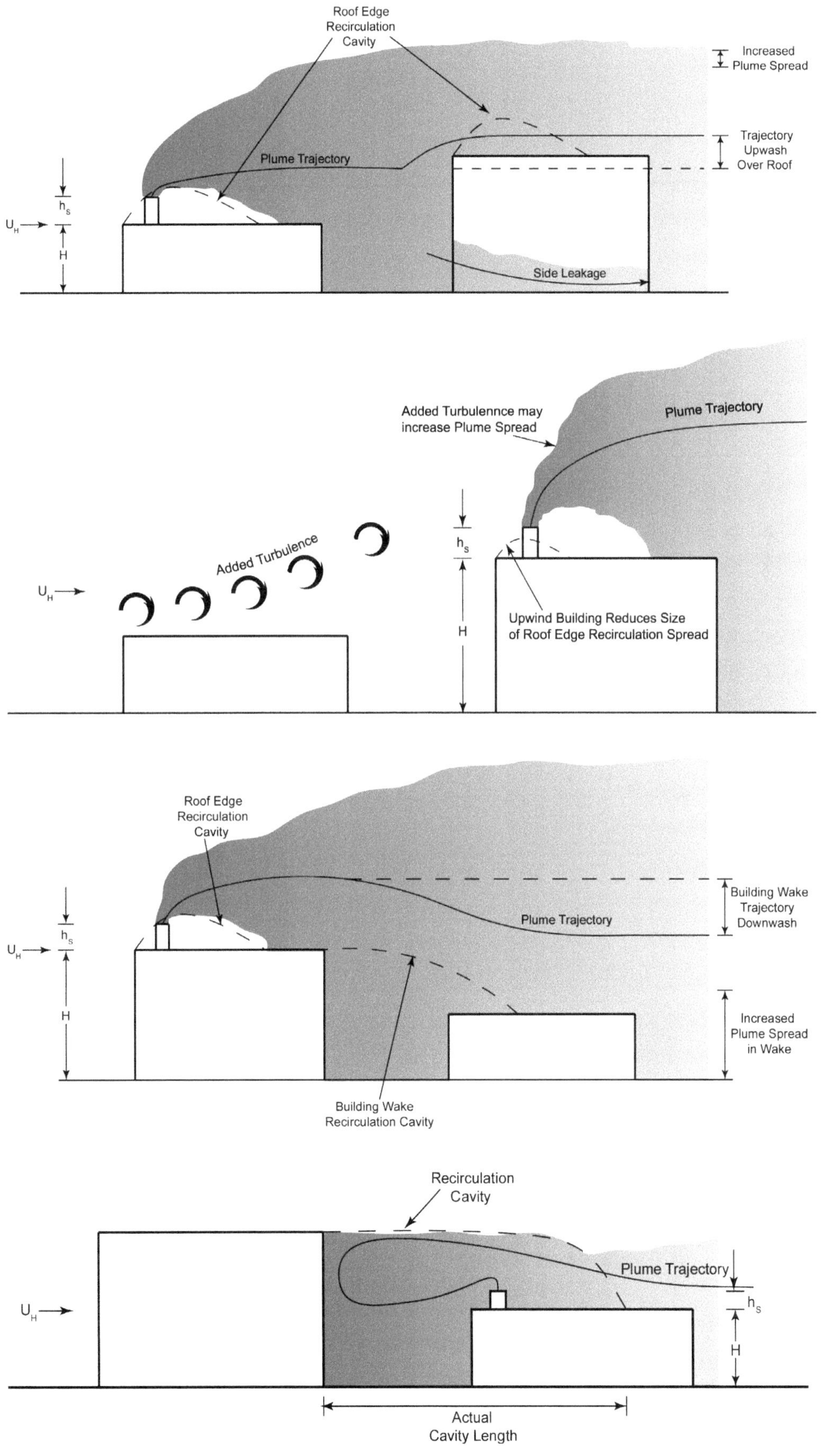

Figure 9-6
Exhaust Plume Impacting the Side Wall of a Neighboring Downwind Taller Building

(Courtesy of CPP, Inc.)

Figure 9-7
Exhaust Plume Caught in the Wake of a Neighboring Upwind Taller Building

(Courtesy of CPP, Inc.)

- When an adjacent building is as high as or higher than the emitting stack, designers should maximize the separation distance between the emitting stack and the adjacent building. If this is not possible, the exhaust system on the emitting building should be designed to create sufficient plume rise to clear the adjacent building.

DESIGN ISSUES

In addition to design parameters, other issues influence the design of the stack:

- Stacks that are highly visible can be considered aesthetically unappealing.
- Energy use is an issue that often causes changed exhaust volumetric flow rates that influence the stack design.
- Noise and vibration can cause problems both outdoors and indoors if not considered.
- Rain can infiltrate exhaust systems with insufficient exit velocities, so it is important to design exhaust stacks for weather control and drainage.

- While rare, mixing of incompatible contaminated airstreams has the potential to create increased hazards.
- Ultimately, the primary function of the laboratory exhaust system is to minimize the reentrainment of toxic and/or odorous emissions; this requires maintaining adequate plume height to clear nearby air intakes and other sensitive receptor locations.

Aesthetics

The required stack height for proper expulsion and dilution of the contaminated air often makes the stack very visible. This visibility is normally unwanted, even though taller stacks generally result in better air quality and lower energy consumption. Therefore, many designers are forced to decide the placement of the stack not only on design parameters, such as location of air intake, recirculating zones, etc., but also on the desire of the owner or architect to make the stack less visible to the surroundings. Fans designed for laboratory exhaust dilution (e.g., induced air systems) may be considered because they typically require less height than conventional stacks and provide the necessary high exit velocities (i.e., high momentum flux). Architectural screens and physical placement of the stack away from the edge of the building can reduce the visibility of the stack. However, when architectural screens are used, care must be taken to ensure that the exhaust cannot enter the recirculating zones the screens create. Additionally, the appearance of the stack can be altered to create a more pleasing appearance than plain metal stacks.

Energy Use

Laboratory buildings are often very high energy consumers due to the need for significant volumes of outdoor air to replace the air exhausted by the fume hoods. To reduce the amount of outdoor air needed, the fume hood exhaust system can be designed to be VAV. However, this also affects the stack design, as the volumetric flow rate out of the stack varies. The requirements for sufficient plume height must be maintained at minimum exhaust air volumes to avoid downwash at the stack and ensure that downwind concentrations remain within acceptable limits. The Maintaining Adequate Plume Height section of this chapter discusses options to accomplish this.

Noise and Vibration

High velocities out of exhaust stacks can cause excessive noise either from turbulence or from the exhaust fan. The noise to the surroundings should be evaluated, and noise reduction equipment should be installed to protect noise-sensitive areas. Local codes and noise level limits should be used as maximum values, with lower values desirable near residential areas. Noise traps and noise screens can effectively reduce noise to acceptable levels but may cause turbulence to effectively lower stack height. Vibration from fans and motors can be transferred to ductwork and cause noise problems from the vibration of the duct against some other object. Further, vibrations cause increased wear and tear on system components, leading to increased maintenance needs and reduced reliability and system life. Vibrations can be reduced with flexible connections to the fan and other vibrating components; the connections must be of a material usable for the contaminants exhausted. Noise and vibration can become important for exhaust velocities above 3000 to 4000 fpm (~15 to 20 m/s).

Weather Control and Drainage

The stack exhaust velocity should exceed 2560 fpm (13.0 m/s) to prevent rain from entering the stack and to prevent condensed moisture from draining into the stack. If a dispersion modeling assessment shows that a lower exhaust speed (and fan energy) is acceptable to avoid fume reentry, or when redundant fans are not in operation, stack designs with internal drains should be installed to prevent water from entering the fan or pooling at low spots in the exhaust ductwork. Caps and coverings or nonvertical exhaust exits greatly reduce the performance of the stack to properly disperse the pollutants and should not be used. Figure 9-8 shows best, good, and poor stack designs for avoiding rain penetration into the exhaust stack.

It is important in cold climates to ensure the stack design prevents freezing by maintaining an internal stack temperature during the minimum design temperature condition. This is typically accomplished by insulating the stack or heating the stack. Freeze protection is especially important for exhaust systems with a washdown system (perchloric fume hoods), where water is used to clean the exhaust air of dangerous pollutants prior to their exhaust. In these systems, the exhaust air is saturated and will quickly freeze up the exhaust stack if allowed to cool and freeze on the stack walls.

Figure 9-8 Best, Good, and Poor Exhaust Stack Design

(ASHRAE [2015b], Figure 2)

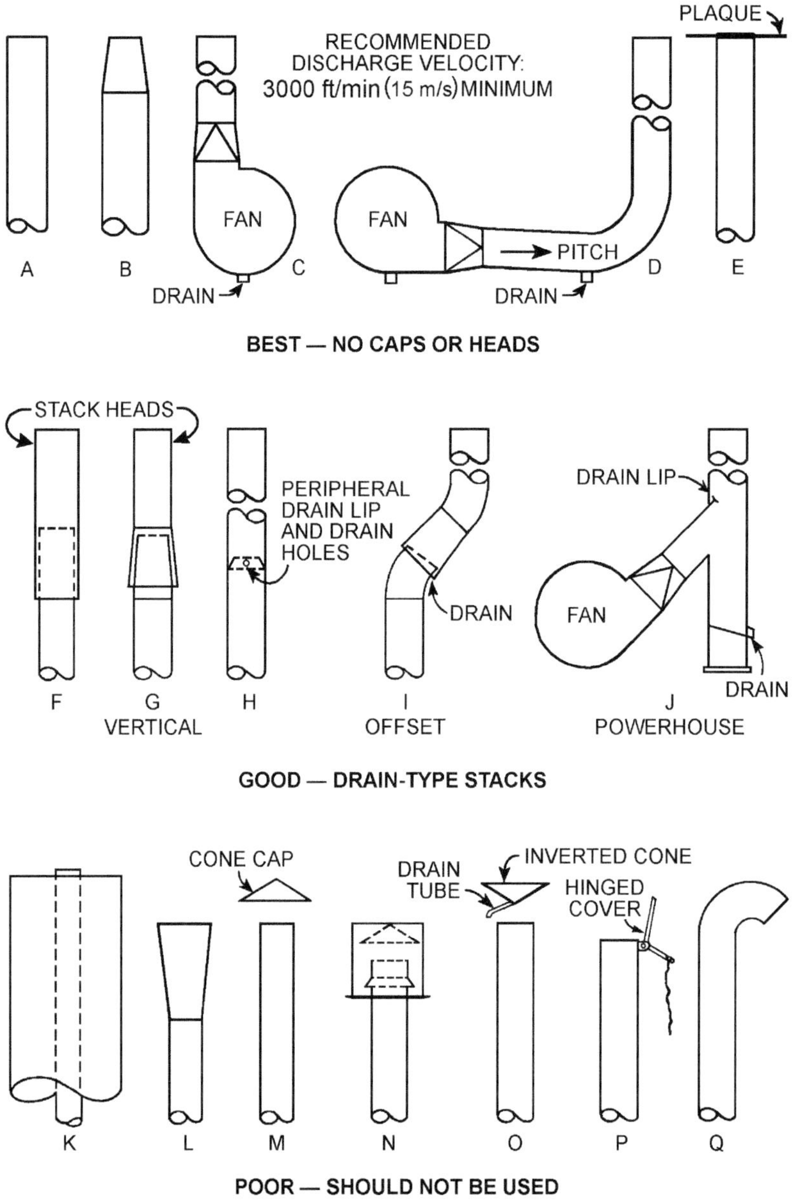

Mixing of Contaminated Airstreams

Section 5.4.3.1 of ANSI/AIHA/ASSE Z9.5 (AIHA 2012) states that laboratory chemical hoods can be ducted into a common manifold with some restrictions for perchloric acid and radioisotope hoods. The guideline further says that, contrary to popular belief, the probability of two or more reagents from different sources combining to produce an explosion or more toxic chemical is very unlikely.

Maintaining Adequate Plume Height

The problem of maintaining adequate plume height at all operating conditions is related to VAV systems, which are primarily designed to save energy by reducing the need for outdoor air to replace the exhausted air from the fume hoods and by reducing the stack velocity. These systems exhaust a varying amount of air, and the air speed out of the exhaust stack subsequently varies. ASHRAE (2015b) recommends designing the stack to meet the requirements at the minimum volume of air exhausted. The requirements at higher volumetric flow rates will thereby be met, but noise and vibration should be considered if the exhaust velocity exceeds 3000 to 4000 fpm (15 to 20 m/s) at the maximum volumetric flow rate.

A dispersion modeling assessment can define the minimum exhaust velocity and volumetric flow rate needed to avoid fume reentry at a specific stack height. If this assessment shows a higher exhaust velocity and/or volumetric flow rate is needed, there are other methods of maintaining adequate plume height:

- Variable-flow geometry (see the cautions in the Stack Design Parameters section on the use of variable-geometry exhausts)
- Bypass air
- Multiple staged fans on a common inlet plenum
- Control system and on-site weather station with reduced velocities set during low winds and high velocities set during high winds
- In-situ chemical monitoring in the exhaust plenum

CAUSES OF PROBLEMS

Many typical problems occur with exhaust stacks that result in significant performance degradation. These include the following:

- Release in turbulent airflow
- Insufficient plume rise
- Stack-tip downwash
- Contamination of elevated receptors

Release in Turbulent Airflow

Release of the exhaust flow into a separated and highly turbulent airflow near the roof, caused by the physical structure, can result in the accumulation of contaminants and prevent the exhaust plume from leaving the building. The turbulent zone may contain critical zones, such as air intakes and entrances that can be polluted even if they are physically far away from the exhaust but are in the same turbulent and recirculating zone. The structures that create turbulent and recirculating zones are architectural screens, roof-mounted air-handling units, penthouses, etc. Turbulent zones are also created behind all "leading edges" on a building, as shown in Figure 9-9. Release in these turbulent and recirculating zones must be prevented. Either the stack must be moved out of this area or the stack height must be increased to release the exhaust above the building. All wind directions should be considered when determining where turbulent and recirculating zones will be created.

Wilson and Winkel (1982) suggest that the dimensions of the turbulent zones be calculated using Equation 9-11:

$$R = B_S^{0.67} \cdot B_L^{0.33} \quad (\text{If } B_L > 8B_S\text{, use } B_L = 8B_S) \tag{9-11}$$

where

R = scaling length, ft (m)

B_S = smaller of upwind building face dimensions, height or width, ft (m)

B_L = larger of upwind building face dimensions, height or width, ft (m)

For buildings with different roof levels, only the height under the roof and width of the roof in question should be used to calculate R, as long as the distance to other obstructions is at least B_S. However, the shape and size of the turbulent zone is not constant, because both the height and the length of the turbulent zone are smaller upwind on the same edge (see Figure 9-10). The turbulent zone at a downwind wall (L_r) extends approximately 1.0R. For a flat-roofed building, the approximate extension of the turbulent zone

Figure 9-9
Turbulent and Recirculating Zones on a Building
(ASHRAE [2013], Figure 3)

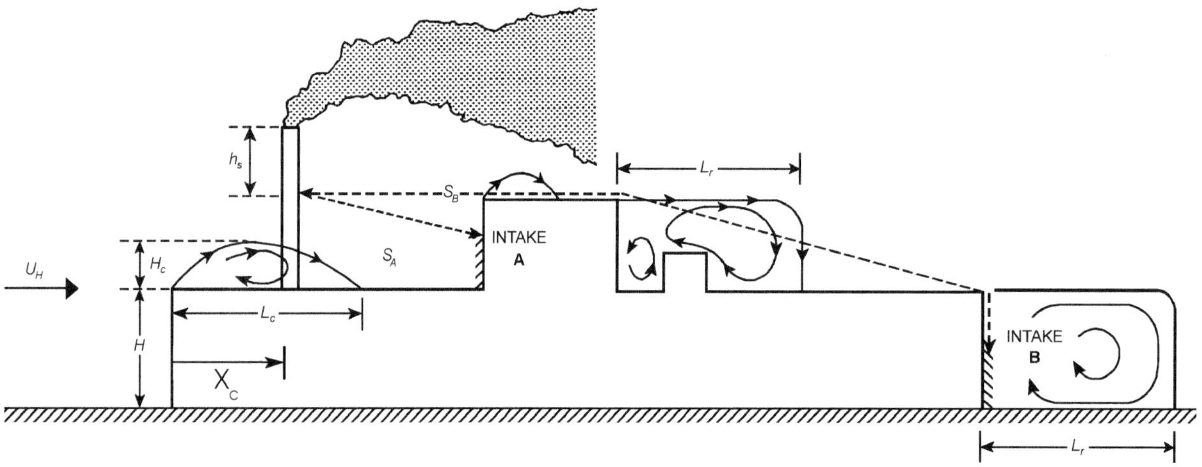

Figure 9-10
Flow Patterns around Two Buildings

(ASHRAE [2015b], Figure 5)

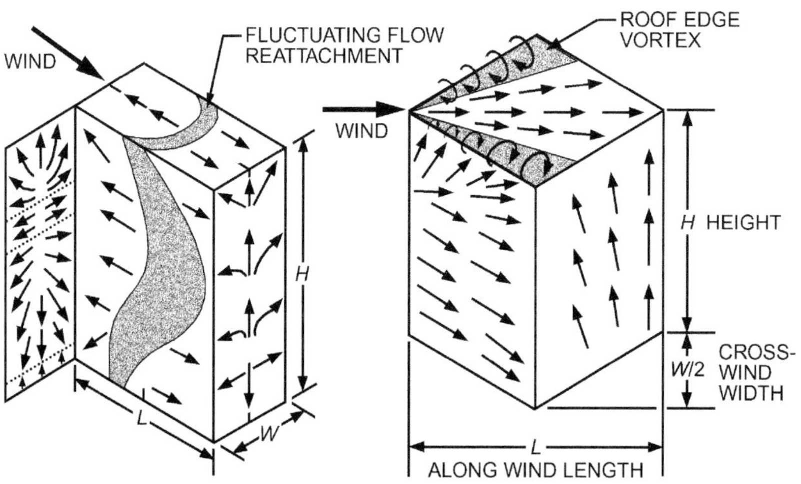

is a maximum height (H_c) of 0.22*R*, a position (X_c) of 0.5*R* behind the leading edge, and to a length (L_c) of 0.9*R* behind the leading edge.

Insufficient Plume Rise

The plume has to rise high enough to avoid contact with critical areas, such as air intakes or entrances to the building. These critical areas can be contained in building recirculation wake (turbulent) regions. Poor design of the exhaust stack (e.g., using a rain cap) or a low exit momentum can reduce or eliminate the plume rise. A conservative design of the stack height does not take into account any plume rise above the height of the stack. This design approach provides a safety margin in case of unfavorable atmospheric conditions.

Stack-Tip Downwash

The exhaust may get caught in the wake region on the leeward side of the stack, reducing the effective stack height. This occurs when the stack exit velocity is insufficient to escape the stack wake. The effect of stack-tip downwash can be diminished through the use of tapered stacks, which increase the exit velocity while reducing the size of the wake. Otherwise, the reduction in plume rise that occurs with stack-tip downwash can be counteracted by increasing the physical height of the stack by the amount specified in Equation 9-7.

Contamination of Elevated Receptors

The exhaust gas can contaminate elevated receptors if their height is within the exhaust plume. To avoid contaminating the receptors, the receptor height can be decreased, the stack height can be increased, or the stack or receptor can be moved so that the receptor is not inside the plume or is far enough away to ensure enough dilution so that the contamination is below the required level.

DISPERSION MODELING

Dispersion modeling is used during design and system troubleshooting to determine the optimum stack design given local conditions. To understand how dispersion modeling works, the characterization of a pollution release must be accomplished including considerations of accidental and steady-state releases. Dispersion modeling can also be used to help design energy-efficient systems by minimizing the required exhaust and bypass air. Advanced systems can be designed such that exit velocities and volumetric flow rates are only increased when infrequent high winds occur and/or when chemicals are present in the exhaust stream (Carter et al. 2005).

EMISSIONS CHARACTERIZATION

In addition to fume hoods, other pollutant sources may also be of concern in laboratory design. These could include emergency generators, kitchens, vivariums, loading docks, traffic, cooling towers, and boilers. The emissions from each source should be characterized to determine their potential impact on air quality at nearby intake locations. Three types of information are needed to characterize emissions: 1) a list of the toxic or odorous substances that may be emitted and the storage quantities, 2) the health limits and odor thresholds for each emitted substance, and 3) the maximum potential emission rate for each substance.

A laboratory must comply with 40 CFR 355 and 40 CFR 370 (GPO 2015a, 2015b) if it stores chemicals listed in the regulation in quantities greater than those specified. Parts 355 and 370 detail how facilities are required to plan for accidental releases of hazardous chemicals. They also stipulate how to respond and when to report the incident to the authorities. Because few laboratories store the listed chemicals in sufficient quantities, most laboratories are not governed by these regulations.

Recommended health limits are based on ANSI/AIHA/ASSE Z9.5, *Laboratory Ventilation* (2012), which specifies air intake concentrations no higher than 20% of acceptable indoor concentrations for routine emissions and 100% of acceptable indoor concentrations for accidental releases. Acceptable indoor concentrations are frequently taken to be the minimum short-term exposure limits (STELs) from the American Conference of Governmental Industrial Hygienists (ACGIH), the Occupational Safety and Health Administration (OSHA), and the National Institute for Occupational Safety and Health (NIOSH), as listed by ACGIH (2015a, 2015b). ACGIH (1989) also provides odor thresholds.

For laboratories, emission rates are typically based on small-scale accidental releases—either liquid spills or the emptying of a lecture bottle of compressed gas—in fume hoods or in rooms. Evaporation from liquid spills is computed from equations in EPA's *Workbook of Screening Techniques for Assessing Impacts of Toxic Air Pollutants* (1992a) based on a worst-case spill within a fume hood or a room. The equations for compressed-gas leaks typically assume the time for emptying of a fractured lecture bottle to be one minute. For other sources, such as emergency generators, boilers, and vehicles, chemical emission rates are often available from the manufacturer.

For general laboratory design purposes, Chapter 16 of *ASHRAE Handbook—HVAC Applications* (2015a) provides an example emission characterization (i.e., design criterion). A 15 cfm (7.5 L/s) chemical emission rate (i.e., one due to a liquid spill or lecture bottle fracture) is assumed along with a limiting concentration of 3 ppm (mg/kg) or less at an intake. For dispersion modeling purposes, the emission characterization can be expressed in SI units as 400 $\mu g/m^3$ per g/s, or dilution units of 1:5300 per 1000 cfm of exhaust flow. ASHRAE (2015a) includes the following disclaimers regarding this design criterion: 1) laboratories using extremely hazardous substances should conduct a chemical-specific analysis based on published health limits, 2) a more lenient limit may be justified for laboratories with low levels of chemical usage, and 3) project-specific requirements must be developed in consultation with the safety officer.

The ASHRAE criterion may be put into perspective by considering the "as manufactured" and "as installed" chemical hood containment requirements outlined in ANSI/AIHA/ASSE Z9.5 (AIHA 2012); that is, a concentration at a manikin outside the chemical hood of 0.05 ppm (mg/kg) or less for "as manufactured" and of 0.10 ppm (mg/kg) or less for "as installed" with a 0.14 cfm (4 L/min) accidental release in the hood as measured using the ASHRAE Standard 110 test method (ASHRAE 1995). The "as manufactured" requirement is equivalent to a design criterion of 750 $\mu g/m^3$ per g/s (1:3000 dilutions per 1000 cfm), and the "as installed" requirement is equivalent to a design criterion of 1500 $\mu g/m^3$ per g/s (1:1500 dilutions per 1000 cfm). Hence, the ASHRAE criterion for a manikin (i.e., a worker outside the chemical hood) is 1.9 to 3.8 times less restrictive than that for the air intake or other outdoor locations. That the air intake has more strict criteria seems reasonable since the worker at the chemical hood can shut the hood or walk away to avoid adverse exposure. Also, the Standard 110 test is not necessarily a worst-case exposure scenario for the worker.

DISPERSION MODELS

Dispersion models are intended to help the designer investigate how the pollutants will be distributed in the atmosphere, around the building, and around adjacent buildings and areas. Understanding this helps identify potential problems that could result in exceeding acceptable pollution concentrations by air intakes, entrances, or other sensitive areas that can be easily corrected during design. These problems can be corrected by changing design parameters such as exhaust exit velocity, volumetric flow rate, location of stack, height of stack, etc. Identifying possible problems during the design phase allows for a less expensive, more efficient solution than correcting a real problem after the building is completed, when, for example, changing the location and/or height of the stack can be very costly. Many models have been suggested, and most of these are available as computer programs. The dispersion models presented in this section include the following:

- EPA models
- ASHRAE dilution/concentration equations
- Wind tunnel modeling
- Computer simulations using computational fluid dynamics

EPA Models

In late 2005, the EPA (Cimorelli et al. 2005) promulgated AERMOD as the replacement for Industrial Source Complex (ISC2) (EPA 1992b). The new model (EPA 2004) includes state-of-the-art boundary layer parameterization techniques, convective dispersion, plume rise formulations, and complex terrain/plume interactions, as well as a building downwash algorithm. AERMOD can be used to calculate short-term (hourly) exposure and long-term (monthly and annual) exposure. Both the short-term and long-term models are divided into three source classifications: point source, line source, and area source. For exhaust stack design, the point source is the model of interest. The EPA guideline (2004) also describes short- and long-term dry deposition models. AERMOD uses the Gaussian equation to calculate the concentration of the contaminant downwind of the source. The models consider the wind speed profile, use plume rise formulas, calculate dispersions factors (which take into consideration different landscapes, building wakes and downwash, and buoyancy), calculate the vertical distribution, and consider decay of the contaminant. More information on AERMOD and other EPA models can be found on EPA's Technology Transfer Network Support Center for Regulatory Atmospheric Modeling website at www.epa.gov/scram001. It should be emphasized that the EPA models are primarily designed to predict concentration (or dilution) values downwind of the building on which the exhausts are located. For predicting the impact at building intakes, operable windows, and entrances, alternative modeling methods are required, such as the ASHRAE dilution/concentration equations, wind tunnel modeling, or computational fluid dynamics (CFD).

ASHRAE Dilution/Concentration Equations

The dilution/concentration equations from Chapter 45 of *ASHRAE Handbook—HVAC Applications* (2015b) can be used to calculate the concentration of contaminants at air intakes on the building roof and/or sidewalls. These equations are based on the Gaussian dispersion equation with simplifications based on ASHRAE-funded research. These equations have been designed to produce conservative (high) concentration estimates, which can result in taller stacks, higher volumetric flow rates, and/or higher exhaust velocities. However, because these dilution equations do not take into account adverse impacts from neighboring and/or complex structures, they may not be valid in these environments.

DilutionCalc, a program that calculates downwind dilutions/concentrations from an exhaust stack using the equations from Chapter 45 of *ASHRAE Handbook—HVAC Applications* (2015b) is available with this Guide at www.ashrae.org/LabDG. Although this calculator is not specifically designed to fully evaluate a stack design, it can be used to spot-check dilution/concentration calculations from a given stack and receptor combination at a specified wind speed. A more elaborate program may be useful for evaluating dilutions/concentrations from an emission source at all nearby receptor locations over a wide range of wind speeds to determine the greatest potential for reentrainment associated with a specific stack design.

Wind Tunnel Modeling

Wind tunnel modeling is often the preferred method for predicting maximum concentrations for stack designs and locations of interest when energy and equipment optimization is desired. It is the recommended approach because it gives the most accurate estimates of concentration levels in complex building environments. A wind tunnel modeling study is like a full-scale field study, except it is conducted before a project is built. Typically, a scale model of the building under evaluation, along with the surrounding buildings and terrain within a 1000 ft (305 m) radius, is placed in an atmospheric boundary layer wind tunnel. A tracer gas is released from the exhaust sources of interest, and concentration levels of this gas are then measured at receptor locations of interest (i.e., air intakes, operable windows, etc.) and converted to full-scale concentration values. Next, these values are compared against the appropriate health or odor design criteria to evaluate the acceptability of the exhaust design. ASHRAE (2015b), Petersen and Cochran (2008), and Snyder and Lawson (1981) provide more information on scale-model simulation and testing methods.

Wind tunnel studies are highly technical, so care should be taken when selecting a dispersion modeling consultant. Factors such as past experience and staff technical qualifications are extremely important.

Computer Simulations using Computational Fluid Dynamics

Computational fluid dynamics (CFD) models are used successfully to model internal flow paths within areas such as vivariums and atriums, as well as in external aerodynamics for the aerospace industry. The aerospace CFD turbulence models, however, are ill suited for modeling the atmospheric turbulence in complex full-scale building environments because of the differing geometric scales. This is exemplified in the conclusions of Castro's (2003) evaluation of applying CFD to the built environment, which have more recently been confirmed by Stathopoulos et al. (2011):

> Despite considerable effort over the last two decades, there is no agreed modeling approach which will automatically yield accurate results for the surface pressure field on and/or the flow field around buildings in the wind. Only large eddy simulation (LES) techniques genuinely have the potential to yield adequate mean and fluctuating data, but these have yet to be fully developed for complex bluff body flows. (Castro 2003, p. 6)

Based on the current state of the art, CFD models should be used with extreme caution when modeling exhaust plumes resulting from laboratory pollutant sources. At this time, research indicates that CFD models can both overpredict and underpredict concentration levels by orders of magnitude, leading to potentially unsafe designs. If a CFD study is conducted for such an application, supporting full-scale or wind tunnel validation studies should be carried out.

A variety of commercial software packages are available for CFD-driven airflow analysis. Most have advanced user interfaces and resulting visualization capabilities as well as sophisticated computational models and solver options. Usually, commercial software includes advanced technical user support provided by vendor specialists. A number of open-source research codes are available as well, but they require a much greater user insight into the underlying solution methods and hardware platforms, and usually no user support or problem-specific validation data are available. It must be emphasized that, regardless of the software package choice, obtaining an accurate numerical solution requires expertise, training, and understanding of the fundamental aspects of CFD algorithm construction and implementation.

There is also a tendency when using CFD for dispersion modeling to significantly limit the number of wind speeds and wind directions included in the assessment. Because each approaching wind condition requires a unique CFD simulation, it is often impractical to conduct sufficient simulations to fully define the worst-case approach condition. Rather, CFD-based assessments typically focus on either predominant wind conditions or assumed worst-case wind conditions. This information can be valuable as a preliminary assessment of the stack design but shouldn't be considered adequate for the final design unless sufficient simulations have been conducted to verify that the worst-case conditions have been adequately modeled.

REFERENCES

ACGIH. 1989. *Odor thresholds for chemicals with established occupational health standards*. Cincinnati: American Conference of Governmental Industrial Hygienists.

ACGIH. 2015a. *2015 Guide to occupational exposure values*. Cincinnati: American Conference on Governmental Industrial Hygienists.

ACGIH. 2015b. *2015 TLVs® and BEIs®*. Cincinnati: American Conference of Governmental Industrial Hygienists.

AIHA. 2012. ANSI/AIHA/ASSE Z9.5-2012, *Laboratory ventilation*. Falls Church, VA: American Industrial Hygiene Association.

ASHRAE. 1995. ASHRAE Standard 110-1995, *Method of testing performance of laboratory fume hoods*. Atlanta: ASHRAE.

ASHRAE. 2013. Chapter 24, Airflow around buildings. In *ASHRAE Handbook—Fundamentals*. Atlanta: ASHRAE.

ASHRAE. 2015a. Chapter 16, Laboratories. In *ASHRAE handbook—HVAC applications*. Atlanta: ASHRAE.

ASHRAE. 2015b. Chapter 45, Building air intake and exhaust design. In *ASHRAE handbook—HVAC applications*. Atlanta: ASHRAE.

Carter, J.J., R.L. Petersen, and B.C. Cochran. 2005. Designing exhaust systems to minimize energy costs. *ASHRAE Journal* 47(7):18–22.

Castro, I.P. 2003. CFD for external aerodynamics in the built environment. *The QNET-CFD Network Newsletter* 2(2), July.

Cimorelli, A.J., S.G. Perry, A. Venkatram, J.C. Weil, R.J. Paine, R.B. Wilson, R.F. Lee, W.D. Peters, and R.W. Brode. 2005. AERMOD: A dispersion model for industrial source applications. Part I: General model formulation and boundary layer characterization. *Journal of Applied Meteorology* 44:682–93.

Dorgan, C.B., and C.E. Dorgan. 1996. *Ventilation best practices guide*. Palo Alto: Electric Power Research Institute.

EPA. 1992a. *Workbook of screening techniques for assessing impacts of toxic air pollutants* (Revised). EPA-454/R-92-024. Research Triangle Park, NC: U.S. Environmental Protection Agency, Office of Air Quality Planning and Standards.

EPA. 1992b. *User's guide for the Industrial Source Complex (ISC2) dispersion models*. Volume II, Description of model algorithms. Washington, DC: U.S. Environmental Protection Agency

EPA. 2004. *AERMOD: Description of model formulation*. EPA-454/R-03-004. Research Triangle Park, NC: U.S. Environmental Protection Agency, Office of Air Quality, Planning and Standards.

EPA. 2013. *AERSURFACE*, Version 13016. Washington, DC: U.S. Environmental Protection Agency, Technology Transfer Network Support Center for Regulatory Atmospheric Modeling.

GPO. 2015a. *Code of federal regulations*. Title 40, Part 355, Emergency planning and notification. Washington, DC: U.S. Government Publishing Office. www.ecfr.gov/cgi-bin/text-idx?tpl=/ecfrbrowse/Title40/40cfr355_main_02.tpl.

GPO. 2015b. *Code of federal regulations*. Title 40, Part 370, Hazardous chemical reporting: Community right-to-know. Washington, DC: U.S. Government Publishing Office. www.ecfr.gov/cgi-bin/text-idx?tpl=/ecfrbrowse/Title40/40cfr370_main_02.tpl.

Petersen, R.L., and B.C. Cochran. 2008. Chapter 24, Physical modeling of air pollution. In *Air quality modeling*. Pittsburgh, PA: EnviroComp and Air and Waste Management Association.

Snyder, W.H., and R.E. Lawson. 1991. Fluid modeling simulation of stack-tip downwash for neutrally buoyant plumes. *Atmospheric Environment*, Vol. 25A.

Snyder, W.H. 1981. *Guideline for fluid modeling of atmospheric diffusion*. EPA600/8–81–009. Research Triangle Park, NC: U.S. Environmental Protection Agency, Environmental Sciences Research Laboratory, Office of Research and Development.

Stathopoulos, T., M. Chaves, and A. Bahloul. 2011. CFD approaches to predicting dilution from exhaust stacks in urban areas. Presented at the 2011 ASHRAE Annual Conference, Montreal, Canada.

Wilson, D.J., and G. Winkel. 1982. The effect of varying exhaust stack height on contaminant concentration at roof level. *ASHRAE Transactions* 88(1):515–33.

Wilson, D.J., I. Fabris, and M.Y. Ackerman. 1998. measuring adjacent building effects on laboratory exhaust stack design. *ASHRAE Transactions* 104(2):1012–28.

BIBLIOGRAPHY

Briggs, G.A. 1984. Chapter 8, Plume rise and buoyancy effects. In *Atmospheric science and power production*. DOE/TIC-27601. D. Randerson, ed. Washington, DC: Technical Information Center, Office of Scientific and Technical Information, United States Department of Energy.

EPA. 1995. *Testing of meteorological and dispersion models for use in regional air quality modeling*. Washington, DC: U.S. Environmental Protection Agency.

Labs21. 2011. *Laboratories for the 21st Century: Best practices—Modeling exhaust dispersion for specifying acceptable exhaust/intake designs*. Washington, DC: Laboratories for the 21st Century. www.nrel.gov/docs/fy11osti/52017.pdf.

McQuaid, J., and B. Roebuck. 1984. *Large scale field trials on dense vapor dispersion*. Final report to sponsors on the heavy gas dispersion trials at Thorney Island, 1982–84. Sheffield, UK: Health and Safety Executive Safety Engineering Laboratory.

Petersen, R.L., B.C. Cochran, and J.J. Carter. 2002. Sp*ecifying Exhaust and Intake Systems*. *ASHRAE Journal* 44(8).

Petersen, R.L., J.J. Carter, and J.W. LeCompte. 2004. Exhaust contamination of hidden versus visible air intakes. *ASHRAE Transactions* 110(1).

Petersen, R.L., J.J. Carter, and M.A. Ratcliff. 1999. Influence of architectural screens on rooftop concentrations due to effluent from the short stacks. *ASHRAE Transactions* 105(1).

Petersen, R.L., M.A. Ratcliff, and J.J. Carter. 1999. Influence of architectural screens of on rooftop concentrations due to effluent from short stacks. *ASHRAE Transactions* 105(1).

Turner, D.B. 1994. *Workbook of atmospheric dispersion estimates*, 2d ed. Boca Raton, FL: Lewis Publishers.

Energy Recovery

10

OVERVIEW

Because laboratories often use 100% outdoor air, energy recovery is an important consideration for design. Energy recovery uses the sensible and latent heat in the exhaust air or other source that would otherwise be wasted to pretreat (heat or cool) the supply airstream, thereby saving energy. These savings can be substantial, with lifetime savings and payback periods of energy recovery systems being very favorable. In some projects, energy recovery has been the main contributor to savings from energy-efficient installations, with others involving variable-air-volume (VAV) and diversification considerations (Streets and Setty 1983). Due to the strong influence of the climate on potential energy savings, each energy recovery system must be evaluated based on local variations in climate and in energy costs.

Energy can be recovered from exhaust air, process water, and cooling water from air-conditioning equipment. Energy recovery from water only transfers sensible energy, while energy recovery from air offers the opportunity to transfer both sensible and latent loads. However, care must be taken to avoid contaminating the supply airstream. Appendix A of *NFPA 45* states that the use of devices for energy conservation purposes should consider the "potential contamination of the fresh air supply by exhaust air containing vapors of flammable or toxic chemicals" (NFPA 2015, p. 45-30). Biological contamination is also a concern.

This chapter discusses the following:

- Air-to-air energy recovery
- Water-to-air energy recovery
- Selection parameters

More details on the theory of energy recovery equipment are available in Chapters 9 and 12 of *ASHRAE Handbook—HVAC Systems and Equipment* (ASHRAE 2012a).

AIR-TO-AIR ENERGY RECOVERY

The intent of air-to-air energy recovery systems is to transfer sensible energy only or enthalpic energy (sensible and latent energy) between two airstreams, typically the exhaust air and supply air. Sensible energy recovery equipment includes runaround loops, heat pipes, sensible heat wheels, fixed-plate heat exchangers, thermosiphons, and reverse-flow plate exchangers. Enthalpic energy recovery includes processes such as liquid desiccants and evaporative cooling.

When selecting the type of air-to-air device, several design issues must be addressed, including the following:

- *Relative airstream location.* Some air-to-air systems require the two airstreams to be located adjacent to each other, whereas other systems can have a reasonable distance between the airstreams.
- *Cross-contamination potential.* Unless there is a physical barrier between the two airstreams, contamination of the supply air could occur.
- *Corrosion.* The chemical in the exhaust air can corrode the energy recovery device, either directly or due to the formation of condensate when the warm moist exhaust air comes in contact with the cold heat exchanger surface. Corrosion can lead to cross-contamination by breaking down the barrier between the two airstreams. Perchloric acid fume exhausts should not be considered for energy recovery due to corrosion and explosion hazards.
- *Fouling.* Fouling of the heat exchanger reduces the supply or exhaust air volume, resulting in reduced heat exchanger and system performance, as cleanliness is critical to maintaining system performance.
- *Freeze-ups.* Freeze-ups occur when the exhaust air contains sufficient moisture and the outdoor air is below freezing. The moisture in the exhaust air condenses and freezes. As frost builds up, airflow is decreased until there is no flow, which seriously compromises occupants' health and safety.

When cross-contamination is a major concern, physically separating the two airstreams by using a second media is recommended. Runaround loops, heat pipes, or evaporative cooling should be used. In some instances, the exhaust air is cleaned by a scrubber or washer prior to the heat exchanger to reduce the potential for cross-contamination.

SENSIBLE ENERGY RECOVERY EQUIPMENT

Runaround Loops

The runaround loop is an energy recovery system where the heat from the exhaust air is transferred to a water-based medium (glycol) using a typical coil heat exchanger. It is then circulated through pipes to the supply air, where the water releases the energy recovered from the exhaust air using another coil heat exchanger. The water then returns to the exhaust air coil. Figure 10-1 depicts this energy recovery system, which is usually used in systems where it is uneconomical or not desired to have the exhaust duct and the supply duct next to each other. These systems have a relatively high initial cost and low performance compared to other energy recovery systems. Runaround loops have a sensible heat transfer effectiveness of 45% to 65% (ASHRAE 2012b).

Runaround loops are excellent energy recovery systems for laboratory applications where it is critical that no cross-contamination occur. To achieve increased effectiveness, the exhaust coil is typically a closely spaced fin coil that requires filtration upstream to keep dirt from accumulating inside the coil. The runaround loop system has exhaust filters upstream of the exhaust coils to minimize cleaning. These systems require periodic cleaning; thus, coil static pressure drop should be monitored to ensure that the coil has acceptable pressure drop and that the laboratory exhaust has acceptable airflow. To clean the exhaust coil, high-pressure water is usually used, which can push particles deeper inside the coil. This forces coil replacement to maintain airflow. Therefore, unless a long-

Figure 10-1
Runaround Loop Recovery System

(ASHRAE [2012b], Chapter 26, Figure 14)

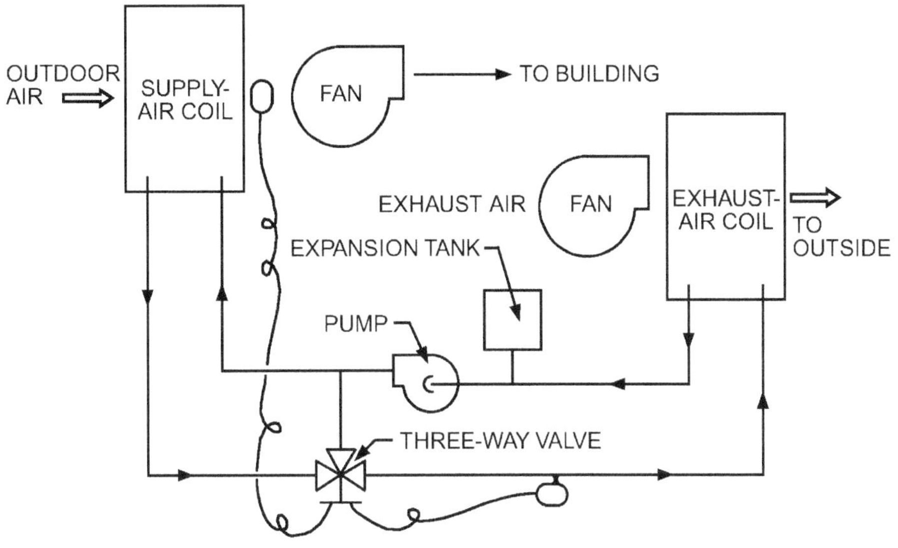

term maintenance plan is developed to maintain the runaround loop, with effective pre-treatment of the incoming exhaust air, this cleaning method should not be used.

Heat Pipes

Heat pipes rely on refrigerant migration within sealed pipes with half of the pipe in the exhaust airstream and the other half in the supply airstream. The heat pipe transfers heat from one airstream to the other by having the refrigerant evaporate at the hot side of the tube (cooling the airstream) and condense at the cold side (heating the airstream). The refrigerant is then returned to the warm side. Figure 10-2 shows the configuration of a heat pipe recovery system, and Figure 10-3 depicts a single heat pipe's operation.

The heat pipe is usually made of copper pipes with aluminum fins. The heat transfer fluid must be selected to work for the operating temperatures to avoid decomposition and deterioration of performance. Because heat pipes can have the supply and exhaust ducts physically separated, there is no cross-contamination. The effectiveness of sensible heat transfer for heat pipes is between 45% and 65% (ASHRAE 2012b).

Sensible Heat Wheels

A sensible heat wheel consists of a cylinder with a large internal surface where one half is in the exhaust air while the other half is in the supply air. The cylinder rotates and heats up on the warm side and transfers this heat to the cool side. Similar to a sensible heat wheel, an enthalpic wheel transfers sensible heat but also transfers humidity using a desiccant coating or molecular science technology. However, the risk of transferring pollutants increases with increased humidity transfer. Figure 10-4 depicts a heat wheel recovery system.

There will always be some cross-contamination with sensible heat wheels. The cross-contamination can happen by carryover or leakage. Carryover is air that is entrained in the sensible heat wheel as it moves from the exhaust to the supply. Leakage occurs because of a pressure difference between the exhaust and supply sides. The amount of cross-contamination is dependent on the pressure of the exhaust relative to the supply airstream at the sensible heat wheel and the design of the sensible heat wheel. The normal recirculation rate is from 1% to 10%. With a purge section, the cross-contamination can be reduced to below 0.1%. SEMCO_Heat_Wheel_Purge.exe, available with this Guide at www.ashrae.org/LabDG courtesy of SEMCO, shows the principle of the purge

Figure 10-2
Heat Pipe Recovery System

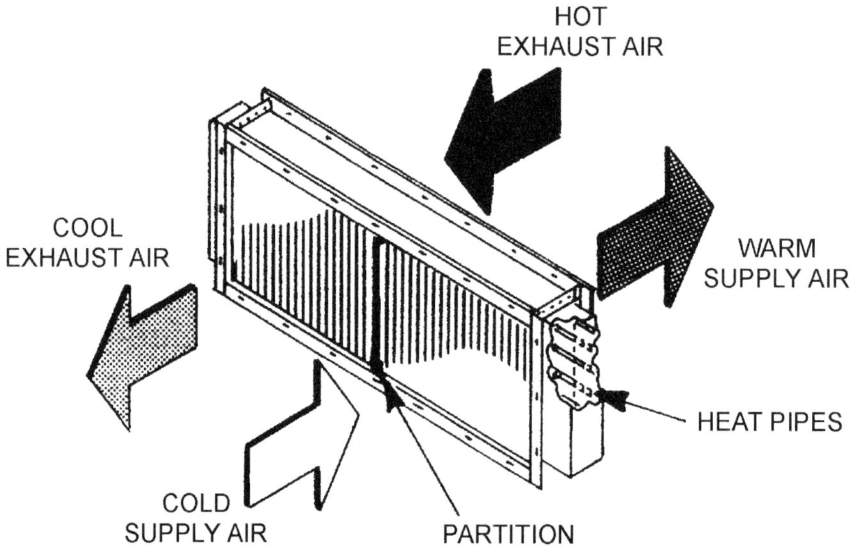

Figure 10-3
Heat Pipe Operation

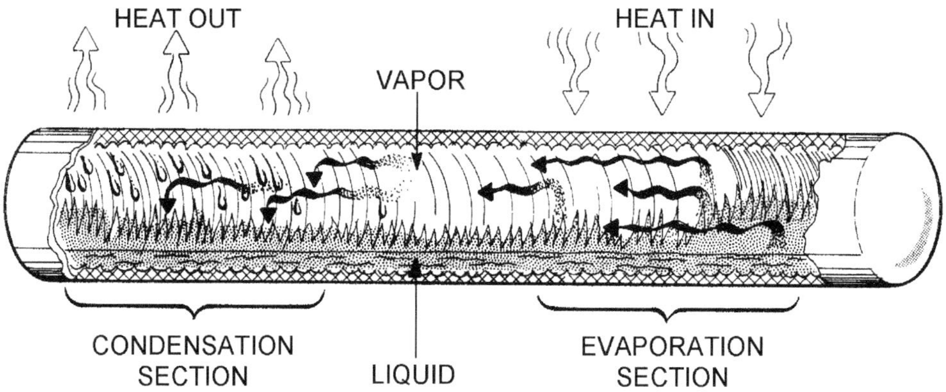

Figure 10-4
Heat Wheel Recovery System

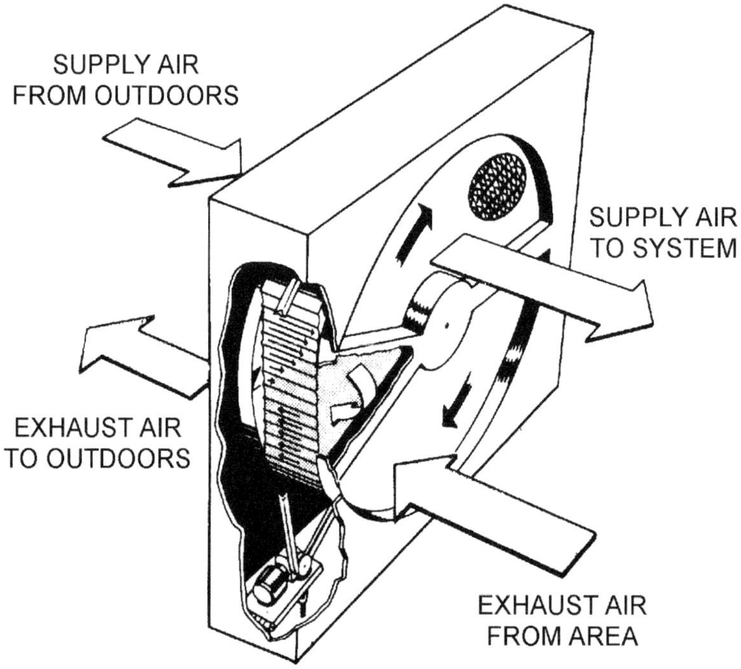

in a sensible heat wheel. The effectiveness of sensible heat transfer is between 50% and 85% and for sensible and latent heat transfer is between 50% and 85% (ASHRAE 2012b).

Fixed-Plate Heat Exchangers

Fixed-plate heat exchangers (depicted in Figure 10-5) have alternating layers of exhaust and supply airstreams separated by a plate. The primary benefit of the fixed-plate system is that there are no moving parts. To maximize heat transfer, the airflow is typically arranged in a cross-flow pattern. Even though a plate separates the airstreams, latent energy can be transferred between the airstreams by using a hygroscopic material for the plate. Some plates built with hygroscopic materials can transfer water vapor.

Though fixed-plate heat exchangers are often considered to have no cross-contamination, the risk of contamination is always present, as the thin plates are all that separates the exhaust and supply. Therefore, fixed-plate heat exchangers usually have between 0% to 5% cross-contamination from leakage. Corrosion, freezing, and cleaning of the fixed-plate heat exchanger can damage the plates and can increase the leakage rate. The effectiveness of sensible heat transfer is typically between 50% and 80% (ASHRAE 2012b).

Thermosiphons

A thermosiphon is used in two energy recovery system designs—a sealed-tube design and a coil tube design. The sealed-tube design is illustrated in Figure 10-6 and is essentially the same as heat pipes, with the exception of the wicker material in the heat pipes. This thermosiphon is dependent on gravity to get the condensate back to the warm side.

The coil loop design is shown in Figure 10-7 and is a thermosiphon coil that is essentially the same as the runaround loop energy recovery system. However, this thermosiphon system does not require any pumping. Thermosiphons typically require a significant temperature difference to initiate boiling. The effectiveness of sensible heat transfer for thermosiphons is between 40% and 60% (ASHRAE 2012b).

Reverse-Flow Plate Exchangers

In a reverse-flow plate exchanger, the exhaust air flows between plates that accumulate energy while the outdoor air flows in another series of plates and recovers energy from these plates (see Figure 10-8). After a period of time, the flows are interchanged from one series of plates to another. This system has a sensible efficiency in the order of 90%. In winter, the water that condensates and frosts on the plates is partly recovered by the enter-

Figure 10-5
Fixed-Plate Heat Exchanger Recovery System

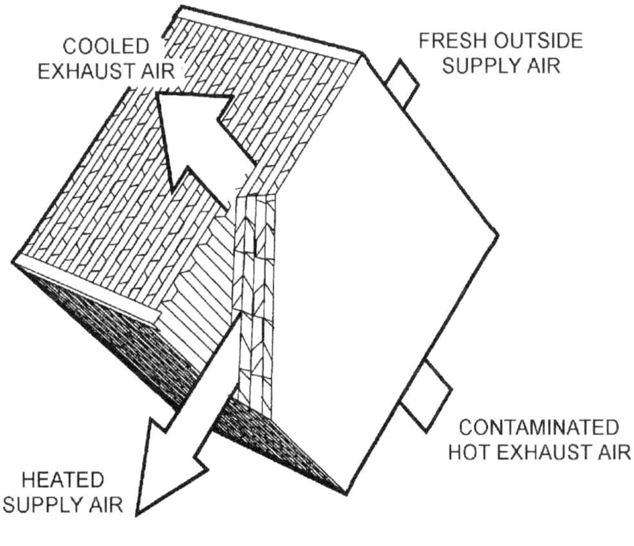

Figure 10-6
Sealed-Tube Thermosiphon Recovery System

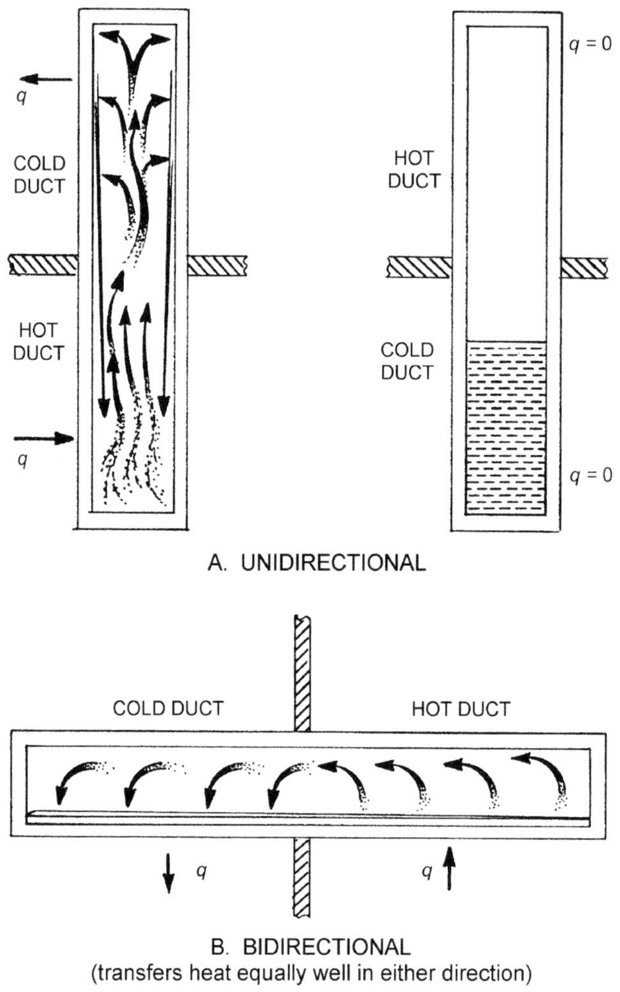

ing air. This is the only device in which there is almost no frosting problem. But with this system there is a little cross-flow contamination during the reversing flow process.

ENTHALPIC ENERGY RECOVERY PROCESSES

Liquid Desiccants

A liquid desiccant energy recovery system consists of a liquid desiccant that is circulated between the supply and the exhaust airstreams. The desiccant transfers sensible and latent energy by direct contact with the air, using contactor towers in both the supply and exhaust. The exhaust air must be filtered to remove particles such as animal hair, food, etc., to avoid contamination of the liquid desiccant. However, any particulates captured in the exhaust air do not typically pose a contamination threat, as the particle cannot be separated from the liquid intake supply airstream. The primary concern is the potential for limited gaseous cross-contamination, which is dependent on the solubility of the gas in the sorbent solution. Tests with sulfur hexafluoride as a tracer gas have shown a typical gaseous cross-contamination rate of 0.025%. The cross-contamination rate should be investigated for a system with large amounts of gaseous contaminants, as well as the gaseous effect on the sorbent solution. The effectiveness of sensible heat transfer is typically between 40% and 60% (ASHRAE 2012b). The capital and operating costs of these systems are high.

Figure 10-7 Coil Loop Thermosiphon Recovery System

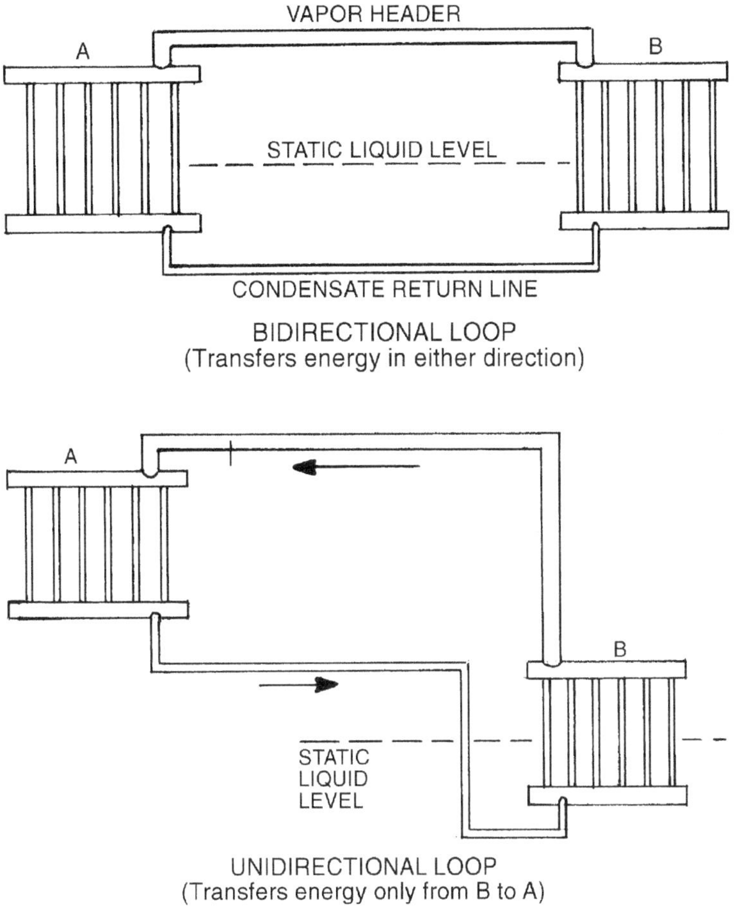

Evaporative Cooling

Evaporative cooling is not really an energy recovery method but a method of using the evaporation of water to cool the air temperature. The cooling occurs when water is evaporated in the air, converting the sensible heat in the air to latent heat and reducing the temperature of the air.

Evaporative cooling may be accomplished directly or indirectly. With direct evaporative cooling, air is cooled by direct contact with the water. This is done either by a series of sprays or by an extended wet surface material. In very dry climates, the supply air can be humidified directly to reduce the air temperature without resulting in excessively humid supply air. Direct evaporative cooling does not cause cross-contamination. However, growth of algae, slime, fungus, or bacteria must be prevented in direct evaporative-cooling systems.

With indirect evaporative cooling, either the exhaust air or the outdoor air is humidified and the dry-bulb temperature is decreased. The cooled air or water can then be used to cool the supply air with a sensible heat exchanger. A combination of both direct and indirect evaporative cooling can also be used. Cross-contamination in indirect evaporative cooling depends on what kind of heat exchanger is used.

Figure 10-9 shows a psychrometric chart that illustrates what happens when air is passed through a direct or an indirect evaporative cooler. Chapter 41 of *ASHRAE Handbook—HVAC Systems and Equipment* (2012a) and Chapter 52 of *ASHRAE Handbook—HVAC Applications* (2015) explain in detail how evaporative cooling is accomplished.

Figure 10-8
Reverse-Flow
Plate Exchanger
Recovery System

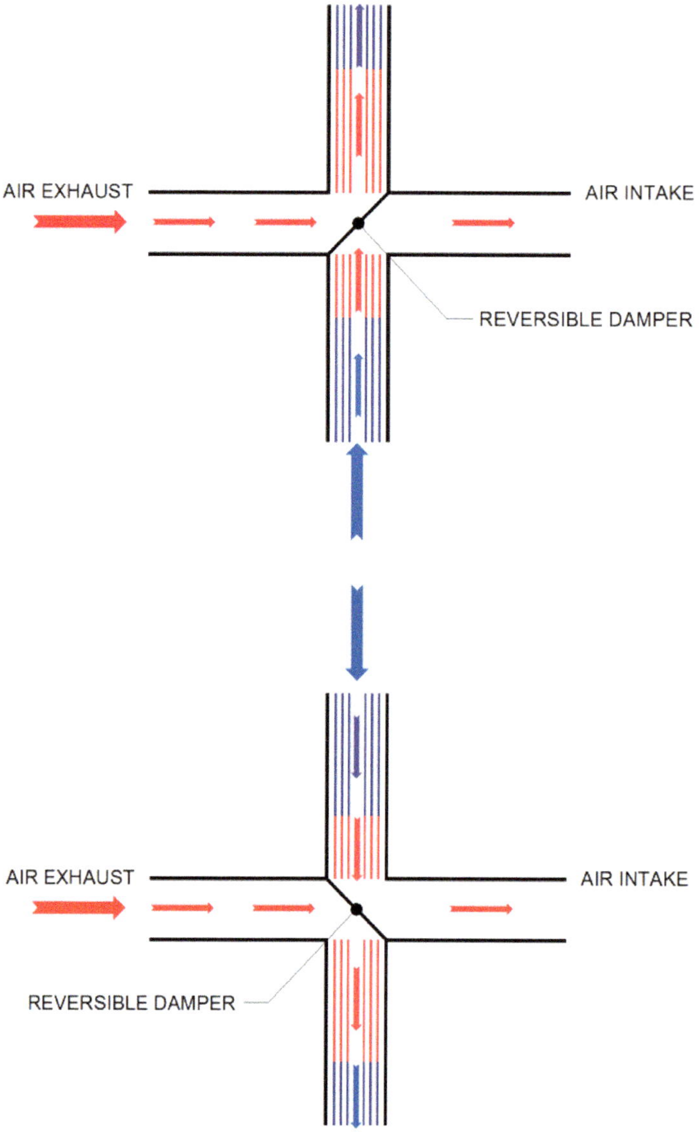

WATER-TO-AIR ENERGY RECOVERY

To overcome the concerns with cross-contamination using air-to-air energy recovery systems, water-to-air energy recovery systems can be used. These systems require piping to the air supply duct and a heating coil that transfers the heat from the water to the air. Potential systems include the following:

- Refrigeration machine energy recovery
- Condenser water energy recovery
- Hot-water waste energy recovery

REFRIGERATION MACHINE ENERGY RECOVERY

Refrigeration machine energy recovery uses a dual air-cooled condenser to heat an airstream. This type of system is used when there is a cooling load (interior zone) and a heating load (perimeter zone) at the same time. Instead of the condenser heat being rejected to the outdoors, it is rejected to the perimeter-zone supply airstream. In some

Figure 10-9
Psychrometrics of Evaporative Cooling

(ASHRAE [2015], Chapter 52, Figure 1)

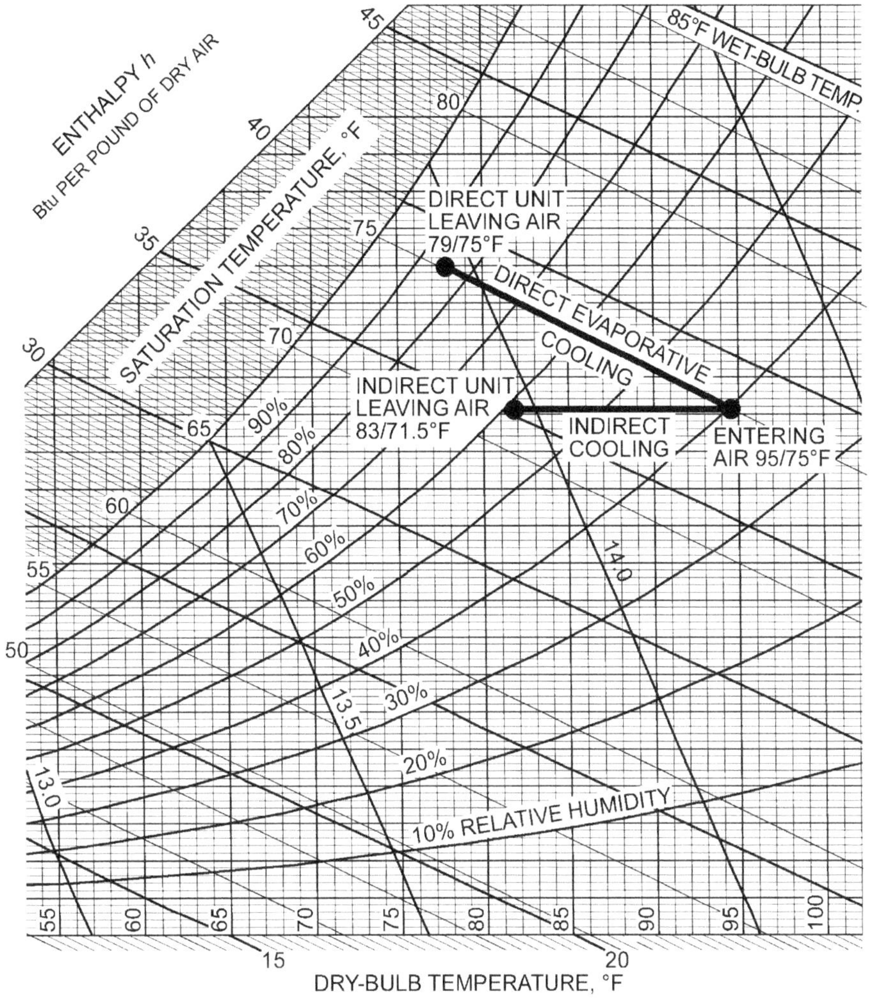

instances, a heat pump is placed across the exhaust and supply air to economically transfer heat from one to the other.

CONDENSER WATER ENERGY RECOVERY

Condenser water energy recovery uses the heat typically rejected by a chiller to a cooling tower. A double-bundle or split-bundle condenser system has historically been the most common configuration used. It comprises two condensers or a single condenser shell with two independent water loops. One of the water loops is connected to a cooling tower to reject excess heat and the other is connected to the heating load equipment.

However, double-bundle condensers are not as widely available in recent years because there have been changes in using newer refrigerants. Instead, dual-condenser chillers for energy recovery are more readily offered by chiller manufacturers due to their greater ease of manufacture. Figure 10-10 illustrates this system configuration. A dual condenser has two distinct condenser shells. These have a water loop connected to the cooling tower and a water loop connected to the heating equipment (e.g., heating coils). Superheated refrigerant gases that result from the compression process flow into both condenser shells, and the heat is rejected to one or both of the condenser water loops.

The double-bundle and dual-condenser arrangements are constructed differently but operated similarly. Figure 10-11 illustrates these condenser arrangements. *Chiller Heat*

Figure 10-10
Dual-Condenser System

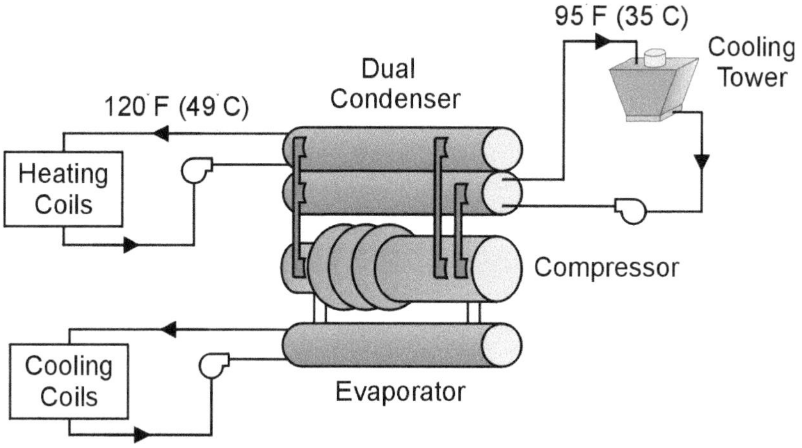

Figure 10-11
Double-Bundle and Dual-Condenser System Arrangement

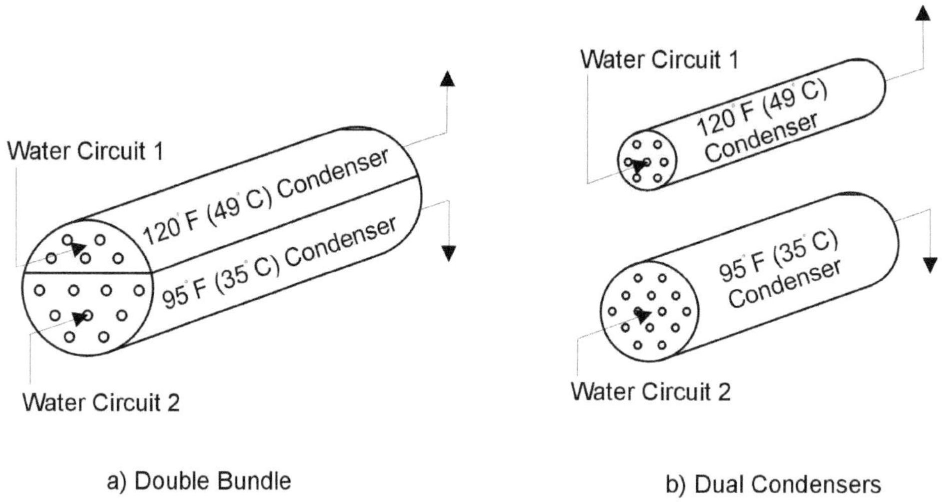

Recovery Application Guide contains details on identifying and sizing double-bundle arrangements (Dorgan et al. 1999).

HOT-WATER WASTE ENERGY RECOVERY

Hot-water waste heat is often available from industrial processes, including high-temperature process cooling water and combustion gases. The temperature is usually very good for preheating the supply air and often high enough and plentiful enough to heat the supply air to the desired temperature. The temperature of hot-water waste energy recovery can be close to the boiling temperature of water. The physical configuration for this type of energy recovery is nearly identical to that of runaround systems.

SELECTION PARAMETERS

The selection of the type of energy recovery system to use is based on parameters such as laboratory requirements, climate, exhaust and supply duct locations, and economics, which are based on the initial cost, expected savings, cross-contamination, and space implications.

Table 10-1 presents a summary of various kinds of energy recovery devices along with their physical features, advantages, and limitations.

Table 10-1 Comparison of Air-to-Air Energy Recovery Devices
(ASHRAE [2012b], Chapter 26, Table 3)

	Fixed Plate	Membrane Plate	Energy Wheel	Heat Wheel	Heat Pipe	Runaround Coil Loop	Thermosiphon	Twin Towers
Airflow arrangements	• Counterflow • Cross-flow	• Counterflow • Cross-flow	• Counterflow • Parallel flow	Counterflow	• Counterflow • Parallel flow	—	• Counterflow • Parallel flow	—
Equipment size range, cfm (L/s)	50 and up (25 and up)	50 and up (25 and up)	50 to 74,000 and up (25 to 35 000 and up)	50 to 74,000 and up (25 to 35 000 and up)	100 and up (50 and up)	100 and up (50 and up)	100 and up (50 and up)	—
Typical sensible effectiveness ($m_s = m_e$), %	50 to 80	50 to 75	50 to 85	50 to 85	45 to 65	45 to 65	40 to 60	40 to 60
Typical latent effectiveness,* %	—	50 to 72	50 to 85	0	—	—	—	—
Total effectiveness,* %	—	50 to 73	50 to 85	—	—	—	—	—
Face velocity, fpm (m/s)	200 to 1000 (1 to 5)	200 to 600 (1 to 3)	500 to 1000 (2.5 to 5)	400 to 1000 (2 to 5)	400 to 800 (2 to 4)	300 to 600 (1.5 to 3)	400 to 800 (2 to 4)	300 to 450 (1.5 to 2.2)
Pressure drop, in. w.g. (Pa)	0.4 to 4 (100 to 1000)	0.4 to 2 (100 to 500)	0.4 to 1.2 (100 to 300)	0.4 to 1.2 (100 to 300)	0.6 to 2 (150 to 500)	0.6 to 2 (150 to 500)	0.6 to 2 (150 to 500)	0.7 to 1.2 (170 to 300)
Exhaust air transfer ratio, %	0 to 2	0 to 5	0.5 to 10	0.5 to 10	0 to 1	0	0	0
Outdoor air correction factor	0.97 to 1.06	0.97 to 1.06	0.99 to 1.1	1 to 1.2	0.99 to 1.01	1.0	1.0	1.0
Temperature range, °F (°C)	–75 to 1470 (–60 to 800)	–40 to 120 (–40 to 50)	–65 to 1470 (–55 to 800)	–65 to 1470 (–55 to 800)	–40 to 104 (–40 to 40)	–50 to 930 (–45 to 500)	–40 to 104 (–40 to 40)	–40 to 115 (–40 to 46)
Typical mode of purchase	• Exchanger only • Exchanger in case • Exchanger and blowers • Complete system	• Exchanger only • Exchanger in case • Exchanger and blowers • Complete system	• Exchanger only • Exchanger in case • Exchanger and blowers • Complete system	• Exchanger only • Exchanger in case • Exchanger and blowers • Complete system	• Exchanger only • Exchanger in case • Exchanger and blowers • Complete system	• Coil only • Complete system	• Exchanger only • Exchanger in case	• Complete system
Advantages	• No moving parts • Low pressure drop • Easily cleaned	• No moving parts • Low pressure drop • Low air leakage • Moisture/mass transfer	• Moisture/mass transfer • Compact large sizes • Low pressure drop • Available on all ventilation system platforms	• Compact large sizes • Low pressure drop • Easily cleaned	• No moving parts except tilt • Fan location not critical • Allowable pressure differential up to 2 psi (15 kPa)	• Exhaust airstream can be separated from supply air • Fan location not critical	• No moving parts • Exhaust airstream can be separated from supply air • Fan location not critical	• Latent transfer from remote airstreams • Efficient microbiological cleaning of both supply and exhaust airstreams
Limitations	• Large size at higher flow rates	• Few suppliers • Long-term maintenance and performance unknown	• Supply air may require some further cooling or heating • Some exhaust air transfer ratio without purge	• Some exhaust air transfer ratio with purge	• Effectiveness limited by pressure drop and cost • Few suppliers	• Predicting performance requires accurate simulation model	• Effectiveness may be limited by pressure drop and cost • Few suppliers	• Few suppliers • Maintenance and performance unknown
Heat rate control (HRC) methods	• Bypass dampers and ducting	• Bypass dampers and ducting	• Bypass dampers and wheel speed control	• Bypass dampers and wheel speed control	• Tilt angle down to 10% of maximum heat rate	• Bypass valve or pump speed control	• Control valve over full range	• Control valve or pump speed control over full range

**Rated effectiveness values are for balanced flow conditions. Effectiveness values increase slightly if flow rates of either or both airstreams are higher than flow rates at which testing is done.*

LABORATORY REQUIREMENTS

The energy recovery device must adequately meet the needs of the laboratory without compromising the safety of the occupants, the functionality of equipment, or the integrity of experiments. Some recovery devices are more appropriate for certain types of laboratories than for others. In general, most energy recovery applications are applicable to most types of laboratories. The key question to ask is: what is the risk of cross-contamination that will threaten the safety of the laboratory occupants, products, or experiments? For example, it is not recommended to use any equipment that expels infectious substances, toxic chemicals, and animal dander, in, respectively, biological, chemical, or animal laboratories, where there is a possibility of cross-contamination.

CLIMATE

The climate determines how much energy is necessary to heat or cool the supply air for a building. The potential energy savings depend on the amount of energy spent. If the supply temperature is close to the ambient temperature for large parts of the year, then the energy savings are often small.

EXHAUST AND SUPPLY LOCATIONS

Most energy recovery systems need the exhaust and supply ducts next to each other. This can cause an increased amount of ductwork unless this is planned from the initial stages of the project. Also, the safety of adjoining contaminated exhaust air and supply air is a concern. Energy recovery systems that allow the exhaust air and supply air to be at different locations, such as runaround systems, can be a solution.

ECONOMICS

Energy recovery is usually aimed at reducing the amount of energy consumed. Most of the savings come from this reduced energy usage. However, energy recovery systems that do not freeze (i.e., that have a lower efficiency during subfreezing temperatures) or have to shut down during subfreezing temperatures can also reduce the component size (or power demand). This demand reduction can reduce or even eliminate the initial cost of the energy recovery system. To evaluate the economic benefits of selected energy recovery systems, an economic analysis can be done using a simple payback analysis or a life-cycle cost analysis. Chapter 15 gives details of these analyses.

REFERENCES

ASHRAE. 2012a. *ASHRAE handbook—HVAC systems and equipment.* Atlanta: ASHRAE.

ASHRAE. 2012b. Chapter 26, Air-to-air energy recovery equipment. In *ASHRAE handbook—HVAC systems and equipment.* Atlanta: ASHRAE.

ASHRAE. 2015. *ASHRAE handbook—HVAC applications.* Atlanta: ASHRAE.

Dorgan, C.B., R.J. Linder, and C.E. Dorgan. 1999. *Chiller heat recovery application guide.* Atlanta: ASHRAE.

NFPA. 2015. *NFPA 45: Standard on fire protection for laboratories using chemicals.* Quincy, MA: National Fire Protection Association.

Streets, R.A., and B.S.V. Setty. 1983. Energy conservation in institutional laboratory and fume hood systems. *ASHRAE Transactions* 89(2B):542–51.

Controls

11

OVERVIEW

Every building requires a properly functioning control system to operate comfortably, safely, and energy efficiently. This is especially true for laboratories, where the HVAC systems protect the occupants from dangerous pollutants and help maintain health and safety. Whereas the physical systems are designed for the peak operating condition, the control system must maintain proper temperature, humidity, air speed, air volume flow, and pressure through all modes of operation. Supply airflows are dictated by one of three criteria: makeup for hood exhaust, thermal cooling demand, or minimum ventilation rate. This is to ensure the safety of the occupants and the accuracy of the experiments and research conducted during all nonpeak periods. In addition, the HVAC system for a laboratory is often subjected to rapid disturbances or changes, such as those caused by opening doors, opening sashes on numerous fume hoods at once, or turning on large pieces of heat-generating equipment. Such disturbances require a fast response from the control system to maintain the precise conditions required in the laboratory. Further, because the materials used in laboratories are commonly hazardous, corrosive, or flammable, the control system components must withstand exposure to such materials.

Laboratories also typically operate using 100% outdoor air, which requires considerable energy to condition before it is supplied to the individual rooms. Care must also be taken to prevent damage to the equipment such as freezing coils, frosted filters, and frosted energy recovery devices. The systems must be able to provide for a wide range of capacities as outdoor temperatures and humidity vary. Therefore, it is very important that a laboratory HVAC control system be designed to minimize energy use by providing the minimum airflow rate that still provides excellent safety through effective hood containment, good thermal comfort, and proper control of laboratory room particulate and chemical contaminants.

Laboratory HVAC controls encompass many strategies and elements that must be engineered properly in order for the system to work, such as the following:

- Constant-volume versus variable-air-volume fume hood control
- Air control devices
- Source containment and exhaust device controls
- Room pressurization control
- Dilution ventilation and minimum ventilation rates

- Room temperature control
- Central system level control
- Emergency modes of operation
- Animal facilities

CONSTANT-VOLUME VERSUS VARIABLE-AIR-VOLUME FUME HOOD CONTROL

Constant-volume systems are the simplest of systems to control. The airflows are manually balanced or controlled by a constant-flow terminal unit to their peak flows needed for makeup air, thermal load, or minimum ventilation rate. Temperature control is typically through the use of reheat coils that are controlled from a room thermostat. However, these systems are very energy intensive and not recommended for modern laboratories.

Variable-air-volume (VAV) systems vary the supply and exhaust volumes based on demand. Typically, VAV fume hood laboratories have hoods that allow the exhaust volumes to respond to sash movement and allow for reduced exhaust requirements to maintain a constant face velocity when the sash is moved to a lower position than the design operating height. With VAV systems, biological safety cabinets (BSCs), canopy hoods, and snorkels are turned off when not needed. The supply air responds to maintain the minimum flow rate needed to meet the highest of the following three different airflow requirements:

- Makeup air for the room's fume hoods and all other exhaust devices to maintain the proper pressure relationship
- Cooling to meet the room's thermal load demands (or at least those that are not being met by a local hydronic cooling approach such as active or passive beams or fan-coil units)
- Supply air needed to provide dilution ventilation to eliminate room-level chemical and particulate pollutants

Dilution ventilation may be achieved through a fixed minimum ventilation or air change rate or, to achieve greater energy savings, a demand-based approach can be used that measures the real-time level of room contaminants and varies the dilution ventilation flow rate as needed to more effectively reduce and eliminate these contaminants. For rooms that are not always dominated by the fume hood or other exhaust device requirements, a room or general exhaust airflow control device is used to exhaust the additional supply air needed for cooling or dilution ventilation. Because the use of outdoor air is typically the single largest driver for energy use in a laboratory, reducing the supply and exhaust airflow volumes can cut the energy use of a laboratory with a VAV control system by 50% or more compared to a laboratory with a constant-volume control system. VAV systems allow for flexibility in laboratory spaces, making it easier to change ventilation rates when equipment and space usage change in the future. As a result, most systems installed today are VAV control systems.

AIR CONTROL DEVICES

Before evaluating control strategies, designers must understand the types of devices that can be used to control and measure the airflow at the room level. Airflow in a duct can be controlled at the room level through the use of blade dampers, pneumatic dampers, or venturi valves. Each device should be evaluated as part of the complete laboratory control system for speed of response, accuracy (at minimum and maximum flow), turndown, material compatibility, operating range, pressure independence, pressure drop, maintenance requirements, and long-term durability. The control accuracy of each airflow in the room affects the ability to maintain a volumetric offset. Required flow control accuracy should be specified with reference to the accuracy needed to maintain required pressure relationships throughout the life of the system.

Blade dampers can be single-blade or multiblade. To provide accuracy of control and pressure independence, the control system must measure the airflow and control the damper to maintain the proper airflow through changing static pressures and heating and cooling demands on the system. Damper construction materials should be compatible with the airstream in which they are located. Blade dampers can be controlled by either pneumatic actuators or electric actuators to provide the speed of response required. Some dampers can be controlled with normal-speed actuators if quick response is not required. Areas in the laboratory without fume hoods may only require a typical 60 to 90 s actuator versus a high-speed actuator with a full stroke of less than 2 s.

One type of pneumatic damper is the bladder damper. These dampers have airfoil blades that expand with pneumatic air to close off the space between them. As with blade dampers, pneumatic damper construction materials should be compatible with the airstream in which they are located. To provide accuracy of control and pressure independence, the control system must measure the airflow and control the damper to maintain the proper airflow independent of changing static pressures across the device.

Venturi valves are airflow control devices utilizing a venturi-shaped body with a cone that rides on a center shaft. The cone is spring loaded to make the valve pressure independent through a mechanical device. As pressure increases in the duct, the spring is compressed and the cone decreases the area between the cone and the valve body, thereby increasing the pressure drop across the valve. Venturi valves can meter the airflow using a characterization curve that relates airflow to the position of the base of the spring in the cone, or effectively the shaft position. The valves are pressure independent over a specified pressure range (valves are available with ranges such as 0.3 to 3.0, 0.6 to 3.0, or 1.0 to 6.0 in. w.g. [75 to 750, 150 to 750, or 250 to 1500 Pa]). Venturi valves typically require a higher pressure drop across the control device than blade dampers or pneumatic dampers to maintain their pressure independence. Valve construction materials should be compatible with the airstream they control. Compared to a blade damper, venturi valves have more precision-machined parts in the airstream. Airflow is controlled by moving the center shaft that moves the base of the spring that is connected to the cone.

A venturi valve can have either a pneumatic or an electric actuator. Airflow can be determined by using an airflow measurement device or by using data that calibrate the valve's airflow to its shaft position characteristic. This characteristic is repeatable, and each venturi valve can be measured and calibrated to determine its characterization curve. This curve correlation can be used to calculate the airflow through the valve based on measuring the position of the shaft that determines the position of the base of the cone, provided that the pressure drop across the valve is within its design specification. The accuracy of a venturi valve based on the position and characterization curves is typically within ±5% of actual airflow. The characterization curves and accuracy of the airflow are dependent on the spring being compressed and the valve operating within its design pressure range.

Upon loss of power or pneumatic air, actuators on the control devices can be provided that fail the device open (full flow), closed (minimum flow), or in the last position. Failing exhaust devices open and supply devices closed provides a more negative space on failure. This prevents contaminants from escaping the space. This can also lead to issues with opening doors and allowing people to leave the space or building. Failing to last position typically keeps the airflow relatively constant. Because the hoods are the primary containment, visual means should be provided to indicate proper operation and alarm functions.

Lab Room Schedule - Sample.xls, available with this Guide at www.ashrae.org/LabDG, allows the engineer to schedule the pertinent data for all the room airflow control devices in one schedule along with the room offset, minimum ventilation rate, and maximum cooling demand quantities that are required for setting up the control system.

AIRFLOW MEASUREMENT DEVICES

To control airflow to a given value, the airflow typically needs to be measured (although some venturi devices control airflow based on calibrated position data). Airflow can be measured with velocity pressure, thermal dispersion, or vortex shedding devices. Each device offers advantages and disadvantages in measuring airflow. When evaluating the accuracy of the devices, the engineer should look at published data sheets indicating accuracy of the complete airflow control system: sensor, controller, actuator, and flow control device. Complete data sheets indicate the airflow range (or velocity range), inlet conditions, and duct pressure loss that correspond to the accuracy data.

Where airflow measurement is used for either control functions or ongoing commissioning, the airflow measurement device should be selected and located carefully to ensure that the resulting system meets the design objectives. Manufacturer's recommendations on straight duct should be followed. Drift and recalibration maintenance requirements should also be considered.

Velocity Pressure

Averaging pitot tubes, flow crosses, and orifice rings are examples of devices that measure and may amplify the velocity pressure. The velocity pressure device is connected to a pressure transducer. A controller reads the output signal from the pressure transducer and calculates the flow rate from the sensed pressure and calibration data. Accuracy of the velocity pressure sensing system depends on the characteristics of the velocity pressure device and the pressure transducer. A design engineer must understand that potentially significant component errors are addressed by product qualification and balancing and commissioning processes.

The velocity pressure is amplified by the static pressure port being in the wake zone on the back side of the sensor instead of on the side as with standard pitot tubes. According to manufacturers, this amplification effect is fairly constant between 500 and 3000 fpm (2.5 and 15.2 m/s). If the amplification factor varies, that appears as a nonlinearity. This is best addressed by examining product performance specifications across the range of interest. Designers should indicate the flow range when specifying airflow control accuracy, and product data sheets from control system suppliers should indicate velocity range or flow range corresponding to accuracy specifications.

Inlet conditions can change the amplification factor. The nominal value published by a supplier might not apply in a particular installation. The construction team adjusts the factor to bring control system readings within a few percentage points of the value measured by the air balancer. This adjustment takes care of the amplification factor and any span error in the transducer.

All airflow sensors are affected by nonuniform airflow into the unit. Velocity pressure sensing devices average and amplify the velocity pressure at the various inlet holes in the devices. Multiple sensing ports make these devices less sensitive to the airflow profile than single-point sensors.

ASHRAE Research Project RP-1353, *Stability and Accuracy of VAV Box Control at Low Flows*, shows that a ±5% accuracy can be read from the flow sensors down to 140 fpm (0.71 m/s) with proper duct configuration (Liu et al. 2012). When reducers are added before the inlet, the response of the velocity pressure device is not linear and amplification is lost at lower velocities within the recommended range for the flow sensor (Klaczek et al. 2006). This is not the accuracy at the controller. Between the airflow measurement device and the controller, a pressure transducer is required to convert the velocity pressure to an input signal to the controller.

At low velocities, accuracy is very sensitive to offset error in the pressure transducer (typically ±0.5% of full scale, which can lead to ±20% of flow). Control system providers select the range accuracy rating of the transducer accordingly. Setup procedures generally include "zeroing" the sensor on site. Some transducers are stable enough to perform after being zeroed once. If the transducer drifts (typically 0.5% per year), some automatic zeroing functioning is appropriate. With an accurate zero, velocity pressure sensing can be very accurate even at low velocities.

The pressure transducers can have a variety of pressure ranges and accuracies. Pressure transducer accuracy is typically expressed in percent of full scale. Pressure transducers have several types of errors that should be considered when determining the overall accuracy of the airflow measurement device. The three common errors are nonlinearity, hysteresis, and nonrepeatability, which is commonly referred to as the root sum squared (RSS) error, which is normally associated with the device's accuracy. Overall error in reading can be from ±2% to greater than ±40% and is very dependent on the pressure transducers, temperature changes, and duct placement (ASHRAE 2013). In addition to the RSS error, zero setting tolerance error, span setting error, stability per year, and thermal effects should be considered when selecting pressure transducers for flow measurement. Auto-zero and frequent calibration of the transducers can help maintain stability over time.

Overall airflow sensing accuracy combines the effects of the probe and the pressure transducer. Each can be characterized with a span error and an offset error. Other errors (nonlinearity, hysteresis) are very small in these components. Table 11-1 summarizes the errors associated with pressure transducers. These errors, along with the flow device error (±3%), should be taken into account at minimum and maximum airflows.

Thermal Dispersion

Thermal dispersion devices measure airflow through the use of thermistors in the airstream. The thermistors are typically used in pairs. The first is used to measure the temperature of the airstream. The second is heated above ambient temperature. The voltage across the heated thermistor is measured and the power dissipated to the airstream is calculated. From the power being dissipated to the airstream, the velocity can be calculated.

Typically, several sets of thermistors are used to create a traverse based on the size of the duct. For small ducts (4 to 16 in. [100 to 400 mm] diameter), typically one or two sets of thermistors are used. Microprocessor-based thermal dispersion devices can measure airflow with an accuracy of ±2% of reading when calibrated against National Institute of Standards and Technology (NIST) traceable standards (ASHRAE 2013). The air velocity can be determined from 50 to 10,000 fpm (0.25 to 50 m/s) with an accuracy of ±3% of

Table 11-1 Pressure Transducer Errors

Sensor Characteristic	Definition	Corrective Measure	Typical for 1 in. (250 Pa) Span Transducer, ±0.4% Accuracy
Nonlinearity	Relationship of calibration curve to a straight line	Select appropriate span and accuracy	±0.38% FS
Hysteresis	Maximum difference from increasing pressure and decreasing pressure	Select appropriate span and accuracy; negligible compared to nonlinearity	±0.10% FS
Nonrepeatability	Difference in readings under same conditions	Select appropriate span and accuracy; negligible compared to nonlinearity	±0.05% FS
Thermal Effect	Change in reading due to change in temperature	Auto zero or keep at constant temperature	±0.03% FS/°F (±0.05% °C)
Long-Term Stability	Change from one period to the next with no recalibration	Auto zero	±0.5% FS/yr
Zero Offset	Factory set 0 compared to actual 0	Auto zero or field set at start-up	±0.2% FS

FS = full scale

reading. The engineer should check the manufacturer's accuracy and speed of response to changes in airflow, as all thermal dispersion devices do not have the same accuracy and speed of response. Some thermal dispersion devices have sensor-to-output errors around 2% of reading with temperature compensation and no drift over time.

Vortex Shedding

Vortex shedding devices measure airflow by measuring the pressure pulses or vortices formed on the leeward side of an obstruction in the airflow and calculating the airflow velocity based on the pulses. Vortex shedding is good for measuring velocities from 450 to 6000 fpm (2.3 to 30.5 m/s) with an accuracy of ±2.5% to 10% of reading (ASHRAE 2013). Below the specified sensing range a vortex shedder turns off and stops communicating data, in contrast to other devices that continue to operate with reduced performance. The devices are usually built with two acoustic sensing elements in the airstream responding to two points in the velocity profile of the terminal. Response time for vortex shedders is typically slower than that of velocity pressure sensors.

SOURCE CONTAINMENT AND EXHAUST DEVICE CONTROLS

Because of the critical nature of the exhaust equipment needed to maintain containment of chemical, biological, and particulate contaminants, the development of a laboratory airflow control system typically starts with the control of fume hoods and other exhaust devices. Maintaining source containment requires proper control of the exhaust volume out of these local containment devices. Though fume hoods are the most common type of exhaust equipment specific to laboratories, there are several other types of exhaust devices that must be considered when designing the control system for a laboratory, including snorkel exhausts, BSCs, and flammable and solvent storage cabinets, as well as direct equipment exhausts (canopy hood or hard ducted). In addition to source containment exhaust, general exhaust from the space is used for maintaining ventilation rate and/

Figure 11-1
Typical Constant-Volume Fume Hood

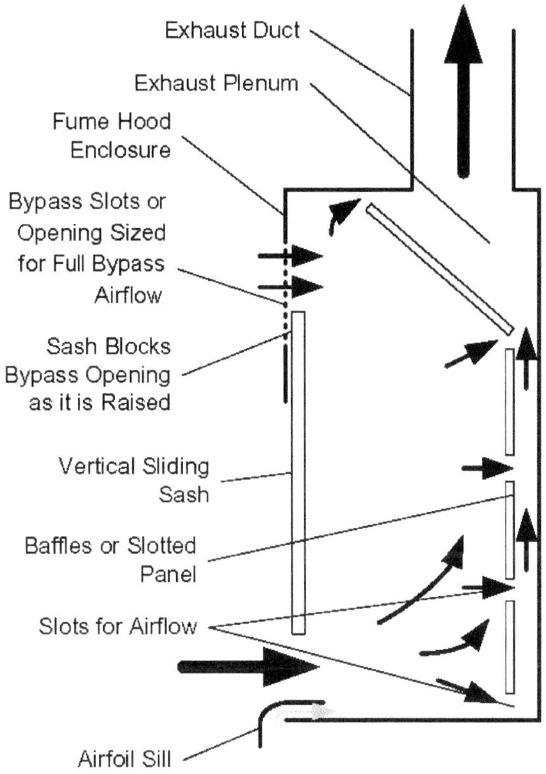

or pressure relationships. Information on fume hoods and other types of exhaust equipment can be found in Chapter 5.

FUME HOODS

Constant-Volume Fume Hood Control

Constant-volume fume hoods or bypass hoods are typically vertical sash hoods with a bypass section located above the sash opening. As the sash opens, it blocks off the bypass opening, allowing the exhaust system to pull air through the face of the fume hood. As the sash is closed, the reverse happens. This provides a manual method of controlling where the exhaust airflow comes from. Constant-volume fume hoods do not provide a constant face velocity because the bypass is typically much smaller than the potential open sash area.

A fume hood monitor that measures face velocity through a sidewall sensor or measures the exhaust airflow can be used to provide a local visual and audible alarm. The alarm setpoint may be for a reduced and excessive flow or face velocity level, such as at 80% and 120% of the desired value.

Constant-volume flow can be accomplished many different ways depending on the type of exhaust system. With a single fan per hood, the flow can be set by balancing the fan with no additional controls required. With a constant-volume manifolded system, a manual balancing damper can be used. Figure 11-1 shows a schematic of a typical constant-volume hood.

When a constant-volume fume hood is connected to a VAV exhaust system, a control system is required. One method of control is to measure the exhaust flow from the hood and modulate a damper to maintain a constant airflow (pressure-independent control). Figure 11-2 shows a typical control schematic for a constant-volume hood.

Figure 11-2
Typical Control Schematic for Constant-Volume Fume Hood

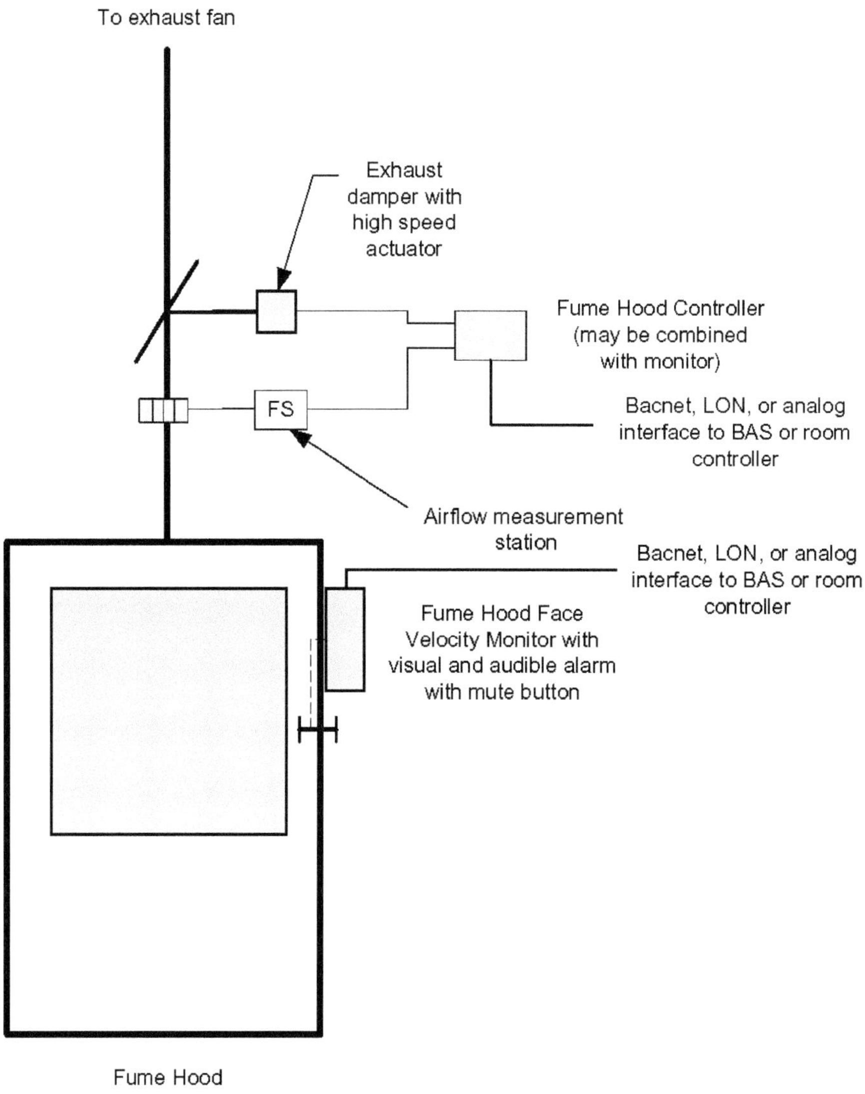

VAV Fume Hood Control

VAV fume hoods vary the exhaust airflow from the hood and maintain a specified face velocity at the hood opening as the sash position is varied. On a manifolded exhaust system, a central exhaust fan with a variable-frequency drive (VFD) or outdoor air bypass dampers is used to vary the total system volume from the building by maintaining a set-point exhaust pressure at some point in the exhaust ductwork. Occasionally, although much less prevalent, a single-exhaust-fan-based fume hood system may vary fan speed directly to control the hood's airflow without a damper or airflow control device.

As with any laboratory hood, the primary goal is to contain pollutants to maintain the health and safety of laboratory personnel. VAV fume hoods achieve this goal by ensuring that the proper hood face velocity is maintained at all times. The face velocity needed for safety can be actively and continually verified. Typically, the hood controller also indicates proper function to the user, with an alarm for high or low face velocity. Face velocity alone does not ensure containment, but face velocity can be an indicator that conditions at a hood that previously maintained effective containment have changed. Sensors should be verified to be operating correctly at least once a year.

An important goal of VAV hoods is to minimize system energy use. By controlling the exhaust airflow needed to maintain safety in the laboratory and not exhausting additional air, VAV fume hoods require less makeup air. This reduces the energy consumption of the exhaust fans and supply fans and the conditioning (heating, cooling, and humidity control) of the supply air. Because laboratories typically use 100% outdoor air and as a result have an HVAC energy consumption that is 5 to 10 times that of a typical commercial building, the energy reduction associated with VAV control is significantly more than that attainable in a typical office building that uses a large portion of recirculated air.

VAV or restricted-bypass hoods are designed to allow the exhaust airflow to be varied as the hood sash is moved. VAV hoods can have vertical sashes, horizontal sashes, or combination sashes. With restricted-bypass hoods, the bypass area is smaller and is typically uncovered only when the sash is fully closed. This allows the control system to vary the exhaust flow from the hood while maintaining a constant face velocity and to maintain a minimum flow when the hood is closed.

There are two ways of controlling the face velocity of a VAV fume hood: directly measuring air velocity and controlling the damper or calculating the required exhaust flow based on the desired face velocity and measured open sash area to control the airflow to the calculated exhaust airflow setpoint.

Measurement of the fume hood face velocity is typically accomplished using a small air velocity sensor (hot-wire or vane anemometer), which is placed in an opening in the sidewall of the fume hood. Controllers using sidewall sensing typically modulate a variable-speed hood fan, damper, or other airflow control device in the fume hood exhaust to maintain face velocity at setpoint. Sidewall sensors have been found to give accurate results with different sash types and positions in hoods up to 8 ft (2.4 m) wide (Zhang and Agarwal 1993). Sensor location is important, and the sensor should be located per the manufacturer's instructions for accurate results. A fume hood monitor and/or controller that measures face velocity through a sidewall sensor also provides a local visual and audible alarm. Figure 11-3 shows a typical control schematic for a VAV hood with a sidewall face velocity sensor.

Measuring the sash position and calculating the open area of the sash to set the exhaust airflow is accomplished by first determining the sash position is using a potentiometer or other variable-resistance means attached to the sash. The sensor reads the position of the sash and calculates the open area of the fume hood face based on the known width. An airflow measurement station or venturi valve in the exhaust duct measures the airflow from the fume hood. The controller calculates the desired setpoint for the exhaust flow rate by multiplying the open area of the fume hood by the desired face velocity. The controller then modulates the exhaust damper or airflow control valve to the airflow desired setpoint. A fume hood monitor can use a sidewall velocity sensor for verifying the face velocity, or the monitor can compare the flow measurement to the calculated setpoint. Figure 11-4 shows a typical control schematic for a VAV hood with a sash position sensor.

VAV Fume Hood Face Velocity Setback

For safety purposes, fume hood sashes should be closed when the hoods are not in use, as the hood sash presents a physical barrier between the chemicals in the hood and the room occupants. However, when there is poor sash management and sashes are left open when the hoods are not in use, face velocity setback can be used to reduce the exhaust flow. Energy is saved by reducing the hood makeup air when it is the driving factor for determining the supply air to the space. This energy-saving measure typically only

Figure 11-3 Typical Control Schematic for VAV Fume Hood with Sidewall Face Velocity Sensor

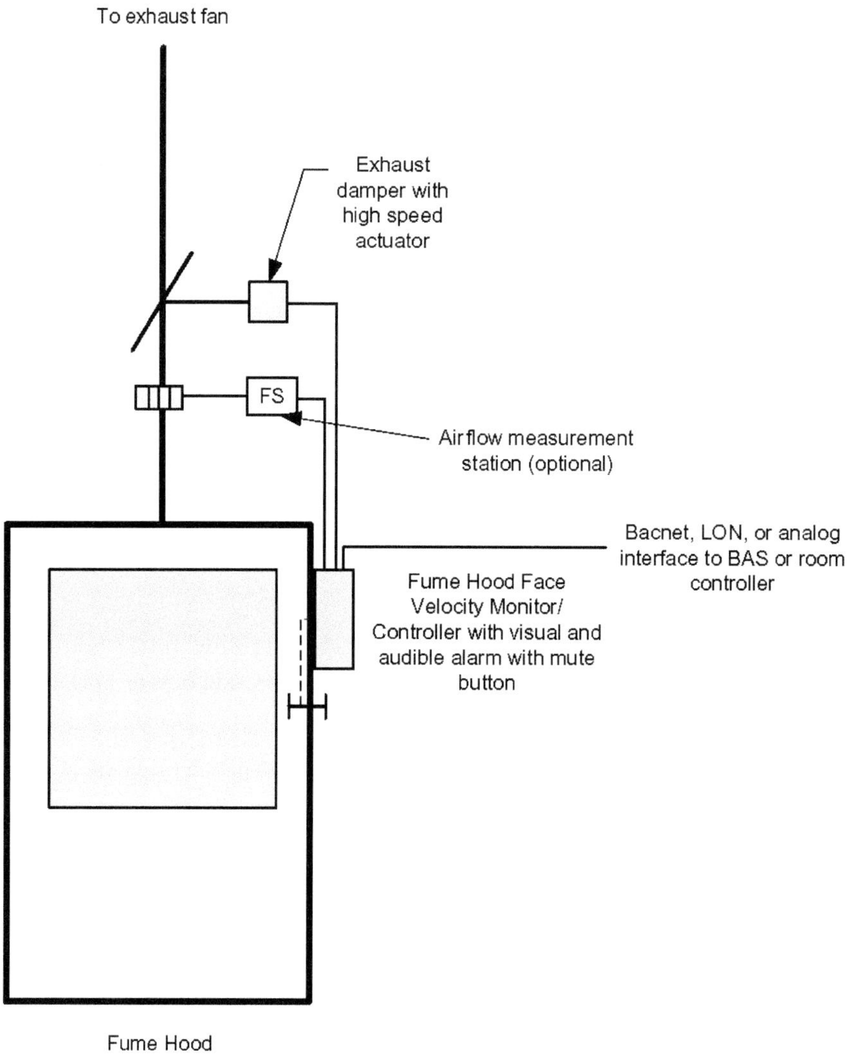

saves energy if the hood sashes are left open; VAV hoods typically exhaust their minimum volume of air when the hood sashes are closed.

Hood face velocity can typically be safely reduced from 100 to 60 fpm (0.5 to 0.3 m/s) when there is no activity in front of the hood. An occupancy sensor aimed at the area directly in front of the hood can be used to set the face velocity setpoint back when the hood is not in use. Alternatively, a room occupancy sensor can be used to set the hood face velocity back when no one is in the laboratory room.

VAV Fume Hood Flow Range

An important set of parameters for a VAV fume hood system is the maximum and minimum flow rates and flow range of the system. The maximum flow rate may be set for the maximum opening of the hood but is often set for the working height of the hood, such as 18 in. (450 mm). In the latter case, if the sash is raised above this point for experiment setup, then the face velocity will be reduced. Chemicals should not be used in the hood if it is open above the height used for design. Typically sash stops are provided to prevent the hood from being raised above the design height (typically 18 in. [450 mm]).

A more important parameter from an energy-efficiency standpoint is the minimum flow setpoint. Previous recommendations from standards such as *NFPA 45* and ANSI/

Figure 11-4
Typical Control Schematic for VAV Fume Hood with Sash Position Sensor

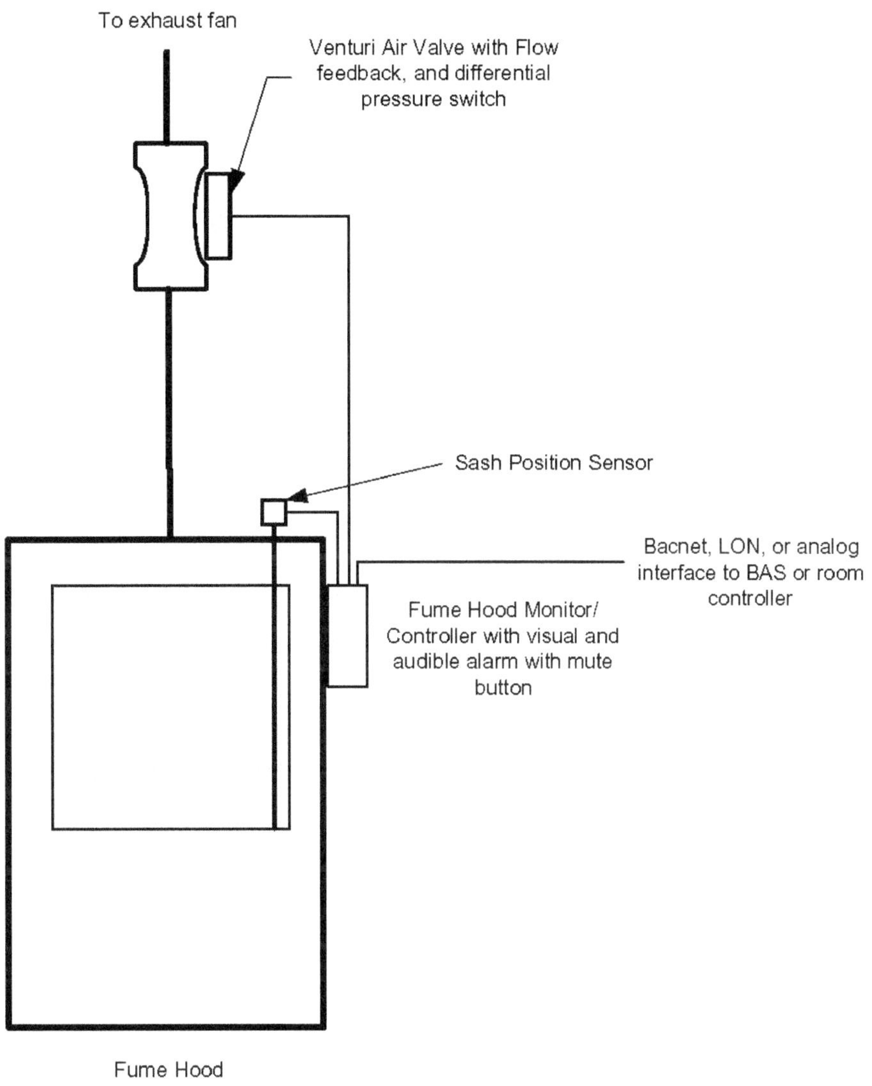

AIHA/ASSE Z9.5 provided guidance of 25 cfm/ft^2 (12 L/s·m^2) of hood bench area, or about 250 cfm for a 6 ft hood (118 L/s for a 1.8 m hood). Over the last 10 years, empirical research studies have shown that this level, which is related to concerns over the lower explosive limit (LEL) for most spills or excessive use of solvents in a hood, is overly conservative. The guidance changed in the 2011 edition of *NFPA 45* and the 2012 edition of ANSI/AIHA/ASSE Z9.5. The latter now requires determining the minimum based on the possibility of an explosion and other considerations. It notes that values in a range of 150 to 375 hood air changes per hour have been used. For a typical 6 ft (1.8 m) benchtop hood, this corresponds to a range of 100 to 250 cfm (47 to 118 L/s), or about 10 to 25 cfm/ft^2 (5 to 12 m^3/h·m^2) of bench area. It is up to the safety professionals and the design team to determine the flow value for a particular installation. Issues to be considered include but are not limited to frequent use of very corrosive acids (such as acid digestion) that could corrode the fume hood liner, control limits of the fume hood airflow control devices, the ability to maintain laboratory offset or pressure, and use of large quantities of highly volatile solvents. This reduced minimum ventilation when the hood sash is closed can provide significant energy savings when the fume hood sashes are closed.

VAV System Response Time

System response time is an important factor for maintaining proper airflow velocity and direction to prevent spillage from the fume hood during disturbances, such as opening the sash from closed to fully open. Careful selection and programming of controllers are important to provide a fast and accurate response to minimize or eliminate fume hood spillage. ASHRAE Standard 110 (ASHRAE 1995) defines an acceptable response time for a VAV hood as the time needed for the face velocity to stabilize within 10% of the design value from the time of the disturbance. This is determined while the sash is opened from a working height of 25% open to fully open at both 1.0 and 1.5 ft/s (0.30 and 0.46 m/s). The response time of the control system should maintain proper safety and pressurization differentials when the system is disturbed. Response times on the order of 3 s or less are typically acceptable (AIHA 2012).

OTHER EXHAUST DEVICES

In addition to controlling fume hoods, the control of other equipment, including snorkel exhausts, BSCs, flammable and solvent storage cabinets, and direct equipment, must be properly integrated into the control of the primary air systems for laboratories to maintain proper room balance. Additional general information on these types of equipment can be found in Chapter 5.

Snorkel Exhausts

Snorkel exhausts are used to provide small, movable exhaust intakes for benchtop laboratory equipment. Though they remove heat and nontoxic particle emissions, they do not provide adequate protection against harmful chemicals. The typical control of these systems is the use of a switchable independent exhaust fan or the use of a manual damper to connect to a manifolded exhaust system. Sometimes the snorkels are provided with an automatic airflow control device to allow one or several snorkels to be opened or closed with a wall switch. See Figure 11-5 for a typical control schematic for a two-position snorkel.

Figure 11-5
Typical Control Schematic for Two-Position Snorkel

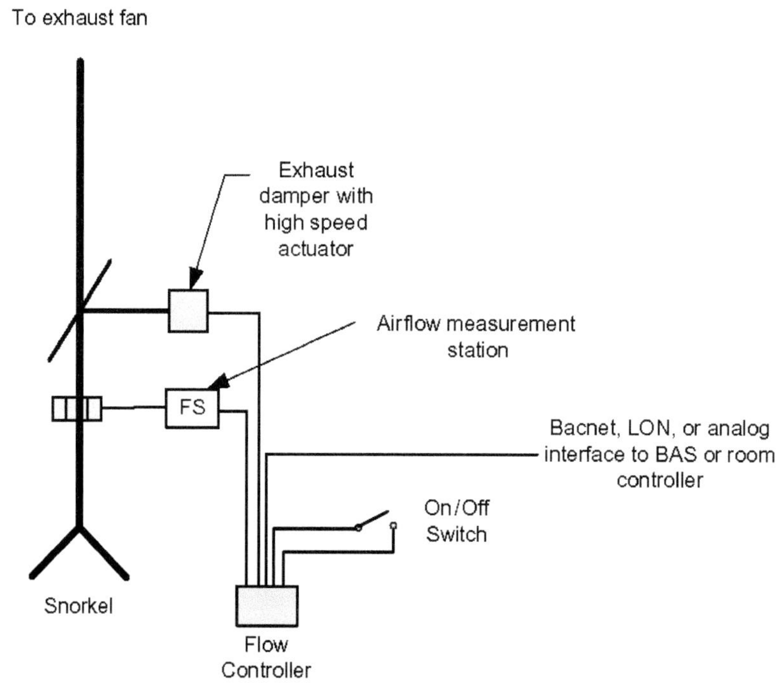

Biological Safety Cabinets

Biological safety cabinets (BSCs) are used to protect laboratory workers and/or the research work being performed from various biological agents. BSCs have higher static pressure drops than fume hoods, and they can add up to 2 in. (500 Pa) of static pressure requirement to the exhaust system. When using a manifolded exhaust system, the extra differential pressure should be considered. Connecting a BSC and a general exhaust to the same duct requires a significant pressure drop through the control damper of the general exhaust device to match that of the BSC. This additional pressure drop can lead to noise and control issues.

As discussed in Chapter 5, there are three levels of safety cabinets—Class I, Class II, and Class III—which provide increasing levels of protection. Some BSCs recirculate some or all of the exhaust air back into the laboratory once it has been filtered through a high-efficiency particulate air (HEPA) filter, while others exhaust all of the air to the atmosphere after being HEPA filtered. Exhaust from BSCs is either through dedicated exhaust stacks or a manifolded system, which only serves biological exhaust sources of similar hazard level. Figure 11-6 shows a typical control schematic for a BSC.

Figure 11-6 Typical Control Schematic for BSC

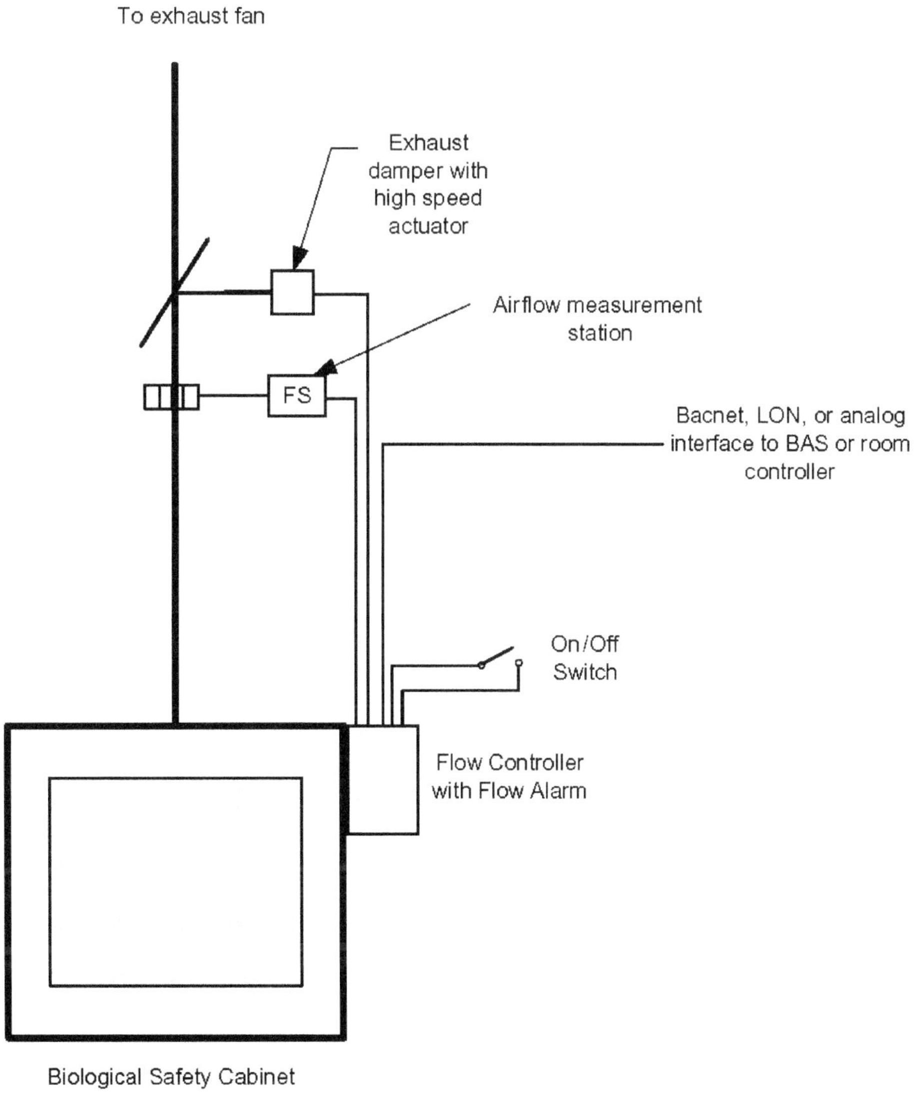

Class I cabinets provide worker protection against low- and moderate- level biological agents. They do not filter or treat the supply air that is introduced to the cabinet and should not be used for high-risk biological substances, because disruptions in the inward airflow to the cabinet can allow airborne particles to escape through the inlet. The exhaust air is HEPA filtered. Because the air entering the cabinet is not filtered, these cabinets do not provide protection for the product. Class I cabinets use constant-volume or possibly two-state control with a fixed opening (no sash).

Class II cabinets are used for moderate-risk biological agents; they use HEPA filtered supply and exhaust air. There are currently four types of Class II cabinets, Types A1, A2, B1, and B2, each which uses a different combination of recirculated air and filtering. Each type of Class II cabinets has its benefits, is used for different types of work, and has a sash that can be opened for performing work. Class II cabinets use constant-volume or possibly two-state control with a movable sash and a filtered bypass opening.

Class II Type A1 cabinets recirculate 70% of the air within the cabinet, and 30% is exhausted through a HEPA filter. Typically these units discharge the air into the room and do not require any external control.

Class II Type A2 cabinets recirculate 70% of the air within the cabinet, and 30% is exhausted through a HEPA filter. These units discharge the air into the room or are exhausted to the outdoors. The control system controls the exhaust to a constant or possibly two-state volume when it is exhausted to an dedicated exhaust fan or a manifolded system.

Class II Type B1 cabinets recirculate 40% of the air within the cabinet, and 60% is exhausted through a HEPA filter. These units are exhausted to the outdoors. The control system controls the exhaust to a constant or possibly two-state volume when it is exhausted to an dedicated exhaust fan or a manifolded system. The control system needs to be able to shut off the airflow to allow for decontamination of the BSC.

Class II Type B2 cabinets exhaust 100% of the air through a HEPA filter. These units are exhausted to the outdoors. The control system controls the exhaust to a constant or possibly two-state volume when it is exhausted to an dedicated exhaust fan or a manifolded system. The control system needs to be able to shut off the airflow to allow for decontamination of the BSC.

Class III cabinets provide a physical barrier between the user and the biological substance. The typical device is a glove box with HEPA-filtered supply air and HEPA-filtered exhaust air. The control system controls the exhaust to a constant volume when it is exhausted to an dedicated exhaust fan or a manifolded system. The control system needs to be able to shut off the airflow to allow for decontamination of the BSC.

Flammable and Solvent Storage Cabinets

Flammable and solvent storage cabinets protect laboratory personnel from the dangers of stored gases and chemicals, including possible explosions and volatile fumes. These cabinets may come with a variety of features, such as fire sprinklers, access openings, and safety controls. Flammable and solvent storage cabinets exhaust air from a high-strength (explosionproof) cabinet, provide supply air (typically room air) through a filtered opening in the cabinet, and are maintained at a slightly negative pressure to prevent fumes from entering the laboratory.

Sometimes these cabinets are vented to the outdoors. When they are, both vents should be piped to the outdoors. Venting only one outdoors will draw air into the laboratory since the room is usually under negative pressure.

An alternative is to provide for active exhaust either through an independent exhaust fan or the use of a manual damper to connect to a manifolded exhaust system. When a manual damper is used, its minimum value is set to provide the design airflow. Depending on the location of the damper, the damper can be opened farther in the case of an emergency, such as a spill inside the storage cabinet. The fan for the main exhaust system is run continuously, because turning off the exhaust to the storage cabinet could allow flammable vapors to build up to dangerous levels. Figure 11-7 shows a typical control schematic for a flammable and solvent storage cabinet.

Direct Equipment Exhaust

Direct equipment exhaust is used for pieces of equipment that generate considerable amounts of heat, humidity, or pollutants. By providing adequate direct exhaust, the heat/humidity gain from the equipment does not contribute a load to the space. However, additional supply air is required to make up for the exhaust air. Equipment that may have direct exhaust includes large furnaces, ovens, centrifuges, autoclaves, and glassware wash machines. Equipment may have either direct duct connection or a canopy hood.

The typical control of these systems is by an independent exhaust fan or by an air control device connected to the manifolded exhaust system that enables the exhaust whenever the equipment is operating. Independent exhaust fans are typically wired to automatically turn on when the equipment is operating, or they have a separate switch to

Figure 11-7
Typical Control Schematic for Flammable and Solvent Storage Cabinet

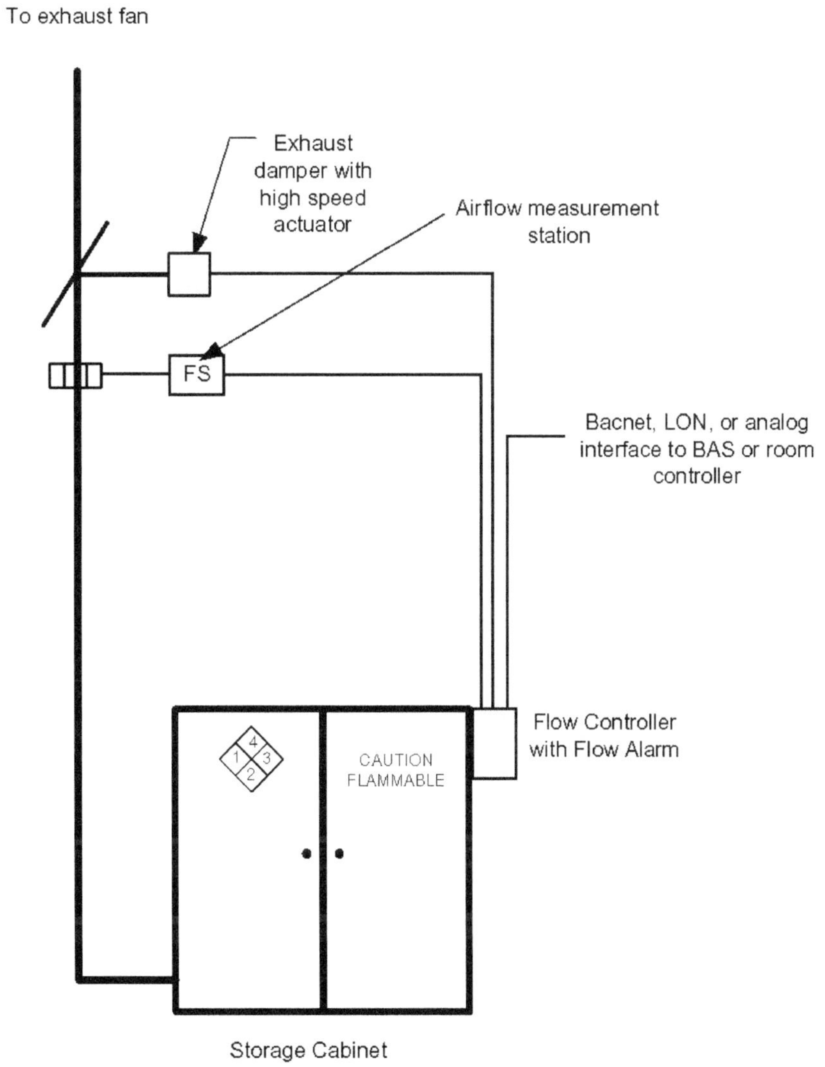

enable the fan. When an air control device is used to connect to a manifolded exhaust system, the device can be either opened manually by the operator or, preferably, automatically by an actuator, which monitors the on-off status of the equipment. The fan for the main exhaust system is run continuously while the equipment requires direct exhaust, and an alarm or interlock device may sometimes be used to ensure that while the equipment is running it has exhaust from the main system. For equipment that stays hot for a considerable time after being turned off, such as a large furnace or autoclave, direct equipment exhaust may use a delay timer to continue providing exhaust for a certain length of time after equipment shutoff. Figure 11-8 shows a typical control schematic for a canopy hood used for direct exhausting.

GENERAL LABORATORY EXHAUST

Whereas equipment exhausts such as fume hoods provide ventilation to a confined area within the laboratory, general laboratory exhaust provides ventilation to the laboratory as a whole. General laboratory exhaust serves two purposes: it allows additional ventilation of the room to remove odors and contaminants that were not captured by the equipment exhausts, and it allows for higher supply airflows to provide the necessary cooling in cases where the cooling load, not the exhaust airflow, dominates the design of the laboratory systems.

Typical control of general laboratory exhaust is through the use of a manifolded exhaust system and balancing dampers for a constant-volume system or airflow control devices for a VAV system. General laboratory exhaust is typically combined with the fume hood exhaust flow in the same manifolded exhaust system. Exceptions to this are when the fume hood exhaust needs to be specially treated, such as with radioisotope fume hoods or perchloric acid fume hoods. Another reason might be the use of an enthalpy wheel, which is typically only used with room exhaust air to reduce the chances of introducing contaminants into the room supply air.

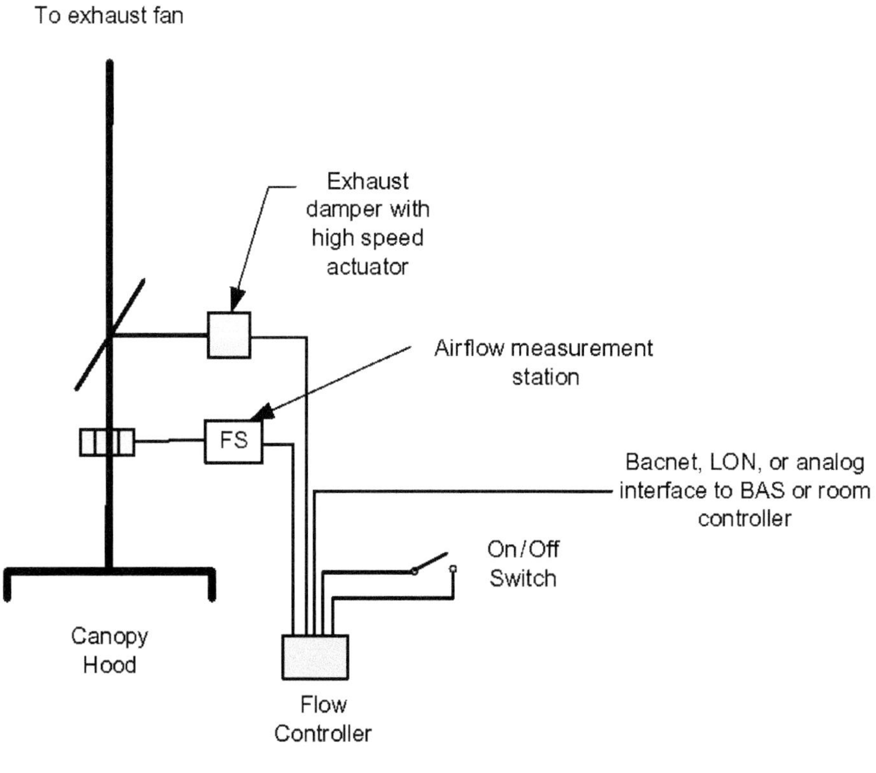

Figure 11-8 Typical Control Schematic for Canopy Hood Used for Direct Exhausting

ROOM PRESSURIZATION CONTROL

Laboratories require accurate control of the pressure differentials between adjoining spaces. Room pressurization is controlled in a laboratory to ensure the safety of the occupants, protect the environment, and protect the research that is being conducted. A rapid increase in exhaust airflow from the room (e.g., opening a sash on a fume hood) requires a similarly large increase in supply air to the room. Pressurization is used to prevent transfer of pollutants to and from unwanted areas. Differential pressure is created by maintaining a differential airflow between the exhaust and supply airflow. Spaces such as cleanrooms and tissue culture and laser laboratories typically must be positively pressurized to prevent unwanted pollutants from entering from surrounding areas, while most other laboratories must be negatively pressurized to prevent airborne pollutants generated in the laboratories from migrating to other areas of the building. Figure 11-9 shows how room pressurization is used to prevent contaminants from spreading.

A typical value for room pressurization differential is 0.05 in. w.g. (12.5 Pa) for stable flow control with a single-stage pressurization control (Sun 2003). This differential pressure has been found to keep contaminants out of cleanrooms as doors open and close. For multiple staged rooms of pressure control, this value may vary from 0.03 to 0.08 in. w.g. (7.5 to 20 Pa). Differential pressure should not exceed 0.30 in. w.g. (75 Pa), as this could make it difficult to open and close typical doors. The maximum delta pressure should be determined by the door opening force and should take into account the force of any automatic door openers.

The room differential pressure can be determined by using the following equation from Chapter 53 of *ASHRAE Handbook—HVAC Applications* (2015a):

$$V = 776CA\sqrt{\frac{2\Delta P}{\rho}} \qquad \text{(I-P)}$$

$$V = CA\sqrt{\frac{2\Delta P}{\rho}} \qquad \text{(SI)}$$

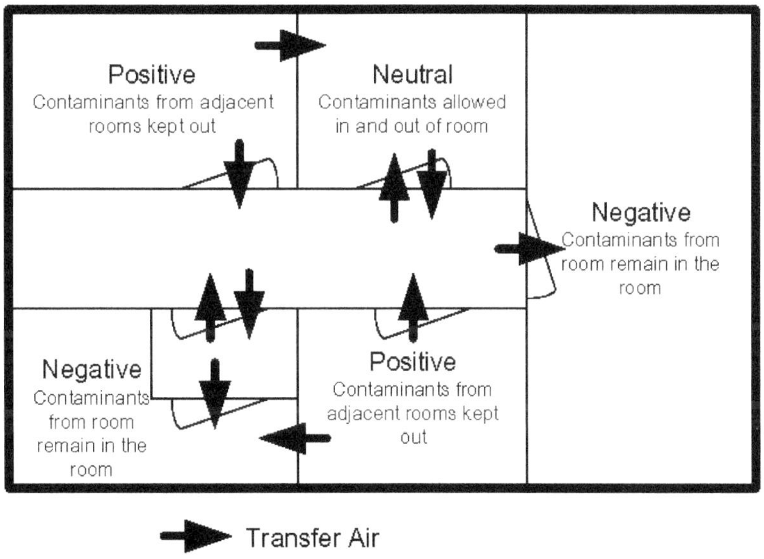

Figure 11-9 Room Pressurization to Prevent Contamination Spread

where

V = volumetric flow rate, cfm (m^3/s)
C = flow coefficient (typically 0.65)
A = flow area (leakage area), ft^2 (m^2)
ΔP = pressure difference across flow path, in. w.g. (Pa)
ρ = density of air entering flow path (0.075 lb_m/ft^3 [1.2 kg/m^3] for standard air)

Figure 11-10 shows a graph of the airflow required to maintain various differential pressures for various leakage areas. In determining the amount of offset air needed to obtain the desired pressure differential, the leakage area is the most difficult number to estimate. But, it is the most basic mechanical parameter in a pressurization design. Designers need some estimate or range of values in order to design the system. Leakage area is dependent on actual building construction. Leakage rates can be in the range from 3.4 to 31.8 in.^{2} per 100 ft^2 (0.0022 to 0.22 m^2 per 10 m^2) of surface area (Geeslin et al. 2008).

To specify a pressurization control system, a designer considers the following:

- Pressurization relationships desired between spaces
- Expected range of leakage values
- Pressurization control sequence for normal operation
- Potential airflow reset strategies for unoccupied conditions
- Desired operation in the event of failures or degraded operation
- Performance specifications of the airflow control components

There are three widely known methods of maintaining room differential pressure: direct pressure control, volumetric offset control, and cascade control. This chapter discusses these methods in the following subsections; see Chapter 16 of *ASHRAE Handbook—HVAC Applications* (2015b) for more detailed descriptions as well as the criteria for selecting a method.

DIRECT PRESSURE CONTROL

Direct pressure control is accomplished by measuring the differential pressure between the room and the reference space (typically the adjacent space) and controlling the exhaust or supply flow to maintain the differential setpoint. Measuring and tracking

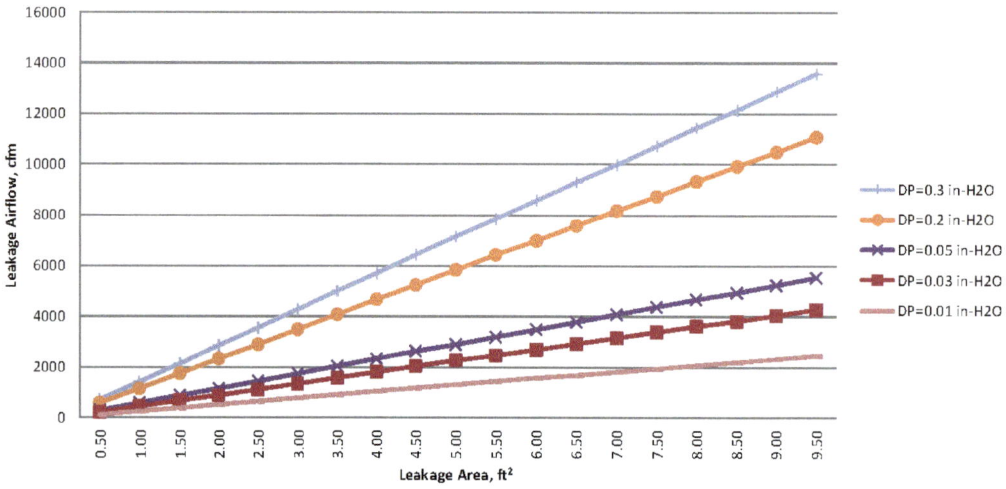

Figure 11-10
Leakage Area versus Flow Rate

all the exhaust and supply airflows is not necessary with direct pressure control. Measuring of the clean supply airflow is needed to maintain minimum ventilation rates. Figure 11-11 shows a typical control schematic for direct pressure control.

Direct pressure control is subject to disturbances of the pressure in the reference space and changes in the leakage area of the room. As doors to the room open and close, the leakage area of the room changes and requires a greater differential airflow to maintain differential pressure. With a door open, the airflow required to maintain the differential pressure may exceed the capacity of the air system. A door switch is sometimes used to freeze the control outputs when the door is open. The system is also subject to changes in the reference space pressure. Opening and closing doors, elevator movements, wind,

Figure 11-11
Typical Control Schematic for Direct Pressure Control

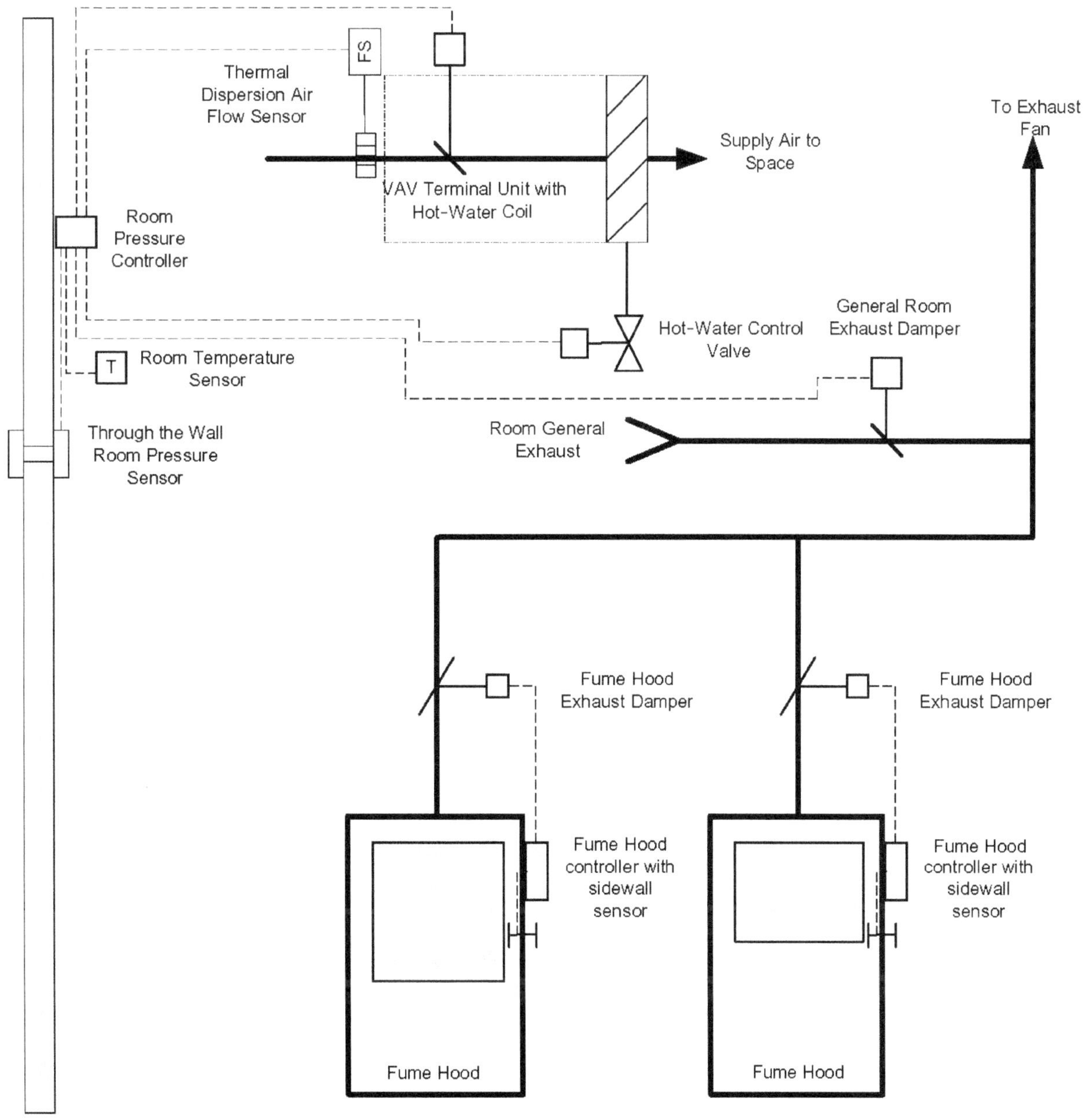

and other factors can affect the reference space pressure. Direct pressure control systems respond to these fluctuations, adjusting airflows to achieve room pressure setpoint.

In a negatively pressurized laboratory, the supply terminal is usually selected as the room pressure control device. This is based on consideration of failure modes in the event that either the supply or exhaust falls short of the setpoint (which is based on minimum exhaust requirements, minimum ventilation rate, or differential pressure). In dynamic, manifolded VAV systems, with energy-conserving strategies implemented at the fans, the possibility of a particular terminal being temporarily starved is very real. If the exhaust falls short, the supply terminal continues to control pressure. If the supply falls short, the room pressure goes more negative than desired. This set of failure conditions is usually preferred to the reverse for safety reasons. Too much negative pressure can lead to issues with opening doors, excess infiltration, or even damage to walls and ceilings.

VOLUMETRIC OFFSET CONTROL

Volumetric offset control is accomplished by measuring all the supply and exhaust flows in the space and maintaining a set differential flow rate between them, thereby maintaining room pressure differential. Measuring and tracking all the exhaust and supply airflows is necessary with volumetric offset control. The offset is often determined based on the number of doors (typical offset values are 75 to 100 cfm [35 to 47 L/s] per door) to adjacent spaces with differing pressurization requirements, or the balancer may adjust the offset to obtain a desired differential room pressure. In determining the offset, the accuracy of control must also be taken into consideration. Figure 11-12 shows a typical control schematic for volumetric offset control.

This method does not recognize changes in the pressures of adjacent spaces. A room pressure monitor with an alarm can be provided to verify room pressure is being maintained. The supply air is controlled to meet the maximum of the thermal demand airflow requirements, the minimum ventilation rate, and the makeup air needed for the hood exhaust and other exhaust devices. The controller maintains the general exhaust at the minimum flow necessary to maintain the desired room offset. The application of the offset depends on the relative pressurization of the room. For example, a negatively pressurized room that has a minimum ventilation rate of 1000 cfm (470 L/s) and an offset of 100 cfm (47 L/s), the offset is typically applied so the total exhaust is 1000 cfm (470 L/s) and the supply is equal to 900 cfm (425 L/s). If the same room were designed to be positively pressurized, the total exhaust would be 900 cfm (425 L/s) and the total supply 1000 cfm (470 L/s). This minimizes the amount of supply air that needs to be conditioned and supplied to the space.

Flow measurement is subject to errors in measurement. If a room has a total supply of 1000 cfm (470 L/s) and a total exhaust flow of 1200 cfm (566 L/s) and a measurement error of ±5%, the nominal differential flow rate is 200 cfm (94 L/s) with a worst-case error of ±110 cfm (52 L/s). This numerical example illustrates that performance of a volumetric offset control system is sensitive to the accuracy of the individual flow control components. ANSI/AIHA/ASSE Z9.5 (AIHA 2012) suggests calculating volumetric offset control performance specifications from pressurization requirements.

The following process, illustrated in Figure 11-13, gives designers a methodical way to specify control components so that the resulting volumetric offset system performs as intended.

1. *Define pressurization relationships between spaces.* The desired pressure relationship, or direction of airflow, must be identified between any pair of adjacent spaces in the vicinity of the laboratory. ANSI/AIHA/ASSE Z9.5 (AIHA 2012) suggests expressing this information directly on a floor plan to clarify

intentions for all contractors, commissioning authorities, and representatives of laboratory users.

2. *Identify supply and exhaust terminals for each space.* The list of the supply and exhaust terminals for each space, and the required range of flow rates, will become the fully specified "box schedule" by the end of the process.
3. *Select pressurization level and accuracy.* The designer selects the nominal value of the offset by considering the expected leakage area and the desired level of pressurization. Then, with reference to the same leakage curve, the designer specifies the allowable control range for that offset value. As the previous numerical example shows, the performance of the control system is

Figure 11-12
Typical Control Schematic for Volumetric Offset Control

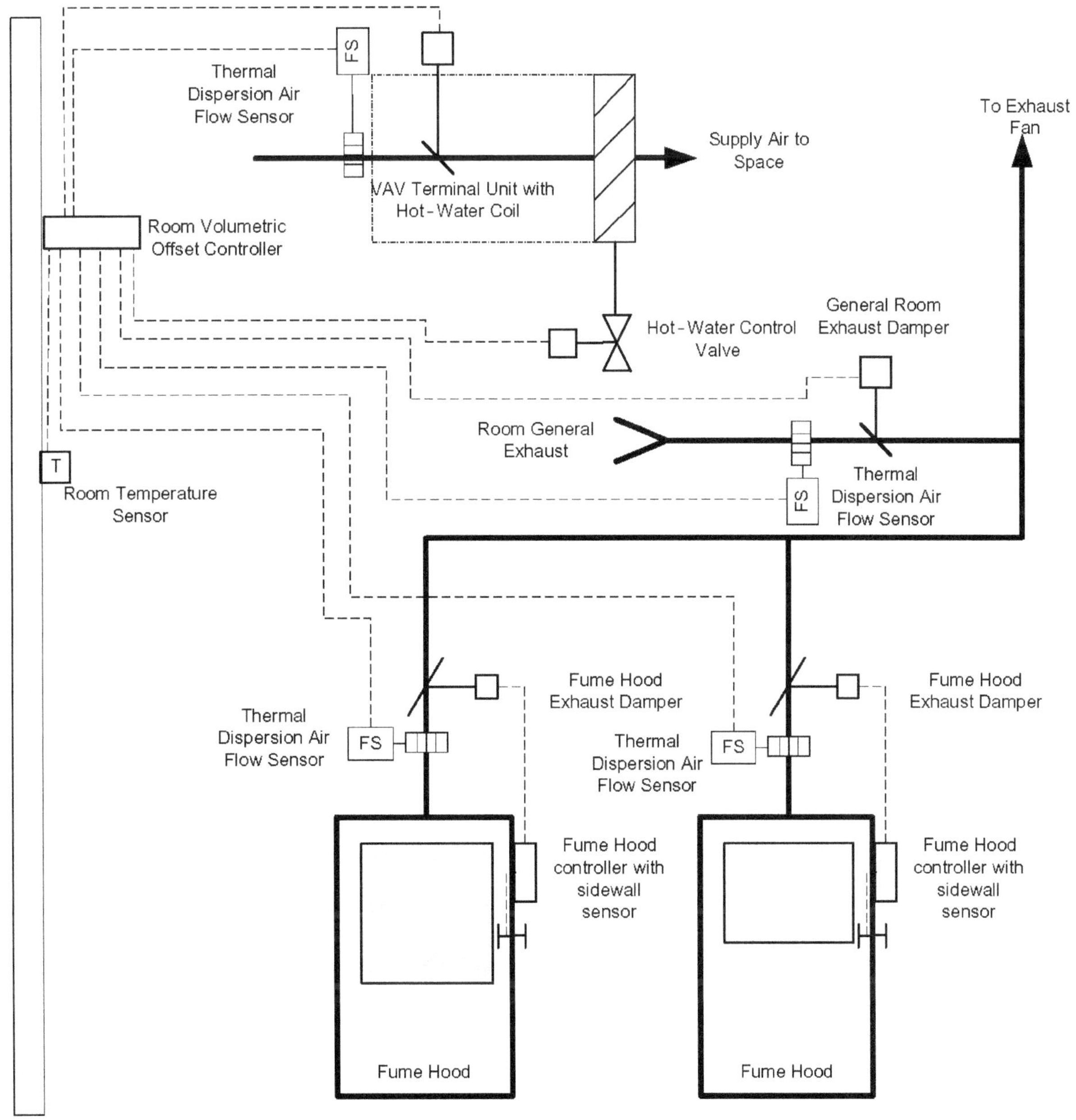

Figure 11-13
Specifying Control Components for Volumetric Offset

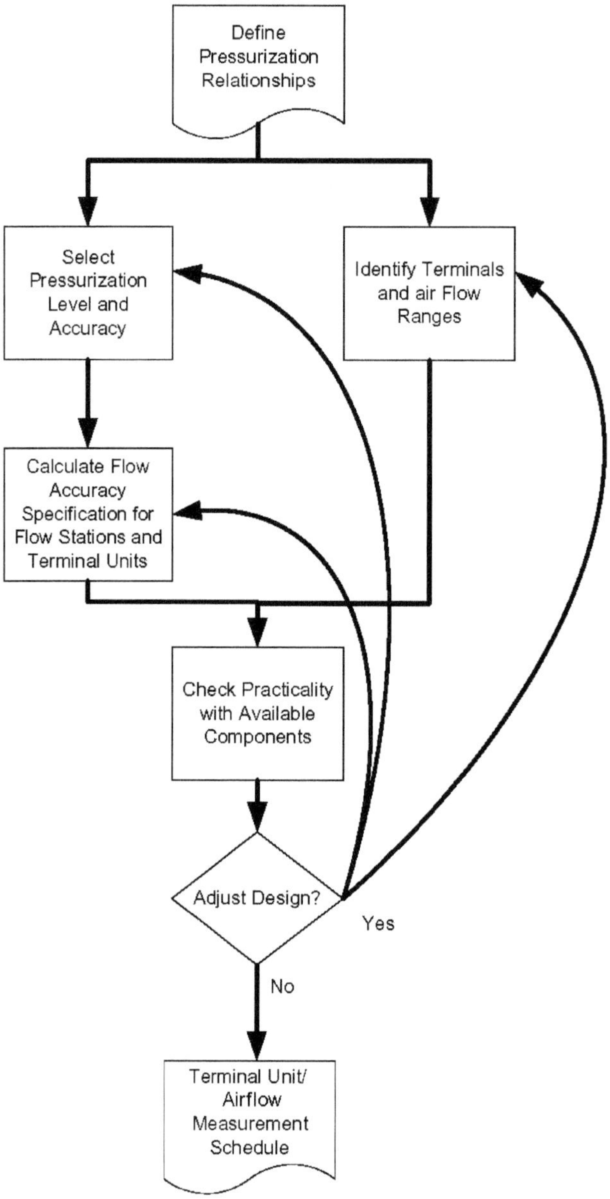

sensitive to the accuracy of the individual flow control components; therefore the designer is advised to allow a fairly wide tolerance on the small offset between two much larger flow values. Tolerance values of 20% to 30% of the nominal offset are often found to be workable.

4. *Calculate flow accuracies.* The chosen airflow tolerance is allocated to each of the airflow control terminals in the room. If they have approximately the same flow range, it makes sense to allow equal tolerance on each one.
5. *Check for practicality.* The designer should check the preliminary performance specifications against what is known to be practically available and within the intended project budget. It is particularly important to check the accuracy specifications at the high and low ends of the flow range for each terminal. If the numbers are reasonable, the job is done. If the accuracy calculated for any of the terminals seems unattainable or impractical, the specifier should adjust the design before putting the project out to bid.

6. *Adjust as needed.* If the designer finds the required accuracy is impractical, there are many possible ways to remedy it, such as the following:
   - Adjust offset and corresponding tolerance.
   - Choose pressure feedback instead of volumetric offset control.
   - Resize terminals to get more accurate flow control.
   - Reselect sensors to increase accuracy.
   - Reduce airflow ranges.
   - Add leakage, move design point.

   The designer can experiment with these modifications and recalculate the flow control specifications to find a workable design.

In a negatively pressurized laboratory, a supply terminal is usually selected as the flow tracking device. This is based on consideration of failure modes in the event that either the supply or exhaust falls short of the setpoint (which is based on minimum exhaust requirements, minimum ventilation rate, or minimum volumetric offset). The designer specifies that if the exhaust flow fails to meet the setpoint, the supply terminal continues to track the actual flow, maintaining the desired pressurization. If the supply falls short, the room pressure goes more negative than desired. This set of failure conditions is usually preferred to the reverse for safety reasons. Too much negative pressure can lead to issues with opening doors, excess infiltration, or even damage to walls and ceilings. (It is possible to design more complex sequences that maintain the desired offset when either fan system lacks capacity. Typically that complexity is only chosen when very special hazards are involved.)

CASCADE CONTROL

For tightly constructed rooms where the leakage can be very small, volumetric offset control cannot typically provide a stable differential pressure because of the small offset flow rates that are needed. For these situations, a cascade control system should be considered. Cascade control can also be used to verify the offset control is working correctly and account for changes in building pressure relationships. A cascade control system is accomplished by combining volumetric offset control and direct pressure control. The differential pressure sensor is used to reset the offset over time to maintain the proper differential pressure. Figure 11-14 shows a typical control schematic for cascade control.

DILUTION VENTILATION AND MINIMUM VENTILATION RATES

Minimum ventilation rates should be established that provide a safe and healthy environment under normal and expected operating conditions. The dilution ventilation provided by this airflow is no substitute for the containment performance of a laboratory fume hood or other primary containment device regardless of the room ventilation rate. The appropriate ventilation rate for clearing a room of fugitive emissions or spills varies significantly based on the amount of release, the chemical's evaporation rate and hazard level, and the ventilation system effectiveness.

Fixed minimum airflow rates in the range of 6 to 12 air changes per hour (ach) when the space is occupied have been used in the past. However, recent university research (Klein et al. 2009) showed a significant increase in dilution and clearing performance when the ventilation rate was increased from 6 to 8 ach, with diminishing returns above 12 ach. Similarly, computational fluid dynamics (CFD) (Schuyler 2009) showed that increasing the laboratory's dilution ventilation rate from 4 to 8 ach reduced the back-

Figure 11-14
Typical Control Schematic for Cascade Control

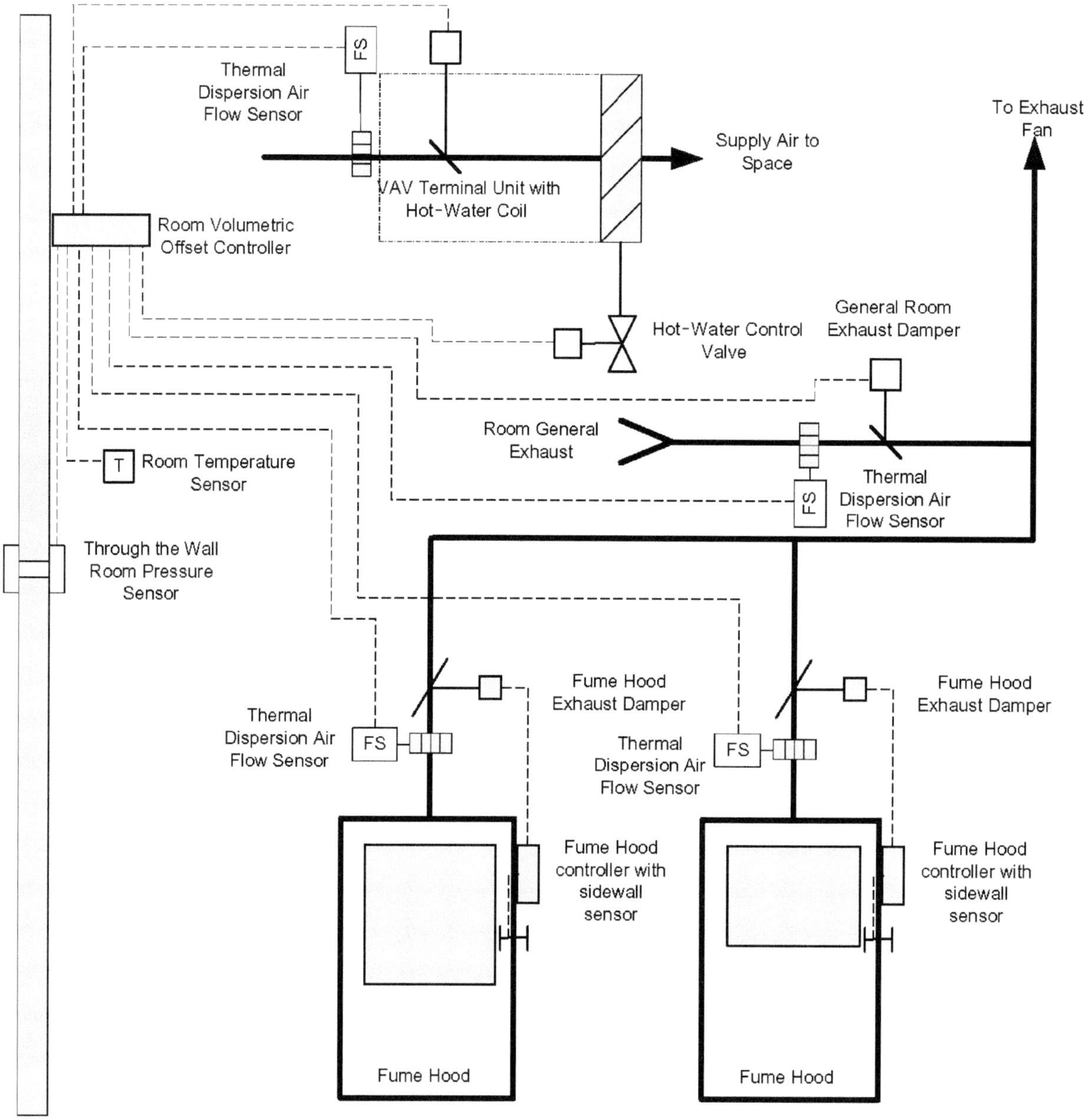

ground contaminant level by greater than a factor of 10. This indicates that minimum ventilation rates at the lower end of the 6 to 12 ach range may not be appropriate for all laboratories. Minimum ventilation rates should be established on a room-by-room basis considering the hazard levels of the materials expected to be used in the room and the operations and procedures to be performed. As the hazard levels, materials, and use of a room change, an increase or decrease in the minimum ventilation rate should be evaluated.

UNOCCUPIED SETBACK CONTROL OF MINIMUM VENTILATION RATES

Some laboratories use a high minimum ventilation rate during occupied periods and when the laboratory is unoccupied, as indicated by a time clock or room occupancy sensors, the minimum ventilation rate is set back to a lower value, reducing the total airflow to the space provided the hood exhaust and thermal demands are also reduced. This approach may save significant amounts of energy, but designers must apply it carefully. The concept behind this method of achieving reduced ventilation is that the laboratory workers are the prime source of contaminants in the room, and when they are not there, working, the strength of the source is greatly reduced, resulting in less of a need for ventilation. There are laboratory rooms where this reasoning is sound; however, there are also cases where it does not apply because of the materials and processes that remain in the room after the workers leave.

This design concept needs to maintain low, safe concentrations of contaminants in the air at all times. It is not part of this concept to allow increased contamination levels on the grounds that the workers are absent. Some of the conditions that preclude use of this concept are laboratory spaces where there is a possibility of fugitive emissions during unoccupied times, the use of chemicals stored outside the hood that may leak into the space, the operation of research experiments and chemically related operations during unoccupied times, or other operations that might release chemical vapors or particulates into the laboratory space that could increase or build up when air change rates are reduced.

For spaces where it is appropriate to use this concept, the following four design steps, adapted from a published process for designing two-state ventilation systems (Coogan 2008), can be followed:

1. *Determine if the unoccupied period presents an opportunity to reduce ventilation.* At this step, the ventilation designer consults with the responsible environmental health and safety professional and a representative for the laboratory users, such as the Principal Investigator or laboratory manager.
2. *Select the minimum flow rates that apply during the occupied and unoccupied periods.* This is the same sort of analysis that applies for the minimum flow rate without a setback. In this case, designers assess the ventilation rate and potential for contamination twice: once with the laboratory in use and once with no one working in the laboratory.
3. *Select a trigger that informs the control system when to switch between the occupied and unoccupied settings.* This is a common building automation system (BAS) function implemented in many familiar ways: schedules, occupancy sensors, manual switches, etc. It is important to select a trigger, or a combination of triggers, that works effectively for the particular space.
4. *Design an indicator for the laboratory users.* It is crucial that anyone in the room or entering the room know that the ventilation system is in the mode that supports their activity. If it occurs that a person is in the room while the system is running at the unoccupied rates, the system must make this fact perfectly evident to the user. It is necessary to train workers on the meaning of the airflow indicator and the procedure to follow if the wrong mode is indicated.

The savings potential that justifies a ventilation setback depends on the amount that the ventilation can be reduced and the amount of reliably unoccupied operating time. Some laboratories have a very consistent weekly work schedule; others are likely to be used at any time of the day or night. The savings potential may also be diminished in laboratories where the hood makeup requirements are greater than the room ventilation requirements.

Note that reducing the air change rate at night or during unoccupied times can be used safely in combination with demand-based control (discussed in the following subsection) since in this case the laboratory is being continuously monitored. As a result, contaminants will be detected and purged before they can build up in the laboratory even during unoccupied periods—no different than the operation during the daytime.

DEMAND-BASED CONTROL

Active sensing of air quality in individual laboratories (Sharp 2010), also known as *demand-based control,* is an alternative approach for dealing with the variability of appropriate ventilation rates, particularly when energy efficiency is important or when less may be known about the hazard level. With this approach, the minimum airflow rate is varied based on sensing the laboratory's actual air quality level or "air cleanliness." Primary contaminants in general laboratories include chemical, particulate, and aerosol contaminants. Appropriate sensors used to determine the laboratory's air quality include, for example, a photoionization detector (PID) type total volatile organic compound (TVOC) sensor to detect a broad range of organic as well as some inorganic contaminants. Some other TVOC or chemical sensors may also be used to pick up other compounds of interest not detected by a PID. A laser-based particle counter is also recommended to detect potential particulate and aerosol pollutants. For special-purpose laboratories that have a dominant potential contaminant, sensors should be evaluated for their ability to detect that contaminant. Carbon dioxide measurement is also often recommended for typical occupancy-related ventilation, particularly where higher people density can occur, such as in teaching laboratories. The engineer should work with the owner's health and safety personnel to determine the chemicals being used in the space and what the maximum permissible exposure level is. The sensors selected should be evaluated for the ability to detect the chemicals present, the ability to measure within acceptable tolerances, and on long-term reliability and recalibration of the sensors.

When air contaminants are sensed in the laboratory above a given threshold, the minimum air change rate is increased proportionally to an appropriate level to purge the room. This purge level depends on the system and airflow control device capabilities but is typically recommended to be in the range of at least 8 ach to as high as 16 ach. When the air is "clean" and contaminants are below the threshold, this approach can safely reduce laboratory air change rates to as low as 2 ach or as determined by the owner's health and safety personnel. A commonly applied approach when the laboratory air is clean is to operate at 4 ach during the day and reduce the air change rate to 2 ach at night, when the hoods are more likely to be closed, the thermal loads are less, and the temperature control can be set back. Based on these airflow reductions and extensive studies of laboratory room environmental conditions that have shown that the air in laboratories is typically clean more than 98% of the time (Sharp 2010), the energy savings from this approach can be very significant. In fact, for many laboratories, such as those that do not have a high hood density, the single largest energy conservation measure that can be applied to the laboratory is the demand-based control of the minimum ventilation rate. For example, Figure 11-15 shows an analysis of the estimated energy savings from reducing the dilution ventilation of a typical laboratory in Boston from a fixed 6 ach to a demand-based control approach operating when the air is clean at 4 ach during the day and 2 ach at night. As can be seen from the figure, the HVAC energy of the laboratory is cut by a total of 51%.

It should be noted that these systems cannot typically be used in biosafety level (BSL) 3 or 4 laboratories, as described in *Biosafety in Microbiological and Biomedical Laboratories* (CDC 2009), for a variety of reasons, such as the facts that no particles can be taken out of these spaces and that it is it practical to mount the sensors themselves inside these spaces. Also, spaces with very high hood densities, such as more than two 6 ft (1.8 m) hoods per a 250 ft^2 (23.2 m^2) laboratory module, are also not good applica-

Figure 11-15
HVAC Energy Use Breakdown

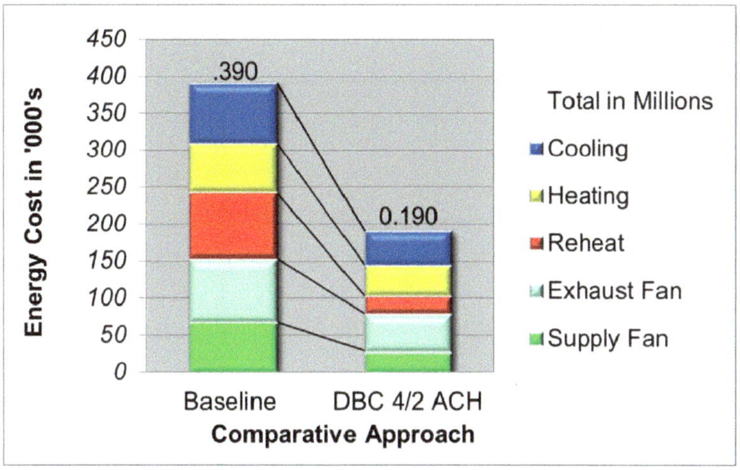

tions for this approach, because the airflow rates are dominated by the fume hood flows even when the sashes are closed.

In terms of implementing this approach, multiple indoor air quality sensors can be individually mounted in each space for continuous monitoring, or else an air sampling system can be used that draws air samples back to a central sensing station to significantly reduce the number of required sensors. It is also recommended to sense both the supply air going into the laboratory and the exhaust air going out of the laboratory and to subtract the supply air levels from the exhaust air levels. The resultant differential measurement eliminates the effect of contaminants in the outdoor or supply air that could create a false positive across the entire building. If an air sampling system is used with this supply-exhaust differential sensing approach, the effects of sensor drift errors can be cancelled out for improved accuracy by using the same central sensor for the exhaust and supply differential measurement. The engineer should review the first cost, energy savings, and life-cycle cost of potential systems, including calibration and maintenance costs of the sensors, to determine the long-term benefits of these systems. The engineer should also review the chemicals being used in the space to determine if the sensors are appropriate for the detection of contaminants, as one sensor cannot typically detect all chemicals. The health and safety personnel will need to review the system when additional chemicals are introduced into the laboratory space.

ROOM TEMPERATURE CONTROL

Room temperature control with constant-volume systems may be accomplished by first fixing the supply airflow rate high enough so that all airflow requirements such as fume hood makeup air requirements, anticipated thermal loads, and the minimum dilution ventilation requirements can be met. A reheat control valve is then used in response to a space temperature sensor to modulate the discharge temperature to prevent the laboratory from being overcooled. With a VAV system, the supply volume can be reduced but the reheat coil is often still needed since the fume hood makeup air requirements or minimum dilution ventilation requirements are often greater than the thermal load requirements for supply airflow.

The temperature sensor location can affect the ability to accurately control the temperature in the laboratory. The sensor should be located where it can read the actual temperature of the laboratory space. The pressure differential between spaces and the possibility of significantly different temperature requirements between adjacent spaces associated with laboratories can pose difficulties in getting the temperature sensor to read the actual room temperature.

For example, if a temperature sensor is located near the door of a negatively pressurized laboratory space, air from the hallway continually flows into the room near the temperature sensor and, depending on the location of the sensor with respect to the door and wall configuration, the temperature sensor can read the temperature of the hallway air instead of the laboratory space. As the temperature differential between the hallway and the laboratory increases, the effect of the hallway air increases.

Another example is a temperature sensor mounted in or over the top of a hole in the wall (or loose electrical box). If the temperature sensor is loose around the hole in the wall, the negative pressure of the laboratory can cause the temperature sensor to read the temperature in the wall cavity.

Temperature sensors should be located in the interior space of the laboratory away from direct sunlight, away from transfer air locations, and away from heat-producing equipment. Care should be taken to locate the sensor where it will not be blocked by shelving and will be representative of the space temperature. In some instances, temperature sensors may be located in general exhaust ducts. Where the exhaust air is filtered at the room level, sensors should be located after the filters.

In terms of estimating the actual versus peak laboratory thermal loads for assessing energy usage and reheat levels, a Labs21 and University of California Davis study (Mathew et al. 2005a, 2005b) found that the average plug and lighting load for a typical laboratory room is 2.5 to 3.0 W/ft^2 (27 to 32 W/m^2), with perhaps only 20% or fewer of laboratories having loads greater than 4 W/ft^2 (43 W/m^2). For these typical average rooms, daytime normal thermal loads are less than 4 ach of conditioned supply airflow, and for nighttime the use of a temperature setback control should push the required supply airflow down to 2 ach or less. Actual airflow turndown should be based on the minimum requirements for the facility.

HYDRONIC ROOM COOLING

In laboratory spaces where the airflow required for cooling exceeds the airflow required for ventilation or hood exhaust, local hydronic cooling approaches such as active or passive beams or fan-coil units can be used for more energy-efficient operation. These approaches are desirable because they decouple the cooling requirements from the airflow requirements so that laboratory airflow can be reduced to the minimum ventilation level, which could be as low as 2 ach, at least when combined with demand-based control. As a result, these approaches may provide lower first or capital costs by reducing outdoor air to reduce the supply airflow conditioning requirements and the size of the supply and exhaust fans. A cost analysis of a combination air/hydronic system versus an all-air system should be performed based on the specific project.

Active or passive beam systems, often referred to as *chilled beam* systems, are a relatively new hydronic cooling coil approach, at least as used in laboratories. These systems are similar to the old constant-volume induction systems. To obtain greater cooling capability, laboratory applications typically use active beams that are provided with conditioned air and that, through the use of multiple high-velocity nozzles, create an induction effect that pulls room airflow through the unit's noncondensing cooling coils as shown in Figure 11-16.

Although the beams can be supplied with constant airflow, a more energy-efficient approach is to vary the airflow at least within a limited range of about 2 to 1 or greater if less cooling is needed at lower airflows. Space temperature is typically controlled by

Figure 11-16
Typical Cross Section of an Active Beam

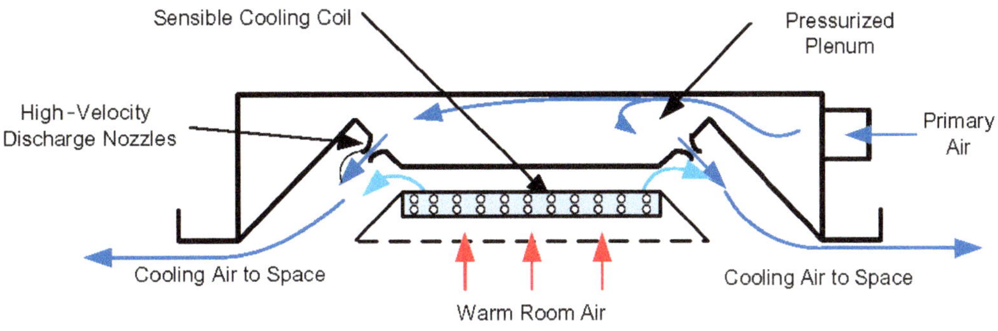

modulating the chilled water serving the beams, and hot-water reheat control valves are still used with the supply airflow control devices to control overcooling. The ventilation air must handle all the latent cooling requirements of the space. The dew point of the ventilation air is controlled at the air-handling unit (AHU) to dehumidify the space. The dew point of the space should be monitored to prevent it from rising above the temperature of the chilled-water in the chilled beam. When the space dew point rises above the chilled-water supply temperature, the chilled water to the beam should be shut off or the chilled-water supply temperature should be raised to above the room dew-point temperature.

When active or passive beams or fan-coil units are used in laboratory rooms having a 6 or 8 ach minimum dilution ventilation flow, the room can often have a large amount of overcooling and thus a large amount of required reheat energy use. Furthermore, the beams or fan-coil units are not often used, as most of the cooling is provided by the main AHU cooling coils, and thus the beams or fan-coils are oftentimes a redundant duplication of cooling equipment. However, if demand-based control is used to bring the room minimum flows down to, for example, 3 to 4 ach during the day and 2 ach at night, then the amount of overcooling and required reheat energy is drastically reduced. If further cooling is required above 2 to 4 ach, the beams or fan-coil units can appropriately provide this peak cooling requirement without impacting the required outdoor airflow. As a result, the HVAC system can be downsized, since the laboratory room's thermal load is decoupled from the airflow, and the air system can be resized to 2 to 4 ach based only on the dilution ventilation and fume hood exhaust requirements.

Occasionally, to try to prevent large amounts of overcooling, "neutral air," or air at about 65°F to 70°F (18°C to 21°C), is provided to the laboratory so that the active or passive beams or fan-coil units provide all the cooling. Reheat still should be provided, although now it must be provided at the supply air handler. However, there are some approaches that use various heat recovery systems to reduce the energy used to provide this reheat, such as dual wheel designs or wraparound coils. In many cases these systems still require some reheat energy, although they can typically provide a significant net reduction in required reheat energy, particularly where a higher density of fume hoods increases the requirements for supply airflow.

On the other hand, the active or passive beam systems or fan-coil units can significantly increase the capital costs and/or complexity of the air handlers and the beams or fan-coils because the room cooling approaches must now provide all the room cooling instead of the supply air providing some of the cooling. When hydronic cooling is used in spaces that require cooling year round, the use of water-side economizers should be considered. Thus, unless the minimum airflow of the room cannot be reduced to less than about 6 ach, the use of neutral air is typically not warranted compared to using approaches to reduce the dilution ventilation requirements to begin with. When demand-based ventilation control is combined with hydronic cooling, the system saves more energy and has less capital cost because the HVAC system and the beams or fan-coil units can be downsized. Laboratories that require decontamination (for example, BSL-3 and BSL-4 laboratories) should not have coils in the space that cannot easily be decontaminated.

CENTRAL SYSTEM LEVEL CONTROL

Control of central air-handling systems in laboratories is similar to control of typical air-handling systems in commercial buildings with a few notable exceptions. Because the exhaust air system typically has airflow control devices throughout the system, the exhaust fans are typically controlled based on maintaining a static pressure setpoint in the exhaust duct. AHUs also typically provide 100% outdoor air, which means the system sees a wide range of inlet air conditions that the unit must respond to and maintain proper discharge air temperatures and humidity for.

To maintain a safe working environment for the laboratory personnel, central systems serving laboratories need to automatically respond to changes in space demand and emergency situations. One of the most important emergency situations is fan failure. This can result in exhaust air not being available for laboratory equipment, such as fume hoods, or supply air not being provided, which can create large negative pressures in the laboratory spaces. When the supply fan fails and the exhaust fan continues to operate on a large system, the building can become extremely negatively pressurized (up to the exhaust fan static). This negative pressure can prevent exit doors from being able to be opened and may cause damage to the building. Fan failures require different control responses for one-fan and two-fan systems.

AIR-HANDLING UNIT

Supply fan controls for VAV systems typically maintain a duct static pressure setpoint with reset based on demand position. Because the units provide 100% outdoor air, there is no return fan control with the unit. Units are typically controlled to maintain a set discharge temperature. Preheat coils should be controlled to prevent freezing of the coils. Where standby fans or units are in place, the controls should provide isolation of units and automatic rotation of the lead unit or fan. When switching from the lead unit or fan to the standby unit or fan, the standby unit should be started, the isolation dampers opened, and the unit brought under control running in parallel with the primary unit, then the primary unit should be isolated and shut down. Opening isolation dampers before the fans start spinning can cause air to flow through the fans in reverse, turning the fans backwards. Turning the fans backwards before starting the fans can cause the motors to overload when trying to reverse the flow and can trip the circuit breakers. Selection of fans with no stall point allows the fan to start before the isolation damper opens. Backdraft dampers can also be used to prevent reverse flow of air.

If a standby unit is provided, when switching to the secondary unit because of fan failure the secondary unit should start immediately and the primary unit should be isolated. Another option is to operate the primary and secondary units in parallel, with each unit carrying half the load. This lowers the static pressure drop through the filters and coils. During a fan or unit failure, the failed unit can quickly be isolated and the other unit ramped up to handle the full load.

Prior to starting the AHU fan, the exhaust fans should be started if the spaces are typically to be kept under negative pressure. This is especially critical in BSL-3 and BSL-4 laboratories. Starting the supply fan first creates a period of positive pressure in the spaces until the exhaust fans are started. This may be acceptable in some spaces, but where containment is critical exhaust fans should start. Fans should have proof of operation points tied to the direct digital control (DDC) system.

Humidification control is important. The typical laboratory air system is 100% outdoor air. Typically, air is supplied to the terminal units at 55°F (12.8°C) year round, with local heating coils providing the final heat. This requires large amounts of humidification to bring the outdoor humidity up to acceptable levels for the laboratories. Supply air temperature reset should be considered for raising the unit discharge temperature when appropriate. The high-limit humidity sensor should be located far enough downstream of the humidifier to ensure the moisture has been absorbed by the cold airstream, which is often a longer distance than is typical. Also, smoke detector placement should follow similar guidelines, because the humidifier fog can trip the smoke detectors.

EXHAUST FANS

In most laboratories, there is an airflow control device in the exhaust ductwork. This device controls the airflow from the space similar to the way a VAV box controls airflow in a typical office building. Because the airflow is controlled at the space level, the exhaust fan is typically controlled to maintain a constant negative pressure in the exhaust ductwork. A static pressure sensor located in the exhaust plenum should be used to maintain a constant negative exhaust plenum pressure setpoint. This can be achieved by modulating the speed of the fan or by modulating the outdoor air bypass damper on exhaust systems designed for constant discharge velocity. The exhaust plenum static pressure sensors can be reset based on room level exhaust demand or damper position. When venturi valves are used in the exhaust system, the reset must maintain a minimum differential pressure across the venturi valves or multiple static pressure sensors near the end of the duct runs. When the bypass damper modulates fully open, a fan can typically be staged off (in multiple-fan systems) if the static pressure becomes too negative in the plenum. When there is insufficient static pressure in the plenum and the bypass damper has modulated full closed, then the next fan should be started. The bypass damper arrangement allows outdoor air to bypass into the exhaust duct near the fan or plenum to maintain constant airflow, discharge velocity, and effective stack height for each fan. In multiple-fan systems, isolation dampers or gravity back draft dampers should be provided for each fan. The fan should start first, then the isolation damper should open. The gravity back draft damper ensures a smooth transition as the fan comes online. This prevents the fan from turning backwards as well as a sudden drop in airflow from the building.

To save energy, VFDs can be added to the exhaust fans. When VFDs are used, the minimum fan speed, discharge velocity, and effective stack height must be considered at both maximum and minimum airflow. A combination system can be used to slow the fans to a minimum speed that maintains minimum discharge velocity and effective stack height, then opens the bypass damper to further lower the exhaust flow from the building.

Additional energy-saving controls for the exhaust fan may include real-time monitoring of either wind speed and direction or contaminant levels in the exhaust plenum (Carter et al. 2011). When the wind speed is low and from the right directions or if the air in the exhaust plenum is already sufficiently diluted or "clean," then the exhaust fan exit velocity and flow maybe reduced because of a lower required stack height and discharge velocity.

EMERGENCY MODES OF OPERATION

When designing the control systems for the laboratory equipment, emergency situations should be addressed in the control sequences at the room level, at the system level, and at the building level. Emergency situations include chemical spills, fan failure, actuator failure, valve failure, network failure, and power failure.

During chemical spills in the space, a manual emergency button on the wall can be used to change the ventilation rate from a minimum value to maximum flow as well as shut down power to the laboratory space. The emergency button should not shut down power to the control system. If demand-based control ventilation is being used, the system can automatically increase the ventilation rate based on the detection of an increased concentration of contaminants.

Failure of a fan should initiate alarms, isolate the fan, and start backup fans if provided. If backup fans are not provided, then if an exhaust fan fails the supply fan should be stopped to prevent positively pressuring the laboratory spaces. If the supply fan fails, the exhaust fan should typically be stopped to prevent depressurizing the building, which can lead to damage and prevent exit doors from being opened. In critical spaces such as a BSL-3 space, backup fans should be provided. Where redundant fans are operated in parallel and one fan fails, the fan should be isolated and the remaining fan(s) ramped up in speed to handle the load.

Power failures stop all systems unless they are on emergency power, in which case they should be restarted based on the start-up sequence. If the supply and exhaust fans are on emergency power, it is important that the entire control system down to the space level is on emergency power in order to maintain control of the system. Power failures should not cause loss of programming or setpoints. A restart sequence should be considered. For example, to maintain negative space pressurization, exhaust fans should start first, then the supply fans.

ANIMAL FACILITIES

In animal laboratories, temperature and humidity often must be maintained at a specified control point as part of the research. Different temperatures may be needed for different types of animals, and ventilation and air-conditioning loads often vary with the type of animal and activity level. Some animals require changing temperatures and humidity levels throughout the day. Also, the types of animals housed in the laboratory may change several times throughout the life of the laboratory. Consultation with laboratory researchers is necessary to ensure that the HVAC system will be able to provide the conditions necessary for the type of work that will be performed in the laboratory during its lifetime. Designing the system for current and future needs in mind includes sizing the heating and cooling systems to account for the maximum anticipated loads and allowing for a wide range of temperature and humidity setpoints, a range of ventilation rates, and a flexible ventilation system that can be easily reconfigured.

Additionally, regarding specific ventilation levels for animal facilities, a guideline of 10 to 15 ach outdoor air has been used for secondary enclosures (animal holding rooms) for many years (ILAR 2011). The control systems used to provide these ventilation levels were typically constant-volume airflow control systems. However, although effective in many settings, these fixed-flow guidelines do not consider the range of possible heat loads; the species, size, and number of animals involved; the type of bedding or frequency of cage changing; the room dimensions; or the efficiency of air distribution from the secondary to the primary enclosure. In some situations, such a flow rate might overventilate a secondary enclosure that contains few animals, wasting energy, or underventilate a secondary enclosure that contains many animals, allowing heat and odor to accumulate.

As such, lower ventilation rates might be appropriate in the secondary enclosure or room provided that they do not result in harmful or unacceptable concentrations of toxic gases, odors, or particles. Active sensing of contaminants in the secondary enclosure and varying the air change rates based on the room environmental conditions is one approach that can be considered to meet these requirements in a more energy-efficient manner. Another approach to reducing the secondary enclosure ventilation rates is to provide ventilated cage racks to enhance ventilation at the primary enclosure level. To implement this manual or automated control of ventilation rates in animal facilities, the eighth edition of *Guide for the Care and Use of Laboratory Animals* (ILAR 2011), by the Institute for Laboratory Animal Research (ILAR) of the National Research Council of the National Academy of Sciences, now recommends the use of VAV for controlling the airflow in vivarium facilities in a more performance-based approach to ventilation:

> Modern heating, ventilation, and air conditioning (HVAC) systems (e.g., variable air volume, or VAV, systems) allow ventilation rates to be set in accordance with heat load and other variables. These systems offer considerable advantages with respect to flexibility and energy conservation, but should always provide a minimum amount of air exchange, as recommended for general use laboratories. (ILAR 2011, p. 6)

It is important to note that *Guide for the Care and Use of Laboratory Animals* is one of the most widely recognized and respected references for the design and operation of animal facilities. It includes data on temperature and humidity requirements along with monitoring requirements.

REFERENCES

AIHA. 2012. ANSI/AIHA/ASSE Z9.5-2012, *Laboratory ventilation.* Falls Church, VA: American Industrial Hygiene Association.

ASHRAE. 1995. ASHRAE Standard 110-1995, *Method of testing performance of laboratory fume hoods*. Atlanta: ASHRAE.

ASHRAE. 2015a. Chapter 53, Fire and smoke control. In *ASHRAE handbook—HVAC applications*. Atlanta: ASHRAE.

ASHRAE. 2015b. Chapter 16, Laboratories. In *ASHRAE handbook—HVAC applications*. Atlanta: ASHRAE.

ASHRAE 2013. Chapter 36, Measurement and instruments. In *ASHRAE handbook—Fundamentals*. Atlanta: ASHRAE.

Carter, J., B. Cochran, and J. Reifschneider. 2011. Saving energy in lab exhaust systems. *ASHRAE Journal* 53(6):26–36.

CDC. 2009. *Biosafety in microbiological and biomedical laboratories* (*BMBL*), 5th Edition. DHHS Publication No. (CDC) 21-1112. Atlanta: Centers for Disease Control and Prevention, U.S. Department of Health and Human Services. www.cdc.gov/biosafety/publications/bmbl5.

Coogan, J. 2008. Constant flow, variable flow and all the space between. Presented at the Labs21 2008 Annual Conference, San Jose, California. www.i2sl.org/elibrary/coogan2008.html.

Geeslin, A., A. Streifel, and G. Nelson. 2008. Air leakage analysis of special ventilation hospital rooms. *ASHRAE Transactions* 114(2).

ILAR. 2011. *Guide for the care and use of laboratory animals*, 8th ed. Washington, DC: Institute for Laboratory Animal Research, Division on Earth and Life Studies,

National Research Council, National Academies Press. http://grants.nih.gov/grants/olaw/Guide-for-the-care-and-Use-of-Laboratory-Animals.pdf.

Klaczek, W., M. Ackerman, P. Fleming, and B. Fleck. 2006. VAV airflow sensor response in relation to "poor" upstream duct geometry. *ASHRAE Transactions* 112(1).

Klein, R., C. King, and A. Kosior. 2009. Laboratory air quality and room ventilation rates. *Journal of Chemical Health and Safety* (9/10):36–42.

Liu, R., J. Win, A. Regnier, X. Zhou, and C. Klassen. 2012. *Stability and accuracy of VAV box control at low flows.* RP-1353. Atlanta: ASHRAE.

Mathew, P., S. Greenberg, D. Sartor, D. Frenze, M. Morehead, and W. Starr, Jr. 2005a. Right-sizing laboratory HVAC systems, Part 1. *HPAC Engineering* 9:24–30.

Mathew, P., S. Greenberg, D. Sartor, D. Frenze, M. Morehead, and W. Starr, Jr. 2005b. Right-sizing laboratory HVAC systems, Part 2. *HPAC Engineering* 10:34–39.

Schuyler, G. 2009. The effect of air change rate on recovery from a spill. Presented in Seminar 26 at the 2009 ASHRAE Winter Conference in Chicago, IL.

Sharp, G.P. 2010. Demand-based control of lab air change rates. *ASHRAE Journal* 52(2):30–41.

Sun, W. 2003. Development of pressurization airflow design criteria for spaces under required pressure differentials. *ASHRAE Transactions* 109(1).

Zhang, Z., and J.K. Agarwal. 1993. Continuous monitoring of fume hood face velocity with a side wall sensor. *ASHRAE Transactions* 99(1).

Airflow Patterns and Testing Procedures

12

OVERVIEW

Proper airflow and room pressure relationships are key to the design and safe operation of laboratories. Important elements of these include the following:

- Airflow patterns and direction
- Testing, adjusting, and balancing
- Laboratory testing requirements

Airflow patterns are used in a laboratory to provide separation between contaminants and occupants to ensure a safe working environment for the laboratory personnel and to provide an appropriate environment to ensure the integrity and accuracy of work performed in the laboratory. Airflow must be maintained in a specific direction and at a specific velocity to protect against harmful substances and is influenced by many aspects of a laboratory, such as equipment placement and traffic flows. Once the desired airflow patterns and direction are determined, testing procedures must be developed and followed so that the airflow patterns are maintained for the life of the laboratory. These procedures include general testing, adjusting, and balancing (TAB) procedures as well as testing requirements specific to laboratories.

AIRFLOW PATTERNS AND DIRECTION

Airflow patterns and direction must be maintained to provide personnel protection and protect research from contamination. Key factors for the protection of laboratory personnel are the hood face velocity and the location and type of the supply air diffuser in the room. Fume hoods must maintain a specified face velocity, typically around 100 fpm (0.5 m/s), to keep the substances used in the hood from escaping to the rest of the room. Improper placement or selection of a diffuser can result in disturbing the airflow near the hood face, resulting in contaminants from the hood being entrained into the room. Nonlaboratory personnel are protected from laboratory contaminants through a combination of physical barriers (walls) and airflow direction (nonlaboratory spaces are typically positively pressurized relative to laboratory spaces).

Protecting the research being conducted in the laboratory from contamination is important for many types of laboratories. It is especially important for cleanrooms, where the research inside must be protected from the less clean air surrounding the laboratory. It is also important for animal laboratories, where the animals also may need to be protected from the less clean air surrounding the laboratory or from contaminants generated by laboratory personnel or other animals. In both cases, loss of this protection can ruin the research being conducted. This is especially true for animal research, where months or years of research are lost when cross-contamination between subjects occurs.

In this section, the following elements of providing personnel protection and protecting research from contamination are discussed:

- Minimizing the spread of contaminants
- Determining the proper airflow pattern
- Air introduction
- Special requirements for critical systems

MINIMIZING THE SPREAD OF CONTAMINANTS

Minimizing the spread of contaminants is achieved by not using recirculated air in laboratory work areas and by maintaining airflow direction from clean areas to dirty areas so that the people and the work performed in the clean areas are not affected by laboratory procedures.

Use of Recirculated Air

A room in a laboratory building can have either once-through airflow or a portion of recirculated air, depending on the uses of the room. Typically, all laboratory rooms, chemical storage rooms, and laboratory support areas use 100% outdoor air or once-through ventilation. This way contaminants are not spread beyond the room in which they are released. However, most buildings that contain laboratories also contain offices, meeting rooms, and other areas where hazardous materials are not found. These nonlaboratory areas can use recirculated air, although it may be desirable to have higher ventilation rates in these areas than if they were in a typical commercial building. Also, it is advisable to have sufficient barriers between areas where recirculated air is used and areas where hazardous materials are used. For example, offices adjacent to laboratory spaces with hazardous materials may not be a good choice for using recirculated air, as a minor spill in the laboratory or a disruption of inward airflow to the nearby laboratory could result in contaminants being transferred to the offices. On the other hand, offices grouped together and separated from laboratory spaces would be an acceptable choice for using recirculated air.

Airflow from Cleanest to Dirtiest

The integrity of the room envelope is critical in providing the separation needed between contaminants and occupants in a given laboratory space. The envelope is the physical barrier between adjacent spaces and, if it is tight and properly sealed, the correct airflow direction can be effectively maintained with a smaller difference in supply and exhaust air volumes.

In laboratories, airflow direction is used to provide a secondary barrier to control the spread of contaminants to other areas. Pressure differentials are maintained to ensure that airflow is from the cleanest areas to the dirtiest areas. Once air has flowed to the dirtiest area, it is exhausted and new clean air is introduced to the clean areas.

For laboratories other than cleanrooms, a negative pressure is maintained to ensure that air flows into the laboratory from the cleaner surroundings, thus preventing dangerous chemicals, which were not exhausted by fume hoods, from migrating to the rest of the building. In these laboratories, the general exhaust for the room will eventually remove or dilute the accidentally released chemicals.

Figure 12-1
Flow from Clean to Dirty

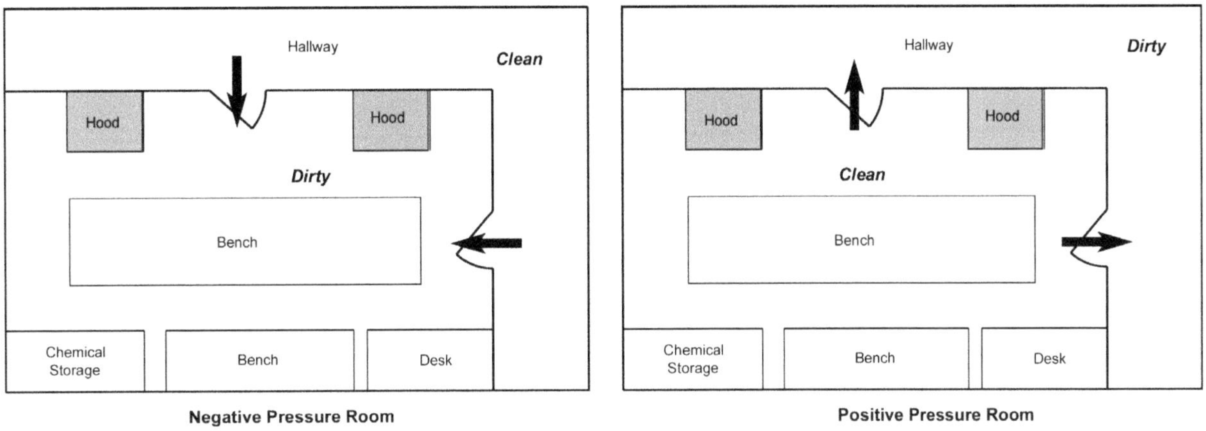

For cleanrooms, which need to be cleaner than their surroundings, a positive pressure is maintained to ensure air flows from the cleanroom to the surrounding areas. In cleanrooms, the contaminants that are contained by airflow direction are in the less clean, unfiltered air from surrounding areas in the building and outside.

Figure 12-1 shows two types of flow from clean to dirty.

DETERMINING THE PROPER AIRFLOW PATTERN

The general goal of maintaining airflow patterns is to provide safe conditions by controlling the movement of hazardous and unwanted substances. Determining the proper airflow pattern needed to maintain the separation between contaminants and occupants in a laboratory building, even during disturbances, is dictated by several variables, including the types of substances handled or created in the laboratory; and the type, size, and number of hoods and protective exhaust systems; and the acceptability of air transfer.

Types of Substances Handled or Created

Early in the planning stage, the types of substances handled or created in each space of a laboratory must be determined so that the proper airflow patterns and setpoints for the laboratory can be chosen to prevent the transmission of contaminants. Areas of similar hazard type and level can be grouped together. However, areas containing substances that could adversely interact must be separated (i.e., not connected to the same manifolded exhaust system) so that exhaust airstreams from the rooms do not mix. Also, the substances that are used in the laboratory may determine the ventilation air changes and relative pressure of the laboratory.

Type, Size, and Number of Hoods and Protective Exhaust Systems

The type of exhaust hood determines many characteristics of the room airflow pattern. First, there is the general choice of a constant-volume or variable-air-volume (VAV) system. Laboratories with VAV systems will have changing airflow patterns as the system changes from minimum flow to maximum flow, and the proper flow directions must be maintained throughout the operating range; constant-volume systems will have a more consistent airflow pattern.

The number of hoods may present challenges if a large number are to be placed in one room, as airflow patterns to each hood need to be maintained so that none of them

Figure 12-2
Conflicting Airflow to Fume Hoods

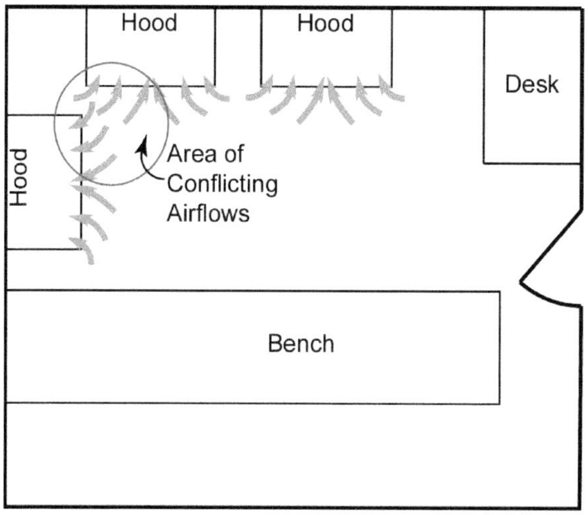

have disturbances in face velocity. Close coordination of the exhaust hood and supply sources during design is needed to ensure that each hood can draw air from the room without drawing in air needed to maintain the proper airflow patterns for a nearby exhaust hood. Figure 12-2 shows conflicting airflow to two fume hoods. For additional information on diffuser location and selection, see the Zone Air Distribution section of Chapter 6.

Acceptability of Air Transfer

Air transfer involves allowing air from one area to be transferred to another area in order to maintain pressure relationships. In some areas of a laboratory, the use of transfer air is acceptable—for example, between a corridor and a laboratory, between two laboratories with similar hazard levels that use compatible chemicals, and from a laboratory that uses low-toxicity materials to a laboratory with moderately toxic materials. In other cases, no air transfer between spaces may be permissible. Examples of cases where air transfer should not be used include from laboratory to nonlaboratory areas (except for cleanrooms), from areas of high hazard level to areas of low hazard, and between two areas of different types of hazard, such as from a biological laboratory area to a chemistry laboratory area. Figure 12-3 shows acceptable and unacceptable cases of air transfer.

AIR INTRODUCTION

Choosing the method most appropriate to the type of laboratory for introducing air into a laboratory space is critical so that the desired face velocity for the exhaust equipment is not disturbed. To obtain proper air introduction into a laboratory, the concept of challenge velocity must be understood, as well as how type, throw, and placement of diffusers and temperature gradients affect turbulence and cross drafts.

Challenge Velocity

Challenge velocity interrupts the airflow across the face of a fume hood or biological safety cabinet (BSC) and causes a disturbance that results in spillage from the hood. Disturbances include the velocity of air from supply diffusers, occupants walking past the hood, or the opening of the door to the laboratory. To prevent hood spillage, the allowable velocity of a disturbance at the hood should be at most 20% of the design face velocity of the hood. For example, if the design face velocity of a hood is 110 fpm (0.56 m/s), the maximum velocity of a disturbance is 22 fpm (0.11 m/s). The throw pattern from the supply diffusers should be less than this value where the pattern intersects the hood face.

Figure 12-3
Acceptability of Air Transfer

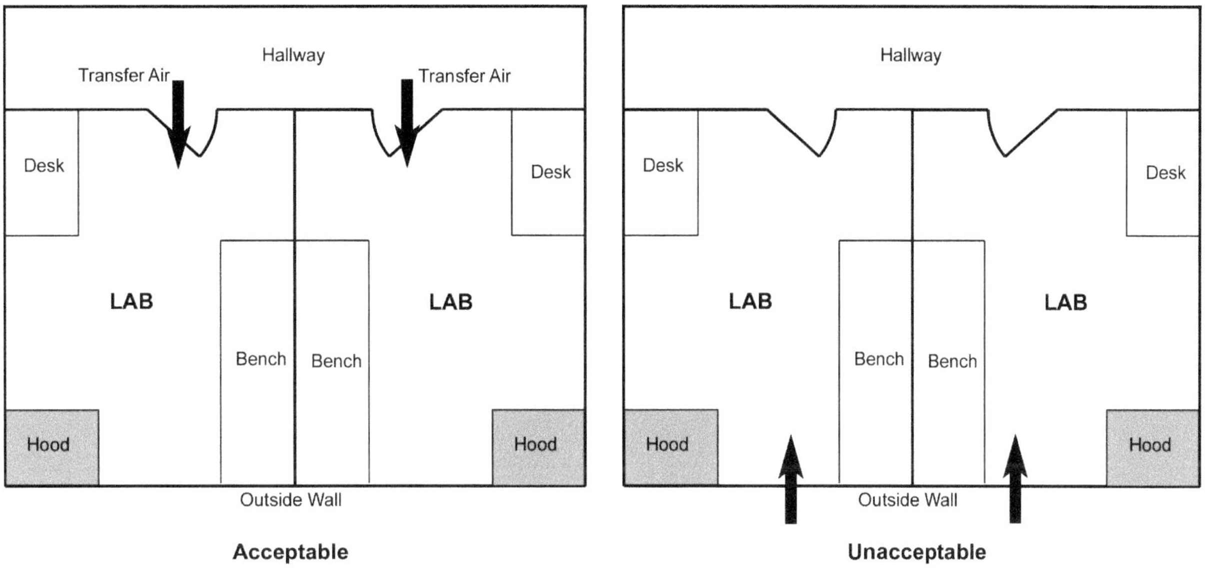

Type, Throw, and Placement of Diffusers

In laboratories, large quantities of supply air must be introduced to maintain pressure relationships while exhaust equipment is operating. However, the large amount of air must be introduced at a low velocity to prevent creating turbulence and cross drafts in the room, as high-velocity airstreams that intersect fume hood and BSC openings could disturb the airflow into exhaust devices. Therefore, the traditional high-velocity blade diffusers used in most buildings should be avoided in laboratory spaces, as their throw pattern would likely create a disturbance in the exhaust equipment airflow. For this reason, non-aspirating diffusers are recommended for laboratories.

Diffuser placement in a laboratory differs from that of a typical building. In typical buildings, equal diffuser spacing is used so that the entire room is covered by the throw from the diffusers and the air in the room is thoroughly mixed. This maintains an even temperature distribution and ensures that all areas of the room are ventilated and the occupants are comfortable. Laboratories, on the other hand, must maintain safety through proper operation of the fume hoods. Diffusers in laboratories should be placed far enough away from exhaust equipment and intakes that they do not cause any interference with the airflow into the exhaust equipment and that their throw pattern does not provide a sufficient challenge velocity to disrupt exhaust airflow patterns under any operating condition. Locations for diffusers should be selected so the velocity profile (both horizontal and vertical) that represents 50% of the face velocity, typically 50 fpm (0.25 m/s), does not intersect the sash area of the fume hood or BSC.

For additional information on locating diffusers with respect to exhaust sources and selecting diffuser types and throws for laboratories, see the Zone Air Distribution section of Chapter 6.

Temperature Gradients

Temperature gradients in a laboratory can create unwanted airflow patterns as the warmer air rises and displaces cooler air. To minimize possible disturbances to the exhaust equipment airflows created as cool air replaces warm air, temperature gradients should be kept to a minimum. This can be achieved by providing direct exhaust for high-heat-producing equipment and proper diffuser selection and location.

SPECIAL REQUIREMENTS FOR CRITICAL SYSTEMS

Critical laboratory systems may have special requirements to maintain airflow patterns for proper control of hazardous materials. These requirements include the use of air locks and multiple-speed fans.

Air Locks

Air locks are frequently used in laboratories where the disturbances created from opening doors could result in undesirable or unsafe conditions and to contain the spread of contaminants. Examples include a laboratory with hazardous materials that can't be allowed to escape without treatment (e.g., a biosafety level [BSL] 3 or 4 laboratory as described in *Biosafety in Microbiological and Biomedical Laboratories* [CDC 2009]), a laboratory with a high pressure differential to maintain, or the need to prevent contaminants from entering the laboratory (e.g., with a cleanroom).

Air locks consist of a series of two doors separated by a small vestibule. When occupants enter or leave the laboratory, they pass through the first door and wait in the area between the doors until the first door is completely closed. Once the first door has closed, the occupants can open the second door and enter or leave the laboratory. Depending on the nature of the hazards that require the use of air locks, either there can be a physical mechanism that locks one door while the other is open or the occupants can be responsible for only opening one door at a time, in which case an alarm sounds when both doors are open. Pressure gages are often provided in the air lock to show that the pressure differential is being maintained.

Multiple-Speed Fans

Multiple-speed fans may be used in critical areas to provide extra airflow in the event of an emergency, such as a large chemical spill. In the event of an emergency, the extra fan capacity can be used to increase the pressure differential of the laboratory and increase the air change rate to remove the spilled chemicals.

TESTING, ADJUSTING, AND BALANCING

The balancing of general HVAC equipment in a laboratory is similar to that for conventional buildings, provided that the equipment is not exposed to hazardous substances from the laboratory and that the laboratory is not in operation. However, if equipment is exposed to hazardous substances, procedures are needed to prevent or control the release of contaminants and ensure the safety of the testing personnel. Also, if the laboratory is in operation, the equipment should only be tested if it will not affect the conditions and safety of the laboratory and if the testing personnel will be safe.

For laboratory health and safety HVAC equipment, there are well-defined procedures that should be followed for proper operation of the equipment. Testing, adjusting, and balancing (TAB) of HVAC systems in a laboratory consists of the following:

- Air and hydronic system balancing
- Ductwork pressure testing
- Equipment balancing
- Conformance to general TAB standards

Designers should taylor the TAB requirements to the needs of each specific laboratory.

AIR AND HYDRONIC SYSTEM BALANCING

The keys to the successful TAB of air and hydronic systems include proper design of airflow differentials, adequate access points to obtain accurate measurements, early selection of the TAB contractor, and clearly defined TAB procedures.

Proper Design of Airflow Differentials

Successful TAB begins during the design stage with the proper selection of airflow differentials (or pressure differentials) needed to contain the spread of contaminants. Because of the critical nature of laboratories, it is recommended that the TAB contractor be required to balance to the following criteria:

- *Supply diffusers.* The listed airflow is the maximum (minimum for cleanrooms) that can be supplied. The actual acceptable value is the listed value to –10% (+10% for cleanrooms) of the value.
- *Exhaust points.* The listed airflow is the minimum (maximum for cleanrooms) that can be exhausted. The actual acceptable value is the listed value to +10% (–10% for cleanrooms) of the value.

When these criteria are used, the pressurization relationship between laboratories is maintained. If these are not used and the typical ±10% of the design value is used by the TAB contractor, there will be many instances of laboratories with airflow in the wrong direction, resulting in loss of contaminant control.

Documenting the TAB criteria during design aids in ensuring the building as a whole works properly. Using the above criteria in a building with 50,000 cfm (23 600 L/s) of exhaust air and 45,000 cfm (21 200 L/s) of supply air to laboratory spaces, the design airflow differential is 5000 cfm (2400 L/s). However, during TAB, the total exhaust airflow could be as high as 55,000 cfm (26 000 L/s) and the supply airflow as low as 40,500 cfm (19 100 L/s), resulting in the building being negatively pressurized by 14,500 cfm (6900 L/s). Conversely, using the incorrect TAB criteria could result in the building being positively pressurized by –4500 cfm (–2100 L/s). This example shows how laboratory buildings can become so negatively pressurized. Though the system was designed to compensate for 5000 cfm (2400 L/s) negative building differential, it can be balanced to a 14,500 cfm (6900 L/s) negative differential. The design of the central systems (chillers, boilers, coils, fans, and heat recovery systems) should have been sized for the larger pressure differential and balanced to maintain a positive, or at least a neutral, building pressure.

Access Points to Obtain Accurate Measurements

Another important step in the design of air and hydronic systems, both for laboratories and buildings in general, is to ensure that adequate test access points for making measurements are provided. For air systems, this means providing sufficient lengths of straight ductwork near equipment to make accurate measurements and allowing sufficient space to gain access to the straight length of ductwork. For hydronic systems, this means providing straight lengths of piping with pressure and temperature ports in the necessary locations and ensuring that they are accessible.

Early Selection of the TAB Contractor

It is important to select a TAB contractor early in the construction phase so that he or she has an opportunity to review the design documents and HVAC system installation. Including the TAB contractor early in the project enables him or her to provide input on the field requirements, such as required balancing devices and proper test port locations, needed to be able to successfully and accurately balance laboratory systems. This way any modifications needed for balancing the systems can be made in the construction phase rather than waiting until construction is complete before discovering whether the TAB contractor can actually balance the laboratory systems.

Clearly Defined TAB Procedures

The selected TAB contractor must submit the TAB plan, which should include how it will specifically meet the requirements of the project specifications. This TAB plan must detail the step-by-step procedures to be followed so that the results can be verified and the system can be retested and balanced in the future (e.g., yearly) by the same or a separate firm. Review and input from the architect, engineer, laboratory personnel, and commissioning authority are very valuable in conveying the requirements of the laboratory that must be verified and the reasoning behind the requirements to the TAB contractor.

DUCTWORK PRESSURE TESTING

Ductwork pressure testing is necessary to ensure that high-quality, low-leakage ductwork was constructed, particularly for exhaust ductwork. Generally, the design of ductwork includes an anticipated leakage rate, which must then be compared to the actual leakage rate to determine if the necessary airflows can be provided to the appropriate spaces.

Because laboratory exhaust contains hazardous materials and exhaust equipment often needs a relatively high static pressure at the hood or cabinet to operate properly, tightly sealed exhaust ductwork is extremely important to physically and economically provide a high static pressure at the piece of equipment. Also, since large quantities of supply air are used for once-through ventilation, significant energy is needed to condition and move the supply air. Tightly sealed supply ductwork in a laboratory helps minimize energy losses more dramatically than in a typical building that uses recirculated air.

For both supply and exhaust ductwork testing, care should be taken to prevent damage to the ductwork during the pressure testing (e.g., it develops leaks or overpressurizing reduces its strength). Details on how to pressure-test ductwork and the typical acceptable leakage rates can be found in the Duct Construction section of Chapter 6.

One last consideration in the TAB of exhaust systems is the sealing of test holes downstream of any exhaust fan. At these locations the ductwork is positively pressurized, and inadequate sealing of the test ports may result in contaminating the space. In some instances, such as BSCs, the test ports should be welded shut to eliminate any possibility of duct leakage. In other systems with low-toxicity chemicals, high-quality plugs that will not react with the exhaust airstream can be used.

EQUIPMENT BALANCING

Equipment balancing for laboratories is performed on two categories of equipment: general HVAC equipment (such as chillers, boilers, and air handlers) and laboratory health and safety HVAC equipment (such as fume hoods and BSCs).

General HVAC Equipment

General HVAC equipment in a laboratory can be tested in much the same way as that in a typical building, provided a few extra safety steps are followed. For example, testing and balancing should only be performed when the laboratory is unoccupied and all hazardous materials have been safely stored so TAB activities will not disturb them. Also, for their own protection, TAB personnel should be provided with all necessary safety equipment and informed of safety procedures when entering laboratory areas. Finally, TAB personnel need to be aware of the hazards associated with working with general building equipment that is essential for laboratory safety, such as high-efficiency particulate air (HEPA) filters for air handlers and general building exhaust equipment.

Laboratory Health and Safety HVAC Equipment

It is critical for laboratory health and safety HVAC equipment to be properly tested and balanced to ensure that design values are met and laboratory personnel will be properly protected. TAB for laboratory equipment has different requirements than that for typical HVAC equipment. Therefore, procedures have been incorporated into a variety of standards, such as ASHRAE Standard 110 for fume hood performance and velocity testing (ASHRAE 1995) and NSF/ANSI 49 for BSC certification for velocity and filter leakage (NSF 2014), as well as others. In cases where a standard is not available to prescribe a TAB procedure, there are best-practice publications, such as SEFA 1, *Recommended Practices for Laboratory Fume Hoods* (SEFA 2010), which may be used. Also in the case of equipment not covered by a standard, the laboratory designer, safety officer, and occupants should be consulted to determine how the equipment is used, the design values needed for safety, and the procedures to test and balance the equipment to the design values.

Laboratories, as most commercial buildings, rarely operate under full design conditions. However, laboratories have significantly more complex HVAC systems that must operate in unison regardless of the current conditions. It is important to consider TAB activities for nondesign conditions, especially room pressure differentials, fume hood performance/face velocity testing, and vented BSCs.

Many actions can cause airflows in a laboratory room to change and create changes in pressure differentials, including leaving doors open, opening and closing fume hood sashes, turning on direct equipment exhaust, and activities in the adjacent laboratory rooms. Therefore, room pressure differentials should be verified under these conditions during testing and balancing to ensure that negatively pressurized areas will maintain the appropriate negative pressure and that positively pressurized areas will maintain the appropriate positive pressure under all operating conditions.

Fume hoods typically operate either at the minimum open position, at a working height, or at the maximum open position, with brief periods to move between these three heights. Each of these three heights result in different flow characteristics for the fume hood. Therefore, fume hood face velocity testing and fume hood performance testing should be performed at these three heights. Face velocity and performance testing are discussed in the Laboratory Testing Requirements section of this chapter.

GENERAL TAB STANDARDS

There are several standards and documents that provide guidance for general TAB activities and testing of airflow patterns. These include those developed and published by organizations such as National Environmental Balancing Bureau (NEBB), Associated Air Balance Council (AABC), and ASHRAE.

National Environmental Balancing Bureau (NEBB)

The National Environmental Balancing Bureau (NEBB) is a nonprofit organization that establishes and maintains industry standards for testing, adjusting, and balancing of air and hydronic systems and for cleanroom performance testing. They also offer a certification program for TAB firms and have developed several standards that relate to laboratory testing, including the following:

- *Procedural Standard for Testing, Adjusting and Balancing of Environmental Systems* (NEBB 2015). This standard provides an overview of the NEBB program, contains suggested TAB specifications, and defines instrumentation requirements and TAB reporting forms. It also provides procedures for preliminary TAB work, air system TAB, hydronic system TAB, and TAB of outdoor air ventilation rates.

- *Procedural Standards for Certified Testing of Cleanrooms* (NEBB 2009). This publication provides standards for testing procedures as well as informational materials for cleanrooms. Some of the testing procedures included in this publication are airflow test procedures, HEPA filter installation leak testing, room cleanliness classification testing, enclosure pressurization and integrity testing, temperature and humidity testing, and lighting level testing. Information on equipment and instrumentation, fundamentals of cleanrooms, cleanroom design, and laboratory/health facility cleanrooms is also included.

Associated Air Balance Council (AABC)

The Associated Air Balance Council (AABC) has developed requirements for total system balancing and special systems testing. The following publications and documents have been developed by AABC:

- *National Standards for Total System Balance* (AABC 2002). This publication defines the minimum requirements for total system balancing and contains information on special systems testing such as laboratory fume hoods and cleanrooms. AABC defines total system balance as the process of testing, adjusting, and balancing HVAC systems according to the design intent.
- *Technician Training Manual* (AABC 2004a). This manual provides detailed, in-depth information on total system balancing for training new technicians.
- *Test and Balance Procedures* (AABC 2004b). This publication details the minimum procedures required to test and balance air and water systems, HVAC system components, and other specialty systems. It also specifies required instrumentation and test procedures.

ASHRAE

The ASHRAE standard related to TAB is ANSI/ASHRAE Standard 111, *Measurement, Testing, Adjusting, and Balancing of Building HVAC Systems* (ASHRAE 2008). The purpose of this standard is to provide uniform procedures for measurement, testing, adjusting, balancing, evaluating, and reporting the performance of building heating, ventilating, and air-conditioning systems in the field.

LABORATORY TESTING REQUIREMENTS

Fume hoods, BSCs, and room pressure relationships are the main components of a laboratory that need to be tested on a regular basis, typically annually, in order to ensure that the safety requirements in the laboratory are being met. The varied periodic testing of laboratory equipment and systems is diagrammatically represented in Figure 12-4 and includes fume hood face velocity and performance testing, BSC performance testing, and verification of room pressurization.

FUME HOOD FACE VELOCITY AND PERFORMANCE TESTING

Regular testing is needed to ensure that fume hoods properly protect laboratory workers from the chemicals that they use. Testing of fume hoods is specified in ASHRAE Standard 110, *Method of Testing Performance of Laboratory Fume Hoods* (ASHRAE 1995). For more information see Chapter 5 of this Guide.

Figure 12-4
Periodic Laboratory Equipment and System Testing

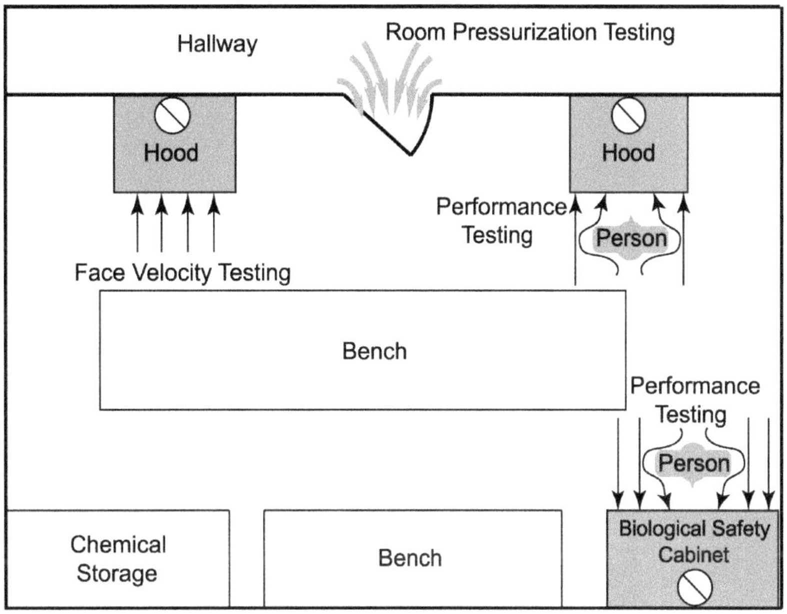

BIOLOGICAL SAFETY CABINET PERFORMANCE TESTING

BSC performance testing is conducted to ensure that cabinets are providing protection from biological aerosols. Testing protocols are specified in NSF/ANSI 49, *Biosafety Cabinetry: Design, Construction, Performance, and Field Certification* (NSF 2014). Depending on the class of BSC, the test parameters may be different because each class of cabinet offers slightly different protective barriers.

The purpose of BSC performance testing is to ensure that the cabinet meets predetermined criteria and listed requirements established by NSF/ANSI 49. Assuming that these criteria are met, a tested unit is determined to have the ability to provide a defined level of protection when made available for use in the laboratory. Following the acceptance testing, the cabinet unit is then listed and bears an NSF 49 seal.

VERIFICATION OF ROOM PRESSURIZATION

Room pressurization is used either to prevent substances not captured by a fume hood from migrating to the rest of the building or to maintain a cleaner environment in the laboratory than the surrounding areas. Supply and exhaust airflows may change over time due to normal wear of the HVAC system. The building envelope and pressure sensor calibration may also change over time and affect room pressurization. Therefore, room pressurization should be tested and verified that it meets the necessary requirements. Both instrument testing and (visual) flow direction testing are used to verify room pressurization.

Instrument Testing

Laboratories must maintain a pressure differential to maintain airflow direction from cleanest to dirtiest. For most laboratories, a minimum of 0.01 in. w.g. (2.5 Pa) is needed to provide this airflow. Laboratories that use highly hazardous materials will likely require greater pressure differentials, up to 0.05 in. w.g. (12.5 Pa) or higher. A pressure differential of 0.01 in. w.g. (2.5 Pa) is intended to provide a 100 fpm (0.51 m/s) velocity through a 0.125 in. (3.2 mm) crack (e.g., under a door), while the actual velocity will vary with the integrity of the envelope construction.

Instrument testing of room pressurization uses a manometer to verify the pressurization of the laboratory. The manometer should be accurate to ±1%, and the precision should be ±0.005 in. w.g. (1.25 Pa). The room pressurization should be tested under the maximum open, minimum open, and typical working positions of the fume hood sashes (ASHRAE 1992).

Flow Direction Testing

Flow direction in laboratories can be visually verified. For example, smoke testing can be used to visually demonstrate that the pressure differential is being maintained for the room and that there is no air leaving a laboratory space and migrating to another part of the building. Flow direction testing can be accomplished by using smoke candles around openings to the laboratory to verify flow direction. The typical openings to be verified under the various operating conditions listed previously are:

- Door (closed and open)
- Electrical outlets
- Windows
- Other penetrations (piping, wiring, duct)

REFERENCES

AABC. 2002. *National standards for total system balance*. Washington, DC: Associated Air Balance Council.

AABC. 2004. *Test and balance procedures*. Washington, DC: Associated Air Balance Council.

AABC. 2004. *Technician training manual*, 2d ed. Washington, DC: Associated Air Balance Council.

ASHRAE. 1992. ASHRAE Standard 41.2-1987 (RA 92), *Standard methods for laboratory airflow measurement*. Atlanta: ASHRAE.

ASHRAE. 1995. ASHRAE Standard 110-1995, *Method of testing performance of laboratory fume hoods*. Atlanta: ASHRAE.

ASHRAE. 2008. ANSI/ASHRAE Standard 111-2008, *Measurement, testing, adjusting, and balancing of building HVAC systems*. Atlanta: ASHRAE.

CDC. 2009. *Biosafety in microbiological and biomedical laboratories* (*BMBL*), 5th Edition. DHHS Publication No. (CDC) 21-1112. Atlanta: Centers for Disease Control and Prevention, U.S. Department of Health and Human Services. www.cdc.gov/biosafety/publications/bmbl5.

NEBB. 2009. In *Procedural standards for certified testing of cleanrooms*, 3d ed. Gaithersburg, MD: National Environmental Balancing Bureau. Available from www.nebb.org/procedural-standards.

NEBB. 2015. *Procedural standard for testing, adjusting and balancing of environmental systems*, 8th ed. Gaithersburg, MD: National Environmental Balancing Bureau. Available from www.nebb.org/procedural-standards.

NSF. 2014. NSF/ANSI 49-2014, *Biosafety cabinetry: Design, construction, performance, and field certification*. Ann Arbor, MI: NSF International.

SEFA. 2010. SEFA 1-2010, *Recommended practices for laboratory fume hoods*. Garden City, NY: Scientific Equipment Furniture Association.

O&M for Ventilation and Exhaust Systems

13

OVERVIEW

Laboratory ventilation systems require regular and appropriate maintenance to ensure proper operation and performance. Proper operation is required to protect people working with potentially hazardous materials and ensure effective and efficient operation over the life cycles of the systems. Poor or ineffective maintenance can lead to premature degradation of the systems and high costs to renovate and upgrade them following degradation. Furthermore, the maintenance helps protect the return on investment.

According to ANSI/AIHA/ASSE Z9.5 (AIHA 2012), all laboratory facilities should develop and implement a laboratory ventilation management program (LVMP). The LVMP should include a description of the systems, define operating specifications, define performance criteria, describe required maintenance tasks, provide a schedule for conducting the tasks, and provide standard operating procedures (SOPs) for all routine test and maintenance methods.

Rather than waiting for major problems to occur, maintenance measures such as monitoring, testing, calibrating, cleaning, replacements, and repairs should be performed on a regular basis (weekly, monthly, and annually). This provides assurance that the laboratory will maintain precise conditions for an extended period of time.

The main elements of operation and maintenance (O&M) that should be addressed for laboratories are as follows:

- Maintenance of equipment and systems
- Decontamination of existing laboratories
- Cost information
- Training

MAINTENANCE OF EQUIPMENT AND SYSTEMS

Routine maintenance measures are needed for various pieces of laboratory equipment and systems, including fume hoods, biological safety cabinets, and ventilation and exhaust systems. Established standards, codes, and guidelines provide requirements and recommendations that should be followed for the safe operation of these equipment and systems. The following subsections summarize and provide resources for assessing these requirements and recommendations.

FUME HOODS

Over time, regular use of fume hoods can cause the flow to vary from the design specifications due to degradation of airflow controls and problems with the operation of the exhaust fans, including belts, fan speed, and operation of outdoor air bypass dampers. Airflow controls are especially critical with variable-air-volume (VAV) fume hoods, as they rely on sensor readings to determine the proper airflow needed to maintain a safe face velocity. If the sensor becomes damaged or goes out of calibration, a VAV fume hood can expose laboratory occupants to potentially hazardous contaminants. Also, if the fume hood flow is higher than necessary, energy will be wasted and the room can become improperly balanced.

All fume hoods should be tested and have their performance verified at least annually. Typically exhaust flow, face velocity, cross-draft test using smoke, and airflow monitor calibration are tested. Many institutions have a standard annual testing protocol, and guidance for fume hood testing and requirements can also be found in the following publications:

- ASHRAE Standard 110-1995, *Method of Testing Performance of Laboratory Fume Hoods* (ASHRAE 1995)
- *National Standards for Total System Balance* (AABC 2002)
- *Procedural Standards for Fume Hood Performance Testing* (NEBB 2009)
- MD 15128, *Laboratory Fume Hoods: Guidelines for Owners, Design Professionals and Maintenance Personnel* (PWGSC 2013)
- ANSI/ASSE Z9.14, *Testing and Performance-Verification Methodologies for Ventilation Systems for Biosafety Level 3 (BSL-3) and Animal Biosafety Level 3 (ABSL-3) Facilities* (ASSE 2014)
- *NFPA 45: Standard on Fire Protection for Laboratories Using Chemicals* (NFPA 2015)

BIOLOGICAL SAFETY CABINETS

Biological safety cabinets (BSCs) protect laboratory users from biological aerosols. These cabinets, like fume hoods, need to be inspected, tested, and recalibrated periodically. NSF International states that the cabinets should be field-tested at least annually (NSF 2012). According to NSF/ANSI 49, "recertification should be performed whenever HEPA/ULPA filters are changed, maintenance repairs are made to internal parts, or a cabinet is relocated. More frequent recertification should be considered for particularly hazardous or critical applications or workloads" (NSF 2014, p. F1).

The following are the tests required for NSF/ANSI 49 field testing:

- Downflow velocity profile test
- Inflow velocity test
- Airflow smoke patterns test
- High-efficiency particulate air/ultra-low particulate air (HEPA/ULPA) filter leak test
- Cabinet integrity test (positive-pressure plenum cabinets only)
- Site installation assessment tests, including alarm functions, blower interlock, and exhaust system performance

For worker comfort and safety, tests on lighting intensity, vibration, noise level, electrical leakage, ground circuit resistance, and polarity may be conducted.

VENTILATION AND EXHAUST SYSTEMS

A laboratory ventilation system should ensure that occupied spaces get the correct amount of outdoor air while maintaining the appropriate temperature and humidity levels for acceptable thermal comfort and experimental quality. The exhaust system expels the air from the laboratory and containment devices to the outside of the building. The exhaust system also works in tandem with the ventilation system to provide the correct pressure relationships for the laboratory spaces. However, without periodic maintenance of the key components of the ventilation and exhaust systems, the requirements for safe and reliable operation will not be met.

The following are some of the items to include in a maintenance plan:

- Fans, motors, and drives
- Supply and exhaust filters
- Dampers
- Airflow monitoring stations
- Humidifiers
- Supply and exhaust air terminal units
- Controls and alarm devices (verify operation and calibration)
- General check of all devices in the airstream

Resources for laboratory O&M requirements and recommendations for ventilation and exhaust systems include ANSI/AIHA/ASSE Z9.5, *Laboratory Ventilation* (AIHA 2012), and ANSI/ASSE Z9.14, *Testing and Performance-Verification Methodologies for Ventilation Systems for Biosafety Level 3 (BSL-3) and Animal Biosafety Level 3 (ABSL-3) Facilities* (ASSE 2014).

DECONTAMINATION OF EXISTING LABORATORIES

During its lifetime, a laboratory will most likely need to be decontaminated. Common reasons for decontamination include the following:

- Changes in the type of work performed
- Renovation or remodeling of the laboratory workspace
- Maintenance of laboratory equipment

Decontamination procedures should be established and documented to assist maintenance staff in providing proper maintenance to the laboratory equipment. Coordination of decontamination amongst maintenance staff and laboratory users is needed to minimally impact the work being performed in the laboratory.

COST INFORMATION

In addition to the initial cost of designing and constructing a laboratory, there are several costs associated with using the laboratory after completion, including the costs of operation, energy, and maintenance. Because laboratory HVAC systems are generally more complex and energy intensive than those of typical buildings and their continued operation is essential for safety, it is important to budget appropriate funds for the O&M of a laboratory, which includes the provision for sufficient qualified personnel. Coordination between laboratory designers, personnel, and maintenance staff can reduce costs by making the laboratory easier to operate and to maintain.

OPERATION COST

The cost of operation includes the costs associated with the operation of the equipment and systems of a laboratory, the salaries of the maintenance staff (or the cost to subcontract maintenance assignments), and other expenses for running the laboratory, such as the costs for non-energy utilities and general building supplies. Energy costs and maintenance costs are major categories of the cost of operation.

ENERGY COST

The energy cost of a laboratory is the cost associated with purchasing electricity, natural gas, and other fuels to operate the laboratory. Because of the large amounts of exhaust air and the typical use of 100% outdoor air for laboratory spaces, energy costs for laboratories are higher than those of typical buildings. The O&M staff should closely monitor the monthly energy costs, as large increases or decreases may identify possible HVAC system problems.

There are many steps that maintenance personnel can take to reduce energy costs, such as regularly changing filters and keeping air passages clean and open. Laboratory users should close fume hood sashes when appropriate. A direct digital control (DDC) system can greatly simplify the optimization of laboratory operation and energy minimization.

MAINTENANCE COST

Maintenance cost includes the costs of repair/replacement parts and materials for the HVAC system. Because laboratories have high concentrations of HVAC equipment, the time and cost to provide maintenance can be significant. Therefore, a carefully planned and executed preventive maintenance program should be used, as it will reduce time and costs for maintenance and ensure a safe working environment in the laboratory. Documentation during design and construction, in the form of the Basis of Design (BOD), as-built drawings, and O&M manuals, is essential to ensuring that maintenance personnel have the necessary resources to maintain the laboratory. Additional costs invested in maintenance can result in significant savings in energy costs.

TRAINING

Training is required throughout the life of any building to keep the maintenance staff and occupants of the building informed of the procedures for safe, energy-efficient, and cost-effective use of the building's HVAC system. Training is especially important for laboratory maintenance staff and occupants because laboratories use many hazardous substances, use more energy than a typical commercial building, and have HVAC systems that can be dangerous or costly to fix if used improperly. Training is also important for laboratories with frequent changes in occupants, such as a teaching or research laboratories.

MAINTENANCE STAFF

To safely and cost-effectively operate a laboratory, the maintenance staff must be properly trained in the operation of building equipment and systems. Maintenance staff should be aware of the hazardous materials used in the laboratory and how they affect maintaining the HVAC system.

Training should be customized to provide the level of detail that the maintenance staff needs, as they may or may not already be familiar with operating the equipment installed in the laboratory. If they have already used the same type of equipment, general background information on the equipment is not needed.

Trainees should be allowed to work with the equipment during training so that they can actually see what the trainer is talking about and have their questions answered. It is unlikely that the maintenance staff will be able to remember or comprehend all of the information that is presented to them during training. Therefore, they should be made aware of the reference materials available to them, such as O&M manuals, and have a chance to use them. Training can be recorded for review and training of new staff.

Finally, maintenance staff should be informed of the preventive maintenance that is required for the laboratory equipment. Maintenance procedures, including work to perform, notification of laboratory users, and documentation of work and testing procedures, should be discussed.

OCCUPANTS

To operate a laboratory safely, laboratory users must know the proper procedures for operating the laboratory equipment to minimize the risk of accidents. Items such as fume hood operation, keeping doors closed to maintain pressure differentials, and limits on where various materials can be used must be discussed. Laboratory users should also be aware of the procedures for reporting a potentially dangerous condition so that maintenance personnel can take corrective action.

Laboratory users should know the proper emergency procedures to minimize the impacts of an accident. Regular reviews of emergency procedures should be performed to sharpen emergency actions, to determine possible improvements, and to develop new procedures when the type of work performed in the laboratory changes. The reviews should also cover use of emergency announcement systems and evacuation procedures.

Finally, laboratory users need to be aware of the impact their actions have on energy usage. For example, they should be aware that leaving fume hood sashes open unnecessarily will cause a significant amount of energy to be used to condition and move air.

REFERENCES

AABC. 2002. *National standards for total system balance.* Washington, DC: Associated Air Balance Council.

AIHA. 2012. ANSI/AIHA/ASSE Z9.5-2012, *Laboratory ventilation.* Falls Church, VA: American Industrial Hygiene Association.

ASHRAE. 1995. ASHRAE Standard 110-1995, *Method of testing performance of laboratory fume hoods.* Atlanta: ASHRAE.

ASSE. 2014. ANSI/ASSE Z9.14-2014, *Testing and performance-verification methodologies for ventilation systems for biosafety level 3 (BSL-3) and animal biosafety level 3 (ABSL-3) facilities.* Park Ridge, IL: American Society of Safety Engineers.

NFPA. 2015. *NFPA 45: Standard on fire protection for laboratories using chemicals.* Quincy, MA: National Fire Protection Association.

NEBB. 2009. *Procedural standards for fume hood performance testing.* Gaithersburg, MD: National Environmental Balancing Bureau. Available from www.nebb.org/procedural-standards.

NSF. 2014. NSF/ANSI 49-2014, *Biosafety cabinetry: Design, construction, performance, and field certification.* Ann Arbor, MI: NSF International.

PWGSC. 2013. MD 15128-2013, *Laboratory fume hoods: Guidelines for owners, design professionals and maintenance personnel.* Gatineau, Quebec, Canada: Public Works and Government Services Canada.

Laboratory Commissioning Process

14

OVERVIEW

Environmental control and monitoring of laboratories are considerably more complicated than for most other buildings because of the stringent requirements of HVAC systems to meet temperature, relative humidity, ventilation rates, directional airflow control and indication, differential pressure control and indication, building equipment dependency integration with the architectural features, and fire and life safety integration. These requirements become more challenging as the scope and complexity of building control systems and associated control strategies expand. Other special needs inherent in laboratory facilities include redundancy, emergency or standby power, energy recovery, and waste treatment, and some laboratories may have additional considerations such as chemical and biological hazards, special filtration, and energy efficiency.

Commissioning has become a necessary methodology for delivering operable laboratories and complying with the requirements of certifying agencies. Commissioning is a quality-assurance-focused process for enhancing the delivery of a project. The process focuses on verifying and documenting that the laboratory and all systems and assemblies are planned, designed, installed, tested, operated, and maintained to meet the Owner's Project Requirements (OPR) (ASHRAE 2013a). The process is most successful when implemented throughout all phases of the project: planning, design, construction, acceptance, and occupancy and operation. The process verifies that at the end of the project the owner receives the operating facility that was expected at the beginning.

This chapter provides a general overview of the commissioning process and guidance for commissioning of the most common laboratory systems, including descriptions of important commissioning tasks for each phase of a typical laboratory project and how the commissioning process can be effectively applied to existing laboratory facilities. The sections include the following:

- Commissioning process
- Predesign phase
- Design phase
- Construction phase
- Acceptance phase
- Occupancy and operations phase
- Commissioning of existing buildings (retrocommissioning)

COMMISSIONING PROCESS

The general phases of the commissioning process are predesign, design, construction, acceptance, and occupancy and operation. While some owners adopt commissioning during the design or construction phases, the greatest benefits of commissioning are achieved by starting in the predesign phase of the project, when the facility owner is defining the project requirements. In so doing, maximum benefits have proven to be achieved throughout the lifetime of the building.

ASHRAE Guideline 0, *The Commissioning Process* (ASHRAE 2013a), describes the commissioning process used to verify that facilities and systems meet the OPR. Guideline 0 is part of the Total Building Commissioning Process as defined by the National Institute of Building Sciences (NIBS). ASHRAE Guideline 1.1, *HVAC&R Technical Requirements for the Commissioning Process* (ASHRAE 2007), provides the technical requirements for commissioning HVAC&R systems in accordance with the process described in Guideline 0. These documents combine to form the basis of commissioning for many types of facilities, including the general process for laboratories. ANSI/ASHRAE Standard 202, *Commissioning Process for Buildings and Systems* (ASHRAE 2013b), was developed as a minimum standard for commissioning.

PREDESIGN PHASE

The goals and expectations of the owner are documented during the predesign phase in a significant document known as the Owner's Project Requirements (OPR). This document describes the functional, operational, and maintenance requirements of the laboratory—how it will be used and operated. The commissioning Basis of Design (BOD) is the engineer's response to the OPR defining how the OPR are intended to be met. The BOD is a living document and must be updated as the project progresses. The commissioning plan, described further in the last subsection of this section, is also drafted during this phase.

ESTABLISHING PROJECT GOALS AND EXPECTATIONS

The building project goals must be stated in clear and concise terms as early in the project as possible to make certain that all the design and construction work performed meets the goals of the owner. While the project goals change during the project, as the owner becomes aware of the implications of the initial project goals, documenting changes and maintaining a current OPR enables everyone involved in the project to work toward the same goals. A number of tools are available to document the updating of the OPR and BOD, including workshops, surveys, and interviews.

In addition to the OPR items described in ASHRAE Guidelines 0 and 1.1, the OPR for laboratories should identify the following:

- Risk assessment or information needed for the risk assessment
- Safety features
- Flexibility, reliability, and redundancy
- Economics
- Functional performance
- Standard operating procedures (SOPs) related to how the users function in the laboratory within the limitations of the systems in the research facility
- Training requirements of the owner and users

Risk Assessment

Ideally the OPR should include a risk assessment for the facility; if one is not provided, the information needed to perform the risk assessment should be included in the OPR. The risk assessment documents the risks associated with the chemicals being used and the procedures being performed in the room that could impact the laboratory personnel, the surrounding areas, and the environment. It also includes how these risks can be mitigated without affecting the research and experiments being performed in the laboratory. The planned usage of the laboratory must be detailed as thoroughly as possible to verify that the designed and constructed laboratory system can meet the requirements. These details can, for example, be the expected use of chemicals (especially perchloric acid), the mixture of chemicals, the efficacy of the biological hazard, the level of cleanliness required, the types of animals involved, the decontamination methodology, operating parameters, etc. More details provided at this stage will help verify that the risks are determined accurately and that the design of the system can mitigate these risks. Chapter 4 discusses risk assessment in more detail.

Safety Features

The owner should list any special safety features required for the laboratory. Based on the risk assessment, additional safety features to handle emergencies such as spills, fires, explosions, malfunctioning of fans or motors, and room pressure problems can be identified. These features help the laboratory systems handle abnormal situations while minimizing risk to the users and the environment. The object of the safety features is to make sure that the users are safe, the environment is protected, and the research is sustained from release of contaminants even in extreme situations.

Flexibility, Reliability, and Redundancy

Flexibility of the design is important for the laboratory to be able to meet changing criteria, as the laboratory's function may change over time. The flexibility of the system may require additional cost, but good planning can considerably ease the process of adding to the system or making a change to the system.

Reliability is an important issue for laboratory systems, because failure of the system can have catastrophic consequences. In critical systems that deal with highly toxic materials or where a shutdown could be extremely expensive, redundant system or component configurations are often necessary. This redundancy ensures that maintenance can be done without shutting down the system and that the system will function even in the case of component failure. The level of reliability must be determined since systems can only approach 100% reliability.

Economics

As in any project, economics is a constraint that determines what can actually be accomplished. The owner always has a limited amount of funds available, and the project must meet this criterion as well as the other criteria. Often, the cost of the requirements exceeds the funds available. While compromises are then made, the safety of the users must be maintained. It is important to update the OPR every time something is compromised and to explain the reason for the compromise so that all project participants understand what is changed and why it was changed to avoid future misunderstandings or discussions (lawsuits) about changes.

Functional Performance

The functional performance to be met by the system must be documented to make certain that all the participants clearly understand what the design and installation must achieve. Functional performance includes the flexibility, reliability, and safety functions with a main focus on the operational requirements. The functional requirements should be as detailed as possible to make certain that the design and construction team can meet the requirements throughout the project.

Standard Operating Procedures

Standard operating procedures (SOPs) must be documented to inform users of the proper procedures for handling chemicals and conducting research and experiments in a safe manner. The SOPs should include limits on the chemicals allowed in the space; instruction on handling of equipment, mixing or dispensing of chemicals, operation of the safety equipment, and operation of the containment equipment (fume hoods, snorkels, etc.); and information about emergency procedures in the event of a spill or accident.

Training Requirements

Training requirements for both laboratory users and the owner should be developed and documented so they can be included in the construction documents. The level of existing personnel training should be documented, as this allows tailored development of training materials for the users and owners.

COMMISSIONING PLAN

The commissioning plan outlines the organization, schedule, allocation of resources, and documentation of the commissioning process. The initial commissioning plan is developed during the predesign phase; it is later ratified by the construction team. Part of the plan includes a review of lessons learned from previous laboratory projects that could be useful in the current project. Additionally, training requirements are identified along with other documentation to identify scope.

DESIGN PHASE

At the beginning of the design phase, the engineer develops the Basis of Design (BOD). During the design phase, the construction documents are developed in accordance with the BOD to meet the OPR and communicate the engineer's design to the contractor and owner.

BASIS OF DESIGN

The BOD is developed in response to the OPR. It is a narrative discussing the systems selected and how they will meet the OPR. Systems that were rejected and the reasons they were rejected are also included. The BOD can be subdivided into two major sections: design intent and design criteria.

Design Intent

The design intent should include an overview of the laboratory control system with general sequences of operation including failure modes. It should also address the risk assessment and how the systems selected will minimize risk. Specific components of the system that should be addressed include the following:

- Airflow and temperature control system selection and operations
- Relative pressure control methodology (e.g., airflow offset, direct pressure control, adaptive offset control, and enhanced differential pressure control)
- Redundancy of equipment (e.g., air-handling units, exhaust fans)
- Emergency power requirements
- Exhaust plume height and dilution

- Materials of construction (e.g., ductwork, grilles, registers and diffusers, airflow control devices)
- Air change rates (occupied and unoccupied)
- Emergency modes of operation
- Failure scenarios
- Energy-saving features
- Testing, adjusting, and balancing (TAB) requirements
- Commissioning requirements
- Instrumentation required for measurement and verification as well as ongoing commissioning
- System maintenance (a review of the differing maintenance requirements for the major equipment selected for the design and the alternatives that would be acceptable during the construction phase)

Design Criteria

As part of the BOD, the design criteria states the codes, standards, guidelines, etc. that are to be followed/complied with in designing the building systems. The BOD may also include a list of manufacturers' names and equipment model numbers for the equipment the system was designed around. Any assumptions used in the design of the HVAC systems should be included. In addition to the general HVAC information, the following specific information concerning laboratories should be included:

- Fume hood airflow rates and sash types
- Biological safety cabinet (BSC) types and airflow rates
- Source capture devices (e.g., snorkels, storage cabinets, canopy hoods) and airflow requirements
- Ventilation rates for occupied, unoccupied, and emergency modes
- Pressure relationships/directional airflow control
- Exhaust reentrainment avoidance for stack heights and intake locations
- Envelope differential pressure requirements and indication
- Corridor and vestibule door interlock scenarios for laboratory ingress and egress
- Alarm notification and user dissemination

CONSTRUCTION DOCUMENTS

Drawings and specifications are the main construction documents developed during the design phase. These documents are used for bidding and the construction of the project and are based on the OPR and BOD. The commissioning requirements for the laboratory building project must be integrated into these documents during the design phase so that all parties involved are aware of the quality process used to achieve the goals and expectations of the owner. The commissioning plan is further developed, along with testing procedures.

Construction documents should be reviewed to verify the system design meets the OPR. Specific items that should be verified include the following:

- Hoods and BSCs are located away from the doors and frequently traveled paths.
- Diffusers are selected and located such that terminal velocities do not interfere with the fume hoods and BSCs.
- Control components and sequences are clearly defined for each mode of operation.
- Fume hood monitoring uses face velocity measurement or flow verification based on calculated values based on sash position.

- Sufficient straight duct has been provided before all air control devices for TAB to verify airflow rates with duct traverses.
- Airflow measurement stations are specified for ongoing commissioning activities.
- Airflow station accuracy, including transmitter and annual drift, meet the system accuracy (±5%) throughout the full flow ranges expected.
- The room pressure control scheme can work with the room envelope construction to maintain the desired room pressure. (In tightly constructed rooms, the amount of differential airflow to maintain the pressure relationship maybe less than the control error on volumetric offset control systems. Direct pressure control or pressure reset of the offset maybe necessary for tightly constructed rooms such as biosafety level [BSL] 3 laboratories as described in *Biosafety in Microbiological and Biomedical Laboratories* [CDC 2009].)
- Test and acceptance procedures are defined to include failure scenarios.
- Ductwork construction and leakage rates are clearly defined.
- Plume heights and exhaust dilution are clearly defined on exhaust fan schedules.
- Reliability and maintenance of major components have been reviewed.
- Redundancy issues have been clearly addressed.

CONSTRUCTION PHASE

Commissioning during the construction phase focuses on the quality of the installations to be in accordance with the OPR as reflected in the construction documents (i.e., drawings and specifications). Though they are similar to those of the design phase, the procedures of the construction phase are different. The emphasis during construction is on accomplishing the installation correctly the first time. Specific requirements for the installers for the installation to meet the OPR should be clearly provided in the construction documents. Commissioning tools and techniques used during construction include mock-ups, construction checklists, review of submittals, and system verification.

MOCK-UPS

The use of mock-ups is one way to ensure quality during construction. The contractor is required to construct a small section of the installation that has to be approved before the remaining construction sections are allowed. This procedure catches installation faults that otherwise would not have been detected until a significant portion of the installation was complete. Mock-ups increase the quality and reduce the cost of the installation for the contractor, as it makes sure the individual workers know what is required to meet the OPR.

CONSTRUCTION CHECKLISTS

The installation of the individual components of the various systems must be verified to meet the OPR. Individual workers accomplish this through continuous spot-checking of work and through the use of construction checklists.

REVIEW OF SUBMITTALS

In reviewing the submittals the following should be verified:

- Systems submitted conform to the construction documents (drawings and specifications) and OPR.
- Equipment and materials conform to construction documents and OPR.
- Maintenance requirements for equipment conform to the construction documents and OPR.

- System integration is clearly defined and identified.
- Airflow measurement and control accuracy conform to the specifications.
- Operation and maintenance (O&M) documentation is complete.

SYSTEM VERIFICATION

Once all systems are installed, performance verification testing (PVT) of the laboratory control system sequences is accomplished. This effort entails the review of control panel wiring and labeling to verify that they match the as-built drawings. The individual control loops are also checked for proper operation during all modes of operation, including failure scenarios. After the control acceptance, all air and water systems can then be tested, adjusted, and balanced. The system TAB is then verified to confirm that the laboratory has the proper flows under all conditions to maintain the health and safety of the occupants and to meet the owner's design intent.

ACCEPTANCE PHASE

The acceptance phase of commissioning focuses on the functional performance of the equipment. Ideally, the commissioning process has eliminated the majority of problems before the functional performance of the system is tested. Minor failures during the test are corrected. Major failures require a retest. The acceptance phase focuses on functional performance tests of HVAC equipment that are critical for laboratories, such as the control system and exhaust hoods and systems.

Functional testing of the equipment verifies that the installed equipment and systems perform as intended. All the normal operating conditions are tested. If an operating condition cannot be tested (e.g., space cooling during the winter season) and there is no way to simulate the necessary conditions, the functional performance test should be scheduled for a later time when it is expected that the necessary conditions are present. In addition to the normal operating conditions, emergency, safety, start-up, shutdown, and all other foreseeable abnormal operating conditions are also tested.

CONTROL SYSTEM

The laboratory control system is the "brain" of the system and integrates the sequences of operation and how the individual pieces of equipment interact. The control system receives information from sensors located at strategic locations throughout the system. A control system can only function correctly if it receives correct information from all the sensors, the control sequences are verified correct, and the equipment performs the tasks the way the control system instructed it.

EXHAUST HOODS AND SYSTEMS

Exhaust systems (fume hoods, snorkels, BSCs, etc.) often have requirements that require complex controls as well as stringent equipment performance requirements. These requirements depend on the use of the laboratory and the specific use of the exhaust hood. Face velocity and proper fume hood containment can be tested by smoke tests under all normal operating conditions. In addition, the fume hoods should be tested per ASHRAE Standard 110, *Methods of Testing Performance of Laboratory Fume Hoods* (ASHRAE 1995). BSCs are tested in accordance with NSF/ANSI 49, *Biosafety Cabinetry: Design, Construction, Performance, and Field Certification* (NSF 2014). The quantity of hoods to be field tested versus factory tested should be determined based on a risk assessment and owner preference. The acceptance criteria for these tests should be determined before projects are bid and included in the construction documents, with the actual test procedures determined when the contractor's submittals have been accepted and the manufacturer determined. The extent of functional performance tests for the laboratory systems include the following:

- Part-load and full-load performance
- Fume hoods
- BSCs
- General laboratory exhaust
- Miscellaneous exhausts
- Failure scenarios

OCCUPANCY AND OPERATIONS PHASE

The last phase of the commissioning process focuses on actions needed to follow up on warranties and making sure the expected performance is maintained for the life of the building. The requirements of documentation and training are addressed in this phase, as well as the need for audits of laboratory systems and ongoing commissioning.

DOCUMENTATION

Documentation should include all the necessary information to safely operate and maintain the laboratory systems. This includes O&M manuals, as-built record drawings, material safety data sheets (MSDSs), and any key references for specific laboratory procedures. The documentation must be well organized and include information on only the installed components; no information on components that were not actually installed should be allowed. The information should be concise and meet the OPR. This allows the O&M staff to have the correct information to perform their required tasks. The documentation should be compiled early during the construction process with details such as the record drawings and the commissioning plan included.

TRAINING

The training of the O&M staff is a key factor that determines whether the laboratory systems are operated as intended when the building is occupied. While the systems may be capable of meeting the owner's and occupants' requirements, the systems will fail if the O&M staff does not understand how to operate the systems and their equipment.

If the maintenance of systems is insufficient, dangerous conditions, as well as shorter lifetimes of individual components, are inevitable. It is an advantage if the training focuses on teaching the O&M staff to use the O&M manual, as such training resembles the way staff will work during day-to-day operation.

The users of the laboratory must also be trained to be able to take advantage of the system's capabilities, know the system's limitations, and understand safety procedures.

AUDITS OF LABORATORY SYSTEMS

The purpose of audits of laboratory systems is to systematically review the existing state of the laboratory facility and determine any code-compliance issues that are related to the safety of personnel and research; energy usage; and the storage, use, and handling of chemicals and biohazardous materials. A report of the findings and recommendations of the audit is an invaluable source for focused improvements. Laboratory personnel should meet to discuss and rank the recommendations of the audit in order of priority. A hazard and operability analysis (HAZOP) style format can be used for these meetings (IEC 2001). Audits of laboratory systems should be performed periodically even if all the steps of a quality control process such as commissioning were followed in the design and construction of the laboratory. See Chapter 4 for a discussion on risk assessment and various other hazard analysis methods.

ONGOING COMMISSIONING

The control system should monitor the system and trend lines developed to monitor the system over time. Sensors (for airflow, pressure, temperature, relative humidity, etc.) should be verified for accuracy and drift and recalibrated as recommended by the manufacturer.

COMMISSIONING OF EXISTING BUILDINGS (RETROCOMMISSIONING)

Commissioning of systems in existing buildings (retrocommissioning) can be completed for two types of work on existing buildings:

- *Retrofitting.* Commissioning to achieve the expected performance from an existing laboratory system.
- *Remodeling.* Commissioning of a project to remodel an existing building that did not have a particular laboratory systems before remodeling.

Retrofitting concentrates on upgrading the existing system to perform as designed or to accommodate slight changes to the existing system. Training of O&M staff, improved documentation, replacement of failing equipment, recalibration of sensors, and TAB of air and water flows are usually the focus of these projects. Laboratory systems can cause significant losses if they do not perform as designed. Therefore, while many of these tasks may be considered maintenance tasks, a more thorough retrofitting will in many cases be required at yearly intervals.

The commissioning process for remodeling of existing buildings usually follows the same process as new building commissioning. However, if certain parts of systems are reused, considerations such as cleanliness, integration with new components, etc., must be considered. Remodeling also requires that contaminated equipment be handled according to safety and health standards and codes during demolition.

REFERENCES

ASHRAE. 1995. ASHRAE Standard 110-1995, *Method of testing performance of laboratory fume hoods*. Atlanta: ASHRAE.

ASHRAE. 2007. Guideline 1.1-2007, *HVAC&R technical requirements for the commissioning process*. Atlanta: ASHRAE.

ASHRAE. 2013a. Guideline 0-2013, *The commissioning process*. Atlanta: ASHRAE.

ASHRAE. 2013b. Standard 202-2013, *Commissioning process for buildings and systems*. Atlanta: ASHRAE.

CDC. 2009. *Biosafety in microbiological and biomedical laboratories* (*BMBL*), 5th Edition. DHHS Publication No. (CDC) 21-1112. Atlanta: Centers for Disease Control and Prevention, U.S. Department of Health and Human Services. www.cdc.gov/biosafety/publications/bmbl5.

IEC. 2001. IEC International Standard 61882, *Hazard and operability studies (HAZOP studies)—Application guide*. Geneva: International Electrotechnical Commission.

NSF. 2014. NSF/ANSI 49-2014, *Biosafety cabinetry: Design, construction, performance, and field certification*. Ann Arbor, MI: NSF International.

HVAC System Economics

15

OVERVIEW

Cost analysis for buildings can be divided into two categories:

- Initial cost
- Life-cycle cost

Generally, budget constraints on a project result in the initial cost of the system being emphasized. Because of the high density of HVAC equipment in laboratories, the initial cost to purchase equipment can be a considerable percentage of the total building cost (between 30% and 50%). Therefore, accurate initial cost estimates are necessary to complete the laboratory within the budget constraints.

First-cost analysis, while important in maintaining the budget, ignores the investment required to operate the building, such as utility costs and maintenance costs, over the mechanical system's life span of 15 to 30 or more years. Because of the large amount of HVAC equipment (and, thus, high energy consumption) in laboratories that must be rigorously maintained for occupant safety, life-cycle cost is highly important in designing a laboratory.

INITIAL COST

The HVAC system choice and layout, and thus the initial cost of the laboratory, is affected by many factors, such as the usage patterns of the laboratory, the substances handled in the laboratory, and the possibility for future expansion and changes in layout.

The usage patterns of a laboratory may dictate which system options are more attractive. For example, a laboratory where most fume hoods are continuously in use may not benefit from the added initial cost of variable-air-volume (VAV) fume hoods, as the need for exhaust is relatively constant.

The substances used in the laboratory may also dictate several design requirements, such as the material to use for exhaust ductwork. Using similar chemicals, which don't interact, throughout the laboratory could allow manifolded exhaust to be used, whereas chemicals that interact require separate exhaust stacks.

The possibility for future expansion and changes in layout may affect sizing factors such as the diversity factor. A laboratory that plans to expand or reconfigure laboratory space in the future may opt to use a higher (or no) diversity factor to allow for additional capacity for future work. In some cases, the diversity factor may equal 1, in that the central equipment has the same capacity as the summed distributed systems.

Each of these factors influence the initial price of the following:

- Central air-handling equipment
- Exhaust system equipment

CENTRAL AIR-HANDLING EQUIPMENT

Central air-handling equipment for laboratories must be able to respond accurately and quickly to changes in load. For constant-volume air handlers, this may mean rapidly adjusting heating or cooling valves to prevent overheating or overcooling the laboratory space. In the case of a VAV air handler, this means making rapid changes in supply air volume when a fume hood sash is opened to maintain room pressure differentials. The equipment selected must be able to perform these rapid changes and not be selected only on lowest cost or familiarity alone. Last, central air-handling equipment may have special filtering and treatment requirements that require additional equipment or add-on options for the equipment.

Air Distribution System

The air distribution system in a laboratory often has special design requirements. For example, diffuser type and location are important in preventing disturbances, which affect fume hood operation. When large quantities of air must be supplied to laboratories to make up for the exhausted air, special methods are needed to prevent temperature variations and air currents that may disturb exhaust equipment.

Constant-Volume Reheat

Constant-volume air-handling equipment with reheat may require less initial cost to purchase the equipment, but it can use significantly more energy in laboratories with fluctuating equipment usage or wide changes in outdoor air temperature throughout the year.

Variable Air Volume

While VAV equipment requires additional initial equipment, such as variable-frequency drives (VFDs) and controls, and thus costs more when compared to constant-volume equipment, this additional initial cost can be offset by the energy cost savings over the life of the building.

EXHAUST SYSTEM EQUIPMENT

To determine the initial cost of exhaust systems, the cost of dedicated versus manifolded exhausts, ductwork material options, and fume hood control options should be considered.

Dedicated versus Manifolded Exhausts

The use of dedicated exhausts can be required for certain substances or if multiple substances that interact are to be used in the laboratory. It is also possible that the building layout makes dedicated exhausts attractive, as in the case of a single-story building, or impractical, as in the case of a tall multistory laboratory, where dedicated exhaust stacks could occupy a significant portion of the floor area on upper floors. The cost-effectiveness of dedicated exhausts is determined by the number of sources to be exhausted (each will need a fan and a stack), the length of ductwork to the exhaust stacks, and the space available in the building for dedicated duct runs and exhaust stacks.

The use of manifolded exhausts can be prohibited for certain substances or if the substances used in different areas of the laboratory will interact. The cost-effectiveness of manifolded exhaust is determined by the grouping of exhaust sources, the building layout, and possibilities for future changes in substances used in the laboratory (there may be a need to convert to dedicated exhausts if different substances that interact are later used).

Ductwork Material Options

As with the choice of dedicated or manifolded exhausts, some chemicals used in a laboratory require special ductwork materials and connection methods. When a choice is available for ductwork materials and connection methods, the material costs and installation time and cost for various options should be considered. Also, the potential for changing the chemicals that are exhausted should be considered, as some ductwork materials limit the types of chemicals that can be handled by the exhaust system.

Fume Hood Control Options

Fume hood controls are either constant volume or VAV. While constant-volume controls are simpler and less expensive for initial cost, constant-volume control may not be economical over the life of the building. VAV controls may be more expensive for initial cost but provide additional features, such as monitoring the face velocity of the fume hood. For additional information on fume hood controls, see the Constant-Volume versus Variable-Air-Volume Fume Hood Control section of Chapter 11.

LIFE-CYCLE COST

To determine the life-cycle cost of a building, both the capital and operating costs associated with the building need to be converted to a common cost per time period using an appropriate interest rate. Typically, the time period is for a 30-year life expectancy. Consultations between the design team and owner are needed to estimate or calculate numerous cost factors during life-cycle cost analysis (LCCA). Often, these cost factors are very project specific; changes in usage patterns of the laboratory or the building location can dramatically affect which cost factor is dominant in determining the most cost-effective system option.

COST FACTORS

The many cost factors that are needed to calculate the cost of a building over its lifetime can be divided into three groups: design factors, economic factors, and performance factors.

Design Factors

Design factors include items selected or determined during the design phase of a project that can greatly affect the materials and equipment used in the laboratory and thus the initial cost. The heat gain from laboratory equipment may be the dominant factor in sizing equipment for some laboratories; for other laboratories, it is the rate of exhaust needed to provide separation between hazardous materials and laboratory occupants.

The climate where the laboratory is located can significantly affect the sizing of the heating and cooling systems for a building. Because of the need to use 100% outdoor air, wide temperature variations throughout the year have a significant impact on laboratory equipment sizing and heating and cooling costs. To compensate for wider variations in temperature, additional design features may need to be considered, such as using VAV fume hoods, improving the building envelope, and using energy recovery.

Exhaust devices such as fume hoods and biological safety cabinets (BSCs) can have significant exhaust air requirements. This results in the treatment of outdoor air to satisfy these requirements. Therefore, during design, the appropriate number, size, and type of devices should be determined. If the laboratory processes can be performed with a smaller fume hood, for example, the energy costs would be reduced significantly compared to the costs for a larger fume hood. Many laboratory processes can use BSCs that, instead of requiring air to be exhausted through the exhaust system, recirculate air back into the laboratory. If this type of BSC can be used, there are reductions in initial costs for exhaust system components and controls as well as reductions in ongoing energy costs for makeup air.

Economic Factors

Several economic factors are used to calculate the life-cycle cost of a laboratory, including interest rate, initial system cost, the cost of maintenance, and operating and energy costs.

The interest rate is used to determine the time value of money, such as determining the current value of electric utility costs over the life of the building. It should be chosen after consultation with the building owner, as organizations often have fixed requirements on interest rates to use for cost analysis.

The initial system cost includes the purchase cost for equipment and materials, the cost of installation, and overhead costs for the installing contractors. The whole building does not necessarily need to be included in the cost analysis. For example, if the building envelope is the same for all options under consideration, its initial cost may be excluded.

The cost of maintenance includes the costs associated with keeping the laboratory equipment functioning as originally constructed. This includes replacement parts for items that normally wear out, such as fan belts and filters, and for items that may wear out prematurely, such as a fan or pump motor. Laboratories have higher maintenance costs than typical buildings because there are many critical systems that require regular, preventive maintenance to ensure that they do not fail and create unsafe working conditions. Typical maintenance costs for commercial buildings should not be substituted for careful consultation between the building owner, operation and maintenance (O&M) staff, and designers to develop a preventive maintenance program and identify the costs of the program.

Operating and energy costs are the costs for items such as electricity, natural gas, and water and sewer. The costs for these items will likely vary in the future, so estimation of an inflation rate may be needed. Reviewing the utility costs for a similar laboratory, if possible, may be useful in estimating energy use. Some building analysis programs used for determining heating and cooling loads can also be used to determine energy usage.

Performance Factors

Performance factors include the diversity factor, service life, average heat gains, and the adaptability of the laboratory. In most laboratories, not all hoods will be used at the same time. If this is the case, a diversity factor, which represents the percentage of equipment that will be used at any one time, may be used to downsize central equipment, such as air handlers, boilers, and chillers. There are some exceptions, though, when diversity factors should not be used. This may include teaching laboratories, which may have students using all of the fume hoods at once during a class. The owner and users of the laboratory should be consulted when determining whether using a diversity factor is appropriate.

The service life for the equipment must be considered when performing LCCA. LCCA requires that a time period be chosen for the analysis. If the expected life of equipment being analyzed is less than the time period for the analysis, the cost of replacing the equipment for as many times as needed to equal the time period of analysis should be included in LCCA.

Average heat gain is the average thermal load supplied to the space, not the connected electric load, as it is rare for all the equipment to be used at once. An average heat gain, however, may not accurately represent a large heat-producing piece of equipment in a laboratory without much other heat-producing equipment. A more accurate method of determining the heat gain in a laboratory is to determine the heat gain from each piece of equipment and use the expected usage, which can be obtained from the building owner, laboratory users, or equipment manufacturers. However, this may also present challenges, as heat gain values from equipment may not be available or the number of pieces of equipment may change.

Because life-cycle cost considers the life span of the building, the amount of adaptability for the laboratory should be considered, as it is likely the type of work performed in the laboratory will change at some point during the life of the building. The building owner and occupants should be consulted to determine the level of adaptability needed to allow for future changes in laboratory uses.

EXAMPLE LCCA CALCULATION

This example is intended to provide guidance on the procedure to use in determining the life-cycle cost of a laboratory.

In the planning stage, the design team suggested two different approaches for the design of the mechanical system for the laboratory. One system emphasized an energy-conservation design and the other was of a more traditional design. The owner wanted to know which system design option had the lower life-cycle cost. The lifetime for both mechanical systems was the same and inflation was not considered in this example.

- Given economic factors:
  - Interest rate = 8%
  - Service life = 20 years
- System 1
  - Mechanical system initial cost = $1,250,000
  - Energy cost = $65,000/year
  - Maintenance cost = $150,000/year
- System 2
  - Mechanical system initial cost = $1,100,000
  - Energy cost = $95,000/year
  - Maintenance cost = $145,000/year

The economic formula that can be used to calculate the life-cycle cost for this scenario is shown in Equation 15-1:

$$LC = IC + AC \cdot \frac{(1+i)^n - 1}{i \cdot (1+i)^n} \tag{15-1}$$

where

LC = life-cycle cost
IC = initial cost
AC = annual cost (energy cost + maintenance cost)
i = interest rate in absolute value (e.g., 8% = 0.08)
n = number of years

System 1 life-cycle cost:

$$LC = 1{,}250{,}000 + (65{,}000 + 150{,}000) \cdot \frac{(1 + 0.08)^{20} - 1}{0.08 \cdot (1 + 0.08)^{20}}$$

$$= 3{,}411{,}000$$

System 2 life-cycle cost:

$$LC = 1{,}100{,}000 + (95{,}000 + 145{,}000) \cdot \frac{(1 + 0.08)^{20} - 1}{0.08 \cdot (1 + 0.08)^{20}}$$

$$= 3{,}456{,}000$$

The energy-efficient System 1 had the lower life-cycle cost with these parameters.

What would the life-cycle costs of the systems be if the interest rate were 12%?

System 1 life-cycle cost:

$$LC = 1{,}300{,}000 + (65{,}000 + 150{,}000) \cdot \frac{(1 + 0.12)^{20} - 1}{0.12 \cdot (1 + 0.12)^{20}}$$

$$= 2{,}905{,}000$$

System 2 life-cycle cost:

$$LC = 1{,}100{,}000 + (95{,}000 + 145{,}000) \cdot \frac{(1 + 0.12)^{20} - 1}{0.12 \cdot (1 + 0.12)^{20}}$$

$$= 2{,}892{,}000$$

The traditional system 2 has the lower life-cycle cost with the higher interest rate.

Microbiological and Biomedical Laboratories

16

OVERVIEW

Microbiological and biomedical laboratories deal with microorganisms that in many cases can cause disease in humans and in some cases cause disease in plants or animals. Such laboratories are often referred to as *biological containment laboratories*, *biocontainment laboratories*, or *containment laboratories*. In these laboratories, the harmful organisms must be handled correctly to protect laboratory personnel as well as to prevent release of the organisms from the laboratory and avoid potential exposure to the general public. The method of handling the organisms greatly depends on the microorganism and its means of causing infection.

This chapter contains information on the classifications of biological contaminants and discusses requirements to provide a basis for the design of microbiological and biomedical laboratories. This information is divided into the following sections:

- Biosafety level classifications
- Users' program requirements
- Engineering considerations
- Special animal considerations

Because it is very important to contain the contaminants within the biocontainment laboratory and laboratory animal areas, issues discussed and illustrated in previous chapters such as space pressurization, system reliability, proper controls, and redundancy apply for microbiological and biomedical laboratories. Specific details about biological safety cabinets (BSCs) used for biological containment are covered in Chapter 5.

INTRODUCTION TO BIOLOGICAL CONTAINMENT

This section introduces the elements, risk assessment, containment barriers, reference standards and design guidelines, and definitions of biological containment, providing a foundation for the more detailed Biosafety Level Classification section that follows.

ELEMENTS

Containment involves safe methods for managing infectious materials in the laboratory environment where they are being handled or maintained. The three elements of containment are laboratory practice and technique, safety equipment, and facility design.

The purpose of containment is to minimize the risk of exposure to potentially hazardous agents. Exposure could involve not only laboratory workers but also other individuals working close by and in the outside environment.

RISK ASSESSMENT

A risk assessment of the procedures to be conducted by the laboratory should be performed by the appropriate laboratory and biosafety professionals to identify the most likely risks associated with procedures involving biohazards in order to determine the appropriate combination of the three elements of containment.

Assistance in the risk assessment process is provided by the risk ranking of biohazards into risk group (RG) classifications. Many organizations have established risk groups for biohazards, and each begin with a ranking of lowest (RG1) to highest (RG4). The risk group classifications of various national and international organizations can be found at the websites or in the publications listed here:

- American Biological Safety Association (ABSA), Risk Group Classification for Infectious Agents, www.absa.org/riskgroups (ABSA n.d.)
- World Health Organization (WHO), *Laboratory Biosafety Manual*, Third Edition, www.who.int/csr/resources/publications/biosafety/Biosafety7.pdf?ua=1 (WHO 2004)
- National Institutes of Health (NIH), *NIH Guidelines for Research Involving Recombinant or Synthetic Nucleic Acid Molecules* (NIH 2013a)
- Government of Canada, Chapter 4, Risk Groups, Containment Levels, and Risk Assessments, of *Canadian Biosafety Standards and Guidelines (CBSG)* (Government of Canada 2013)
- Standards Australia and Standards New Zealand, AS/NZS 2243.3:2002, *Australian/New Zealand Standard™: Safety in laboratories, Part 3: Microbiological aspects and containment facilities* (SA/SNZ 2002)
- European Union, Directive 2000/54/EC (EU 2000)
- Centers for Disease Control and Prevention (CDC), *Biosafety in Microbiological and Biomedical Laboratories* (*BMBL*), www.cdc.gov/biosafety/publications/bmbl5 (CDC 2009a)

The design professional needs to work closely with the laboratory users and owners and compliance officers to understand what facility features are needed for containment and to accommodate the required biosafety protocols based on the risk assessment.

CONTAINMENT BARRIERS

The primary factor in designing a biocontainment laboratory is protection of laboratory workers, the environment, and the public from accidental exposure to potentially biohazardous agents. Primary and secondary barriers must be in place to prevent accidental personnel exposure to or environmental release of contaminated materials. Effective containment requires strict adherence to the following principles regarding the hierarchy of containment barriers:

- *Primary Containment.* Protection of personnel and the immediate laboratory environment from exposure to infectious agents by the use of containment equipment such as the following:
    - BSCs (the primary containment device for the control of infectious splashes or aerosols in the laboratory)
    - Safety centrifuges
    - Sealed sonicators, blenders, and homogenizers

    - Sealed tubes and transport carriers
    - Safe sharps, needle boxes, medical waste bags, etc.
- *Secondary Containment.* Protection of the environment external to the laboratory from exposure to infectious materials provided by a combination of facility design features and operational practices (protocols) such as the following:
    - Specialized ventilation systems to ensure unidirectional airflow from lower to higher hazard spaces
    - Air treatment systems to decontaminate or remove agents from exhaust air
    - Controlled access zones
    - Integrity of the laboratory room envelope (sealing of penetrations through walls, floors, and ceiling)
    - Air locks as laboratory entrances and exits
    - Separate buildings or modules to isolate the laboratory

REFERENCE STANDARDS AND DESIGN GUIDELINES

Adhering to the following nationally recognized reference standards and design guidelines helps reduce the health and safety risks posed by biocontainment laboratory facilities. In the United States, prudent guidelines for working with microorganisms are provided by several sources, the most predominant of which include the following:

- CDC, *Biosafety in Microbiological and Biomedical Laboratories* (*BMBL*) (CDC 2009a)
- NIH, Division of Technical Resources, *Design Requirements Manual* (NIH 2013b)
- United States Department of Agriculture (USDA), Manual 242.1, *ARS Facilities Design Standards* (USDA 2012)
- Federal Select Agent Program (FSAP), CDC and USDA Animal and Plant Health Inspection Services (APHIS), Regulations 7 CFR Part 331, 9 CFR Part 121, and 42 CFR Part 73 (FSAP 2014a)
- CDC, "Primary Containment for Biohazards: Selection, Installation and Use of Biological Safety Cabinets" (Currently published as Appendix A of *BMBL* but planned as a future separate publication) (CDC 2009b)
- Institute for Laboratory Animal Research (ILAR), *Guide for the Care and Use of Laboratory Animals* (ILAR 2011)
- American Committee of Medical Entomology (ACME), American Society of Tropical Medicine and Hygiene, "Arthropod Containment Levels (ACLs)" (ACME 2003)
- Institutional Biosafety Committee (IBC), Rensselaer Polytechnic Institute (RPI), "NIH Risk Group Classifications" (IBC 2010)
- American Industrial Hygiene Association (AIHA), ANSI/AIHA/ASSE Z9.5, *Laboratory Ventilation* (AIHA 2012)
- American Society of Safety Engineers (ASSE), ANSI/ASSE Z9.14-2014, *Testing and Performance-Verification Methodologies for Ventilation Systems for Biosafety Level 3 (BSL-3) and Animal Biosafety Level 3 (ABSL-3) Facilities* (ASSE 2014)

DEFINITIONS

The following are definitions of some of the biological agents that may be used in the facility being designed:

- *Select agents.* Certain "biological agents and toxins have been determined to have the potential to pose a severe threat to both human and animal health, to plant health, or to animal and plant products" (FSAP 2014b). These biological agents have been classified as *select agents and toxins* and are regulated under the Federal Select Agent Program (FSAP) (www.selectagents.gov/index.html).

- *Vector-borne diseases (VBDs).* VBDs are bacterial and viral diseases transmitted by organisms such as mosquitoes, ticks, and fleas. For more information, visit the website of the Division of Vector-Borne Diseases (DVBD) of the CDC: www.cdc.gov/ncezid/dvbd/index.html.
- *Zoonotic agent (zoonoses).* Zoonotic agents are infectious agents that can be transmitted between (or are shared by) animals and humans (Olsen 2000). For more information, visit the Zoonotic Diseases Tutorial website at www.vet med.wisc.edu/pbs/zoonoses.

BIOSAFETY LEVEL CLASSIFICATION

Biocontainment laboratories are designed for researching a wide range of microorganisms from those that are not known to cause diseases in humans (lowest risk) to those that can cause life-threatening diseases (highest risk). In addition to pathogens that cause diseases in humans, there are pathogens that cause animal and plant diseases. Consequently, the laboratories designed for researching these pathogens must be constructed and classified based on the nature and dangers associated with the work and materials housed within the laboratory. *Biosafety in Microbiological and Biomedical Laboratories* (*BMBL*) (CDC 2009a) describes four biosafety levels (BSLs) ranging from the lowest risk (BSL-1) to the highest risk (BSL-4) (see Figure 16-1). The objective of these guidelines is to offer information that will reduce the spread of diseases from laboratories handling microorganisms by recommending and/or requiring practices for the design and use of these facilities. Figure 16-2 provides a comparison of some of the salient features of the four BSLs. Note that not all features are necessarily described for a specific laboratory application.

Note that the selection of a BSL for a specific research program should be made through a risk assessment process conducted by the researchers, clinicians and biosafety professionals responsible for the safety of the program, as stated previously.

The use of "ABSL" in the *BMBL* refers to *animal biosafety level* and applies to animal research facilities (animals used within a laboratory or an animal-specific facility). Generally the animal biosafety guidelines incorporate the same level of secondary barrier precautions (e.g., separation of the laboratory work area from public access), but they also include additional requirements for animal care and husbandry practices set forth in *Guide for the Care and Use of Laboratory Animals* (ILAR 2011). There are many types of animals commonly used in research laboratories. Each species has individual require-

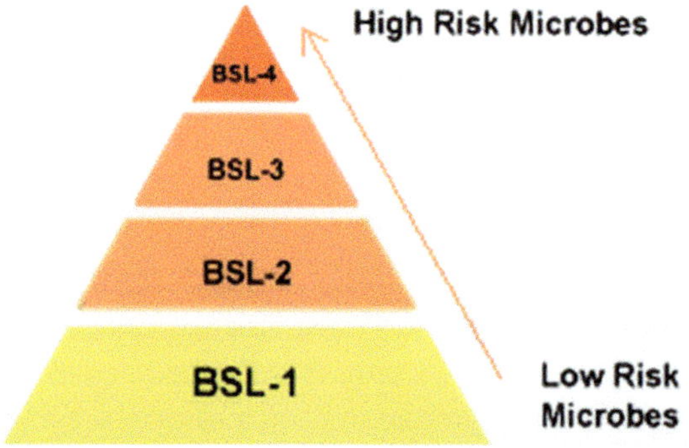

Figure 16-1 Four Biosafety Levels

(CDC n.d., Quick Learn Lesson)

Figure 16-2
Four Biosafety Laboratory Levels
(CDC n.d., 4 Biosafety Lab Levels)

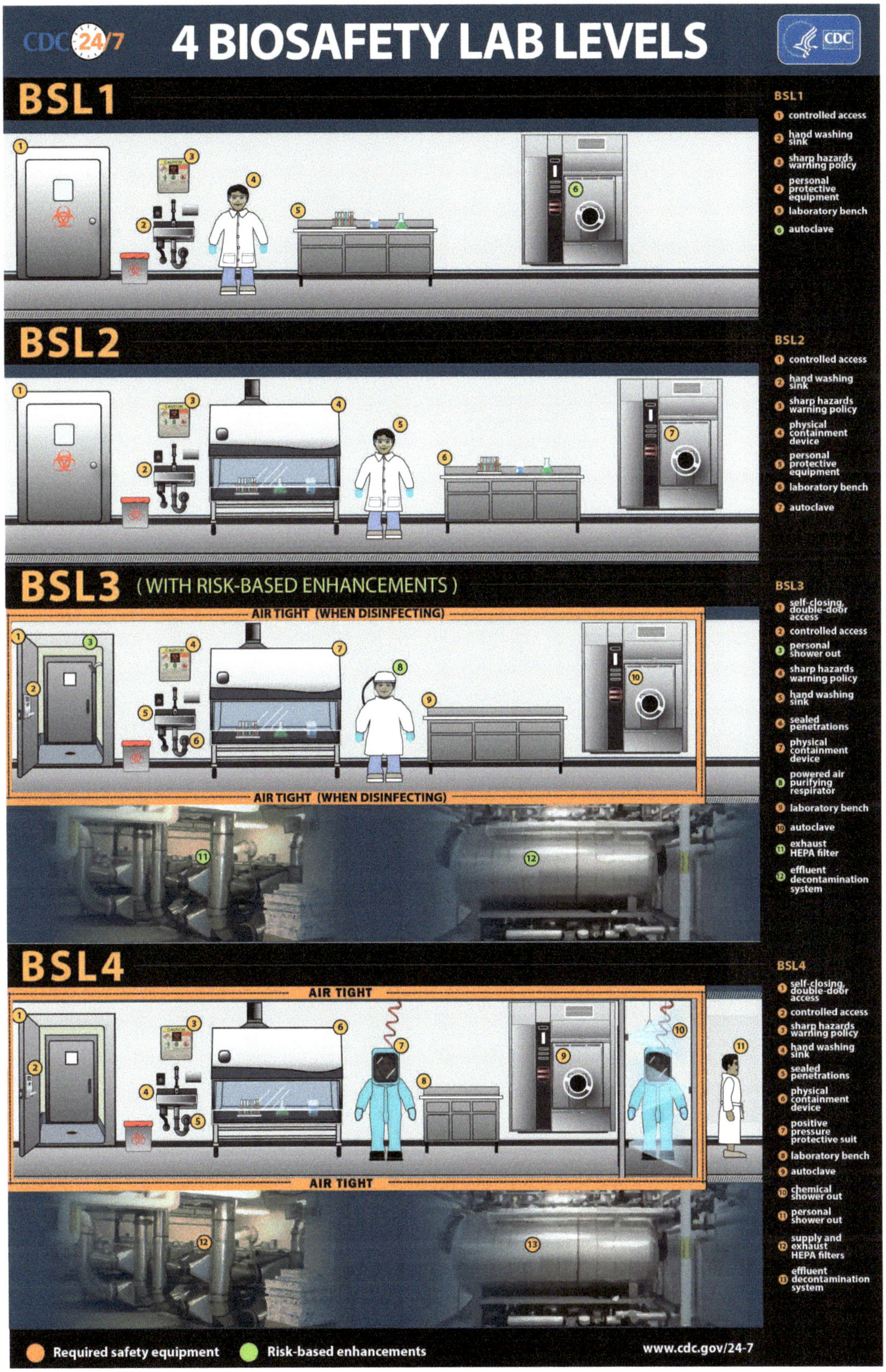

ments detailed in guidelines, standards, and codes. These requirements, including those for temperature, humidity, sound, vibration, and light, are required for healthy and humane housing of animals.

Types of animals used in microbiological and biomedical laboratory research can include the following:

- Invertebrates
  - *Drosophila melanogaster* (fruit fly)
  - *Caenorhabditis elegans* (nematode worm)
  - *Galleria mellonella* (waxworm)
- Vertebrates
  - Amphibians (primarily frogs, toads, and salamanders)
  - Birds
  - Cats
  - Dogs
  - Fish
  - Farm animals (large)
  - Farm animals (small)
  - Nonhuman primates
  - Rabbits
  - Reptiles
  - Rodents (guinea pigs, mice, and rats)

Laboratory animal areas within research facilities have special requirements for maintaining the health and comfort of the animals (specific to the species), for avoiding contamination within and between individual rooms of the facility, and for providing a safe working environment for the laboratory personnel.

(A)BSL-1

Biosafety level 1 (BSL-1) practices, safety equipment, and facilities are appropriate when working with microorganisms that are not known to cause disease in healthy humans. Laboratories for undergraduate, secondary educational training, and teaching are facilities where BSL-1 can be a sufficient measure of precaution. BSL-1 does not require any primary or secondary containment barrier, except a sink for hand washing. Very few biomedical research laboratories operate at the BSL-1 level.

(A)BSL-2

Biosafety level 2 (BSL-2) practices, safety equipment, and facilities are appropriate when working with indigenous moderate-risk agents that are known to cause human diseases with varying severity. Laboratories for clinical, diagnostic, and teaching are facilities where BSL-2 can be a sufficient measure of precaution. Procedures that can cause splashing or aerosolization must be conducted in primary containment enclosures such as certified BSCs. Gloves, face masks, splash shields, and gowns should be used when appropriate. A sink for hand washing and facilities for decontamination of wastes (such as an autoclave or other validated decontamination method) must be available.

(A)BSL-3

Biosafety level 3 (BSL-3) practices, safety equipment, and facilities are appropriate when working with indigenous or exotic agents with a potential for respiratory transmission, which may cause serious and potentially lethal infection. Laboratories for clinical, diagnostic, research, production, and teaching are facilities where BSL-3 can be a suffi-

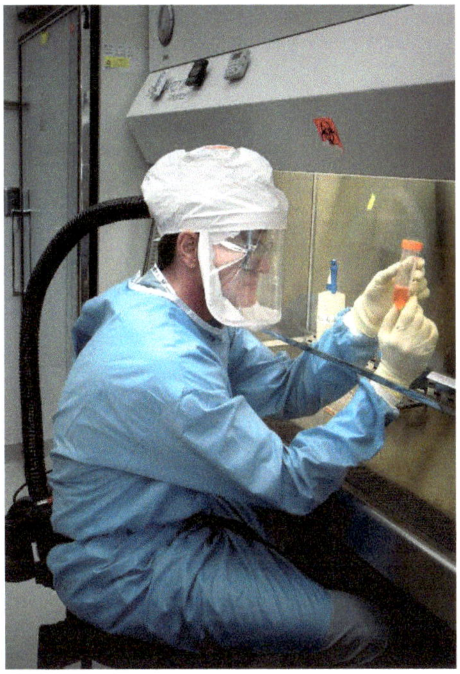

Figure 16-3 Scientist Examining Specimens in a BSL-3 Laboratory

(CDC 2014)

cient measure of precaution. The primary hazards relate to autoinoculation, ingestion, and exposure to infectious aerosols. BSL-3 places emphasis on primary containment barriers, and all laboratory manipulation should be conducted within certified BSCs or similar enclosures (see Figure 16-3). The room ventilation system should be designed and constructed to minimize the release of infectious aerosols from the laboratory should there be an accidental release and the release of limited fugitive emissions from the primary containment device, such as a BSC. Laboratories should have restricted access. Strict attention should be given to the secondary containment barrier construction. These laboratories operate under a defined negative pressure, which must be monitored, and directional airflow must be maintained. Laboratory personnel must be able to verify direction of airflow through a visual monitoring system at the laboratory entry. Visual and audible alarms should be provided inside the laboratory to notify personnel if directional airflow or negative pressurization has been compromised.

BSL-3 laboratories should be provided with single-pass air systems (100% outdoor air) with no recirculation of laboratory air to other parts of the building. Additional best practice guidance added in the fifth edition of the *BMBL* is that "the laboratory shall be designed such that under failure conditions the airflow will not be reversed" (CDC 2009a, p. 43). This requirement should be clearly defined in the design documents, because consideration will need to be given as to how this will be accomplished. Most modern BSL-3 laboratories provide for a single level of high-efficiency particulate air (HEPA) filtration of laboratory room exhaust air in addition to the HEPA filtration of air being rejected from the primary containment device, such as a BSC. The use of these filters should be determined based on a risk assessment, agent summary statement, and local standards. Though it is not specifically mentioned as being best practice in the fifth edition of the *BMBL*, standby power is typically provided to serve critical BSL-3 laboratory systems and equipment. Other codes and standards may specifically require standby power be provided.

Mechanical, electrical, and plumbing penetrations of the secondary containment barrier must be sealed to maintain the integrity of the barrier to allow the airflow controls to work properly and to prevent leakage outside the containment barrier of gaseous agents used to decontaminate the space (e.g., hydrogen peroxide vapor, formaldehyde gas, or chlorine dioxide). It is important to note that when selecting materials that will be exposed to bio-decontamination agents, chemical compatibility should be evaluated in consultation with the laboratory researchers. An anteroom with interlocking doors should be provided between the containment laboratory and adjacent noncontainment spaces. Some facilities require shower-out capabilities, and there must be areas reserved for storage of personal protective clothing, equipment, and devices. To minimize contamination, sinks, lights, and phones should be automatically controlled or operated via hands-free controls and floors should be seamless. Any windows should be permanently sealed, and doors should be installed with automatic closure control and electromagnetic interlocks. All doors should have windows to allow entering personnel to view the laboratory for normal and emergency conditions. It is preferable that, where possible, equipment requiring maintenance be located outside containment rooms in order to facilitate operation and maintenance (O&M) functions. Where possible, duct devices and equipment should be installed on or in uncontaminated duct. Certain control devices that must sense across the containment barrier may require small HEPA filters in the sensing lines. BSL-3 laboratories generally have dedicated exhaust and supply air systems. If building supply air is used, a controlled bubble-tight shutoff supply damper should be considered for space isolation and backflow prevention.

Enhancements to BSL-3 Laboratories

BSL-3 laboratories may need additional measures for the protection of personnel and/or the environment. The enhancements may be required based on the agent summary statement (from the *BMBL*); a risk assessment; the laboratory institution's or funding source's requirements; or applicable local, state, or federal regulations. Enhancements may include the following:

- Single-level HEPA filtration of room exhaust.
- Series and/or parallel exhaust HEPA filters.
- Supply air backflow prevention with a HEPA filter or with quick-acting, bubble-tight, bioseal dampers at supply duct penetrations through the architectural containment barrier. These dampers automatically close if differential pressure sensors indicate a loss in negative pressure control. Bioseal dampers also can aid in the isolation of the HEPA filters when it is time to decontaminate them prior to changeover.
- Additional HVAC redundancy.
- Shower-in/out capability.
- Clean and dirty change rooms.
- Effluent decontamination system (EDS) (chemical or heat treatment)
- Special room sealing for decontamination operations.

When HEPA filters are provided, the housing assembly should include the following:

- Prefilter section, especially if single-stage HEPA filtration is used
- Inlet and outlet
- Gastight isolation dampers
- Inlet and outlet decontamination (decon) ports
- Consideration for bag-in/bag-out arrangement
- Scan section or means for leak testing and validating filters

Most HEPA filters used are rated at 99.97% efficient for 0.3 μm particles, which is considered the filter's most penetrating particle size (MPPS), the particle size at which penetration of particles through the filter is highest. HEPA filters must effectively trap hazardous bacteria and viruses to provide personnel protection. Although an individual virus particle ranges in size from 0.005 to 0.1 μm, viruses generally only survive to travel through the air as part of larger particles (0.3 μm or larger), for example, attached to mucous particles. Because it is difficult to disperse or aerosolize single viral particles and because of the particle collection mechanisms of HEPA filters, particles larger and smaller than a filter's most penetrating size are collected with great efficiency. HEPA filters should be tested and validated annually.

BSL-3Ag (BSL-3 AGRICULTURAL LARGE-ANIMAL FACILITIES)

BSL-3Ag practices, safety equipment, and facilities are appropriate when working with high-consequence livestock pathogens that pose a substantial economic risk to the agricultural industry. Some BSL-3Ag agents also cause disease in humans; these are known as *zoonotic agents*. Where studies involving these high-risk agents are performed using large agricultural animals, the rooms act as a primary containment barrier, much like a BSC in BSL-1, -2, and -3 laboratories. Therefore, additional protocol and operational requirements, as well as additional levels of engineering measures, are needed to protect the environment. Personnel change and shower rooms should be provided to control access to the containment spaces. Doors entering and leaving the containment space should be air pressure resistant (APR) and have either compressed-gasket or inflatable-gasket seals. All penetrations must be sealed and the entire containment boundary should be pressure decay tested as defined by the USDA Manual 242.1, *ARS Facilities Design Standards* (USDA 2012), to verify integrity. The supply and exhaust systems must be dedicated to the BSL-3Ag zone and must be interlocked to prevent a reversal of directional airflow and positive pressurization of the containment space. A visible means of displaying pressure differentials must be provided inside and outside of the containment space to assure workers that it is safe to enter the space. Supply and exhaust air to and from the containment space must be HEPA filtered. Single HEPA filtration on the supply and double (in-series) HEPA filtration on the exhaust are typically required as a minimum for high-risk areas.

Depending on the research program, parallel HEPA filters may be desired on the supply and exhaust. Installing a parallel filter arrangement allows filter maintenance, decontamination, change, and/or certification if continuous operation of the space is required or desired and is based on research program activities and needs.

Higher-risk areas may include areas where large amounts of aerosols containing BSL-3Ag agents could be expected, such as large-animal rooms, contaminated corridors, necropsy areas, and carcass disposal facilities. The risks associated with the agent used may also be a factor. The HEPA filter requirements should be reviewed with the appropriate biosafety personnel for each facility. HEPA filters should be designed to allow in-place scanning of the filters for efficacy testing as well as in-place decontamination. The HEPA filters are to be located as near as possible to the containment space to minimize the length of potentially contaminated ductwork. All ductwork serving the containment spaces between the rooms and the HEPA filters must be welded, tested, and certified as gastight. All liquid effluent leaving the containment space must be treated through an EDS (see Figure 16-4) prior to discharge to the normal sanitary sewer system. Likewise, all solids, such as straw and bedding, must be decontaminated prior to disposal. HEPA filters are to be provided on all plumbing vents from the containment space. If there is a chance of the vent filters becoming wetted, special attention should be paid to the HEPA

Figure 16-4
Heat Treatment EDS Schematic
(Reprinted with permission from Applied Biosafety *[de Kok-Mercado, et al. 2011])*

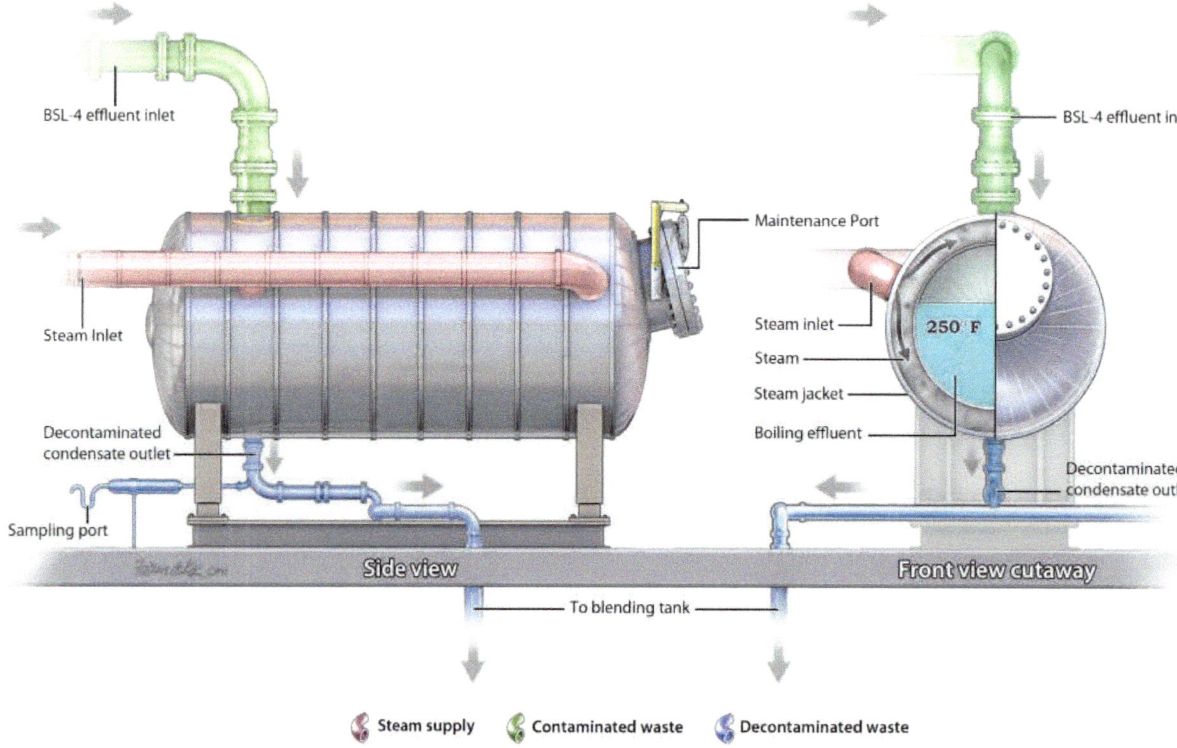

filter materials. A means for safely disposing of carcasses must also be provided. The carcass disposal methodology should be reviewed thoroughly with both the appropriate facility personnel and the local environmental and waste regulatory authorities.

Work with high-consequence agricultural pathogens is done in small animal models or in other primary containment devices such as BSCs, the work may be performed in ABSL-3 facilities with additional enhancements that include the following:

- *Personnel change and shower rooms.* APR doors are not required at the boundary.
- *Supply and exhaust HEPA filters (one filter on the supply and one on the exhaust).* The requirement for these filters should be coordinated with the appropriate facility personnel. Filter housings must be arranged to support the ability to scan the filters in place and to decontaminate the filters prior to removal.
- *Dedicated single-pass supply and exhaust systems that are interlocked to prevent reversal of airflow from containment spaces.* Exhaust ductwork between the containment space and the HEPA filter is pressure decay tested.
- *An EDS.* All liquid effluent leaving the containment space must be treated through an EDS prior to discharge to the normal sanitary sewer system.

BSL-4

Biosafety level 4 (BSL-4) practices, safety equipment, and facilities are appropriate when working with very dangerous and exotic agents that pose a high individual risk of life-threatening disease that may be transmitted via the aerosol route or for which there are no available vaccines or treatment. Agents with a close or similar antigenic relationship to BSL-4 agents should be handled at this level. The primary hazards relate to autoi-

Figure 16-5 Essential Features of a BSL-4 Facility

(Reprinted with permission from Applied Biosafety *[de Kok-Mercado, et al. 2011])*

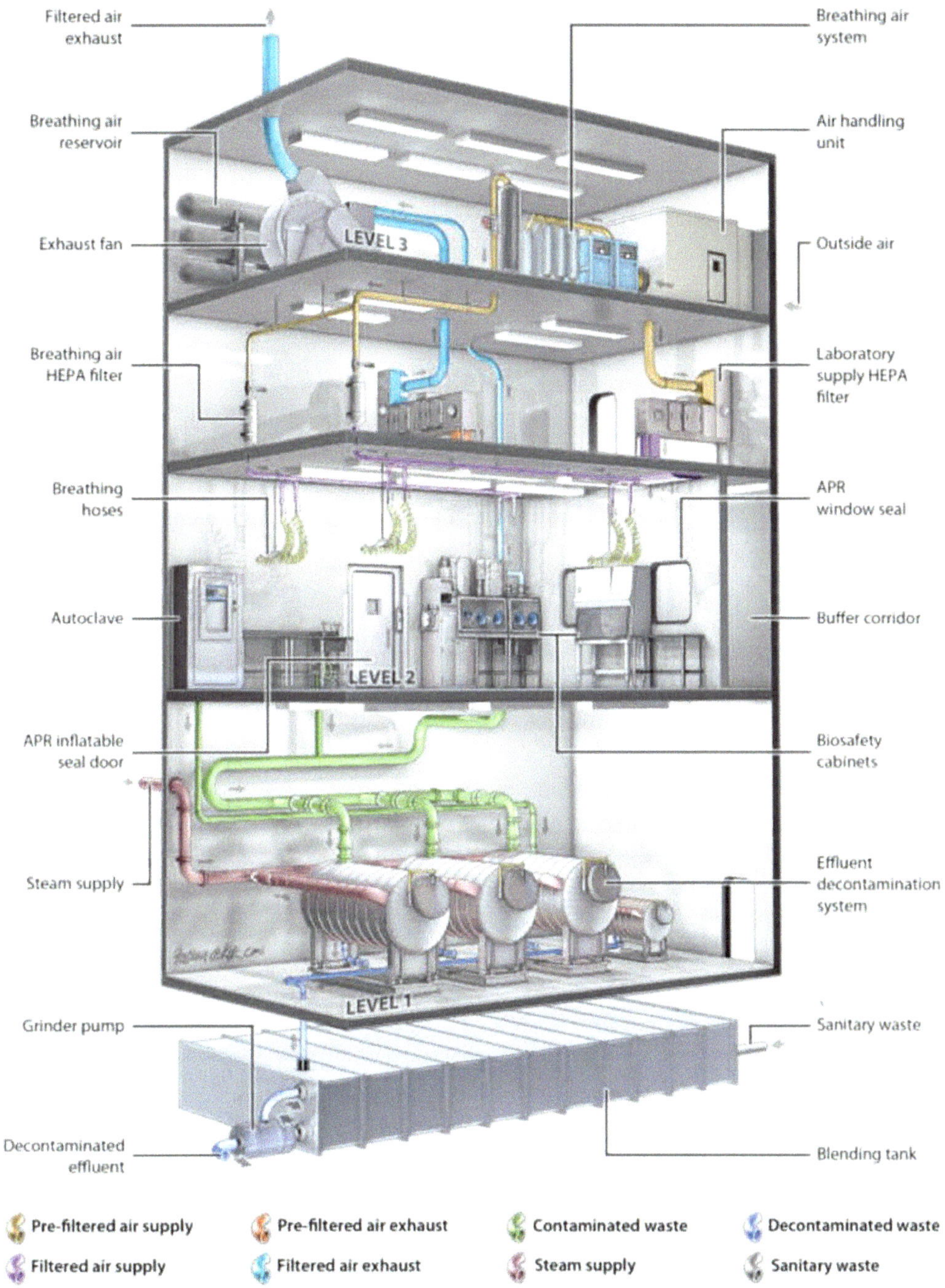

noculation, mucous membrane or broken skin exposure to infectious droplets, exposure to infectious aerosols, and physical contact. All manipulation of BSL-4 agents poses a danger to the personnel, the community, and the environment. Figure 16-5 shows the essential features of a BSL-4 laboratory.

All BSL-3 requirements apply to BSL-4 laboratories. Recommended operating requirements for the design and operation of BSL-4 laboratories are presented in the fifth edition of *BMBL* (CDC 2009a). BSL-4 laboratories should, in general, be located in separate buildings or in completely isolated zones with precisely operated ventilation and waste management systems. In BSL-4 laboratories, as a minimum supply and exhaust ducts should be constructed of Type 304L stainless steel with welded seams and joints to prevent air leakage. Alternative duct material can be used based on risk, ability to withstand pressure and decontamination chemicals, and other considerations.

All BSL-4 facility design criteria and operational procedures must be documented. The facility systems must be tested to verify that the design and operational parameters have been met prior to operation and that all systems are fully functional prior to use. Facility protocols and operating parameters must also be reverified annually. Appropriate communication systems (e.g., voice, fax, and computer) must be provided between the laboratory and the outside. Provisions for emergency communication and access/egress are to be considered.

Upon commissioning and every year thereafter, BSL-4 ducts between the room and the HEPA filters should be pressure decay tested. While the actual testing methods should be approved by the responsible biosafety professional, examples of duct testing requirements are available from the Department of the Army in Table C-1 of Army Pamphlet 385-69, *Safety Standards for Microbiological and Biomedical Laboratories* (U.S. Army 2013), and from the USDA Agricultural Research Service in Manual 242.1, *ARS Facilities Design Standards* (USDA 2012). After the HEPA filters installed at the containment boundaries of BSL-4 spaces, ducts may be galvanized steel.

BSL-4 laboratories typically fall into two design classifications: cabinet laboratories and suit laboratories. Cabinet laboratories have customized glove boxes (certified gastight Class III BSCs that provide the highest attainable level of protection to personnel and the environment) connected in long lines. In suit laboratories, scientists wear impervious, positively pressurized spacesuit-type garments (breathing suits) with a transparent faceplate and breathing air hoses and conduct their work in Class II Type B2 BSCs located in a dedicated containment laboratory.

Both cabinet and suit laboratories share many similar safeguards and detailed levels of operational dependability and support of the environmental systems, and each offers advantages depending on the type of research conducted.

Cabinet Laboratories

A BSL-4 cabinet laboratory either is located in a separate building or consists of a clearly demarcated and isolated zone within a building, such as in a BSL-3 enhanced space. For these laboratories housing Class III glove box lines, laboratory doors must have locks in accordance with the institutional policies supporting biosecurity protocol. Rooms in the facility must be arranged to ensure sequential passage through an inner (dirty) changing area to a personal shower and an outer (clean) change room prior to exiting the room(s) containing the Class III BSCs. (Many times the changing area is designed, constructed, and operated as a BSL-3 enhanced laboratory space).

Class III BSCs are totally enclosed, ventilated cabinets that provide for the work to be performed through the use of attached gloves (see Figure 16-6). These cabinets are gastight enclosures, maintained under a negative pressure, with supply air HEPA filtered and with exhaust air passed through two HEPA filters in series. Transfer boxes or ports allow for passage of materials and equipment into and out of the work zones while maintaining the negative pressure. These Class III cabinets can be assembled in a process line to facilitate research functions.

In a BSL-4 cabinet laboratory, an automatically activated emergency power source must be provided for, at a minimum, the following systems: laboratory exhaust, life support, alarms, lighting, entry and exit controls, BSCs, and door gaskets. Monitoring and control systems for air supply, exhaust, life support, alarms, entry and exit, and security systems should be on an uninteruptible power supply (UPS) of adequate capacity to allow for an orderly exit from the laboratory.

Figure 16-6
Components of a Class III BSC
(Reprinted with permission from Applied Biosafety *[de Kok-Mercado, et al. 2011])*

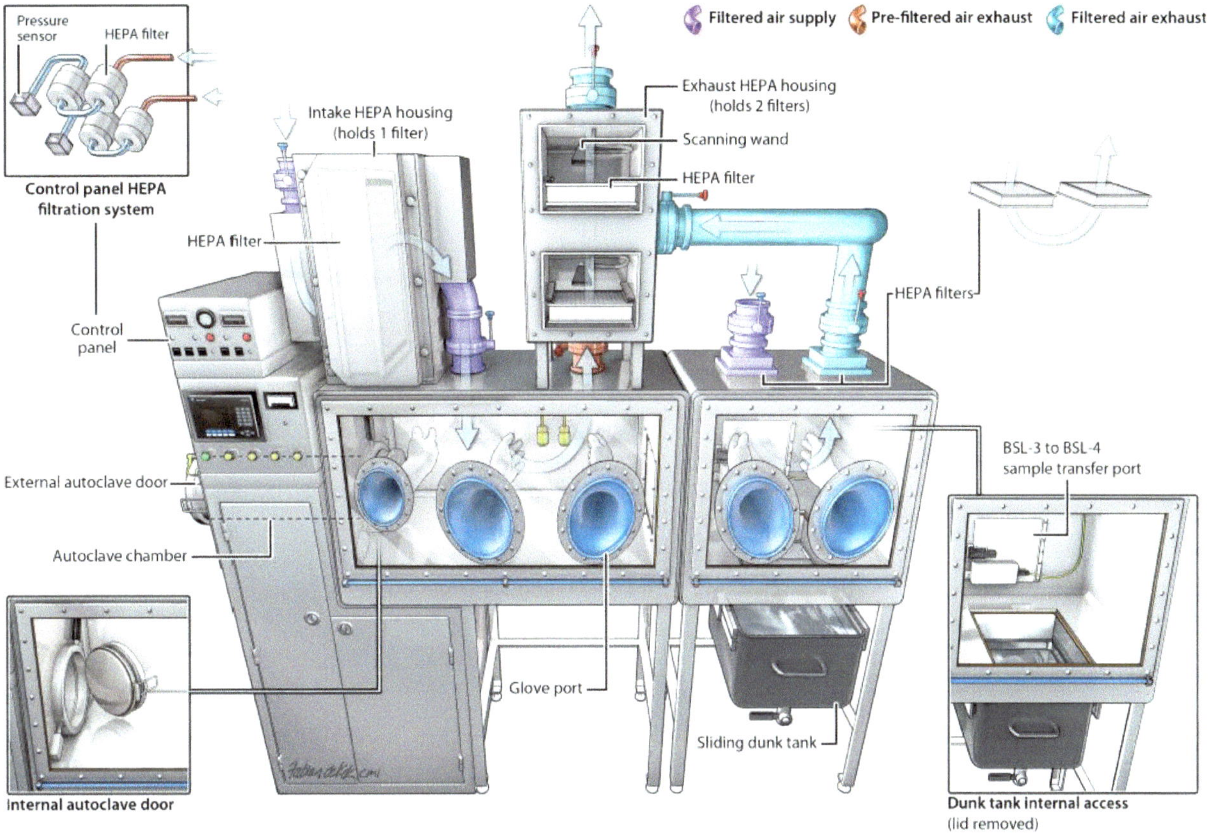

A double-door autoclave, dunk tank, fumigation chamber, or ventilated anteroom/air lock must be provided at the containment barrier for the passage of materials, supplies, or equipment. Liquid effluents from cabinet room sinks, floor drains, autoclave chambers, and other sources within the cabinet room and inner (dirty) change room must be decontaminated by a proven method, preferably heat treatment, before being discharged to the sanitary sewer system.

The walls, floors, and ceilings of the laboratory must be constructed to form a sealed internal shell to facilitate fumigation and prohibit animal and insect intrusion. The internal surfaces of this shell must be nonporous and resistant to chemical liquids, vapors, and gases used for cleaning and decontamination of the area. Floors must be monolithic, sealed, and coved. All penetrations in the internal shell of the laboratory and inner change room must be sealed airtight and gastight. All penetrations must be fitted with two backflow prevention devices in series. Consideration should be given to locating these devices outside of the containment area to facilitate inspection and maintenance. Atmospheric venting systems must be provided with two HEPA filters in series, sealed up to the second filter and resistant to moisture accumulation.

Decontamination of the entire Class III cabinet laboratory must be performed using a validated gaseous or vapor method when there have been significant changes in cabinet usage, before major renovations, or for maintenance shutdowns. Design of cabinet ventilation systems should consider purge of gaseous vapors resulting from the decontamination process. If Class II BSCs are needed in the cabinet support laboratory, they must be installed so that fluctuations of the room air supply and exhaust do not interfere with proper operations.

A central vacuum system is not recommended. However, if used, it must not serve areas outside the cabinet laboratory. Two in-line HEPA filters must be placed near each use point, and the filters must be installed in such a manner as to permit in-place decontamination and replacement.

A dedicated nonrecirculating ventilation system is required for a cabinet laboratory. Only laboratories with the same HVAC requirements (i.e., other BSL-4 and ABSL-4 laboratories) may share ventilation systems, and only if each individual laboratory system is isolated by gastight dampers and HEPA filters. The supply and exhaust components of the cabinets and general space ventilation system must be designed to maintain the cabinets and the laboratory at negative pressures relative to surrounding areas and provide differential pressure/directional airflow between adjacent areas within the cabinet and laboratory sections.

Redundancy level is directly or indirectly related to the capacity N required (needed) for stated operating conditions. The minimum redundancy for a BSL-4 facility is N+1, in which any single point of system failure is backed up by a spare device or piece of equipment ready to start. N+1 redundant supply fans are recommended to enable maintenance to be performed without affecting overall system performance. N+1 redundant exhaust fans are required. Supply and exhaust fans must be interlocked to prevent positive pressurization of the laboratory.

The ventilation system serving the cabinets and general laboratory space must be monitored and alarmed to indicate malfunction or deviation from design setpoints. A visual indicating device, incorporating a magnetically coupled indicating mechanism and a sealed differential pressure diaphragm, must be installed near the clean change room so that proper differential pressures within the laboratory may be verified. Consideration should be given to providing monitoring and/or alarming of pressures within the cabinet laboratory. Supply air to and exhaust air from the cabinet laboratory, inner change room, and fumigation/decontamination chambers must pass through HEPA filters. The air exhaust discharge must be located away from occupied spaces and building outdoor air intakes.

All HEPA (Type 1, Class A) filters should be located as near as practicable to the cabinet or laboratory in order to minimize the length of potentially contaminated ductwork. All HEPA filters must be tested and certified annually. The HEPA filter housings should be designed to allow for in situ decontamination and validation of the filter prior to removal. HEPA filter housings should be constructed of stainless steel with gastight isolation dampers, decontamination ports, and the ability to scan each filter assembly for leaks.

Class III BSCs must be directly and independently exhausted through two HEPA filters arranged in series. Supply air must be provided in such a manner that prevents positive pressurization of the cabinet. It is recommended that supply air and exhaust operation be interlocked to prevent overpressurization of the cabinet in the event of multiple fan failures.

Pass-through dunk tanks, fumigation chambers, or equivalent decontamination methods must be provided so that materials and equipment that cannot be decontaminated in the autoclave can be safely removed from the cabinet laboratory and passed into the lower classified room housing the cabinet line.

Autoclaves that open outside of the laboratory must be sealed to the primary wall using a bioseal gasket that is durable and airtight. The autoclave doors must be interlocked so that only one can be opened at any time and must be automatically controlled so that the outside door to the autoclave can only be opened after the decontamination cycle has been completed.

The entire cabinet line and the supporting facility must be tested to verify that the design and operational parameters have been met prior to operation. Facilities must also be reverified annually.

Suit Laboratories

A BSL-4 suit laboratory is housed in a separate building or in a clearly demarcated and isolated environmental zone within a building. Suit laboratories are typically accessed by passing through a series of interlocking spaces entered through APR boundary doors with security locks meeting the institutional policies supporting biosecurity protocol. The laboratory rooms in the facility must be arranged to ensure exit by sequential passage through a chemical shower, an inner (dirty) change room, a personal shower, and then an outer (clean) changing area (see Figure 16-7). The BSL-4 suit laboratory uses structural containment to provide a static barrier by being constructed of materials that have a low permeability by design and are provided with internal surface coatings that provide the required airtightness and gastightness.

Entry into a BSL-4 laboratory must be through a defined entrance air lock fitted with APR doors (also called *bioseal doors*; see Figure 16-8) and a decontamination shower for cleaning of a suit prior to removal and reuse. There are two types of APR doors: 1) a positive-seal gasketed door that has a turning wheel and lugs to mechanically clamp the door shut and 2) a door with inflatable gaskets that seal the air lock gastight when the door is closed.

Personnel who enter the BSL-4 laboratory area must wear a positive pressure suit furnished with HEPA-filtered breathing air (see Figure 16-9). The breathing air systems must have redundant air compressors, failure alarms, and emergency backup. Because the laboratory areas are fitted with APR doors and the spaces are sealed tight, laboratory pressurization and directional airflow must be considered in conjunction with the HVAC control systems design (i.e., active differential pressure control versus cascade air movement).

A chemical shower that provides full-body coverage spray capability with inflatable gasket seals and double interlocking doors controlled to prevent simultaneous opening must be provided to decontaminate the surface of the positive pressure suit before the laboratory technician enters or leaves the laboratory. See Figure 16-10 for a diagram of a

Figure 16-7
Layers of Containment in a BSL-4 Suit Laboratory
(NIAID n.d.)

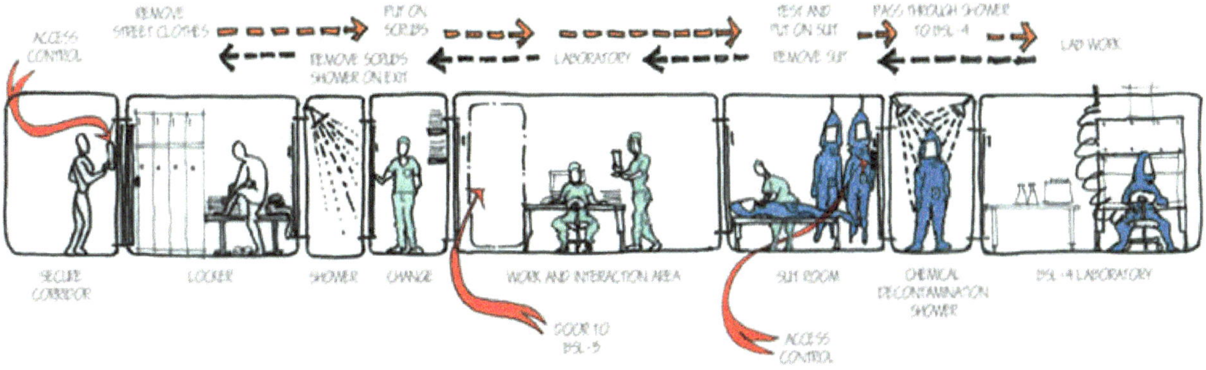

Figure 16-8
Inflatable-Gasket APR Door System
(Reprinted with permission from Applied Biosafety *[de Kok-Mercado, et al. 2011])*

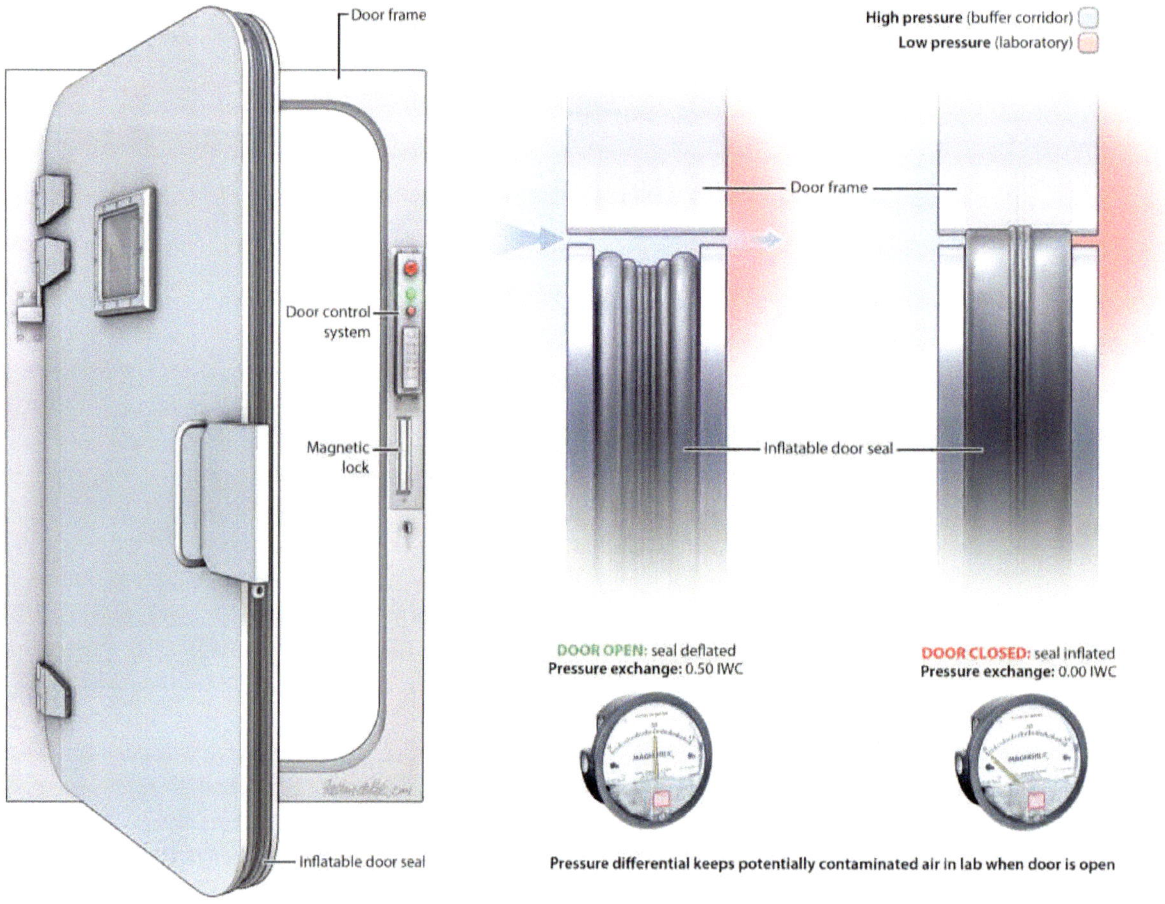

typical chemical shower. A method for decontaminating positive pressure suits as well as the shower room and/or decontamination area itself, such as a gravity-fed supply of chemical disinfectant or a hand-spray mechanism, is required in the event of an emergency exit or failure of the chemical shower system.

The facility must be fitted with an automatically activated emergency power source to ensure full operation of the laboratory exhaust system, life support systems, alarms, lighting, entry and exit controls, BSCs, and door gaskets. Monitoring and control systems for each support system such as air supply, exhaust, life support, alarms, entry and exit, and security systems must be on a UPS, to eliminate transient/transfer upset, and an emergency generator in case of prolonged outages.

Double-door autoclaves, dunk tanks, or fumigation chambers must be provided at the containment barrier of a BSL-4 laboratory for the passage of materials, supplies, or equipment. The walls, floors, and ceilings of the laboratory must be constructed to form a sealed internal shell to facilitate fumigation and prohibit animal and insect intrusion. All penetrations in the internal shell of the laboratory, suit storage room, and inner change room must be sealed. All drains in the laboratory must be connected directly to the liquid waste decontamination system. Atmospheric sewer vents and other service lines must be protected by two HEPA filters, in series up to the second filter, and have protection against cabinet moisture accumulation and degradation and insect and animal intrusion.

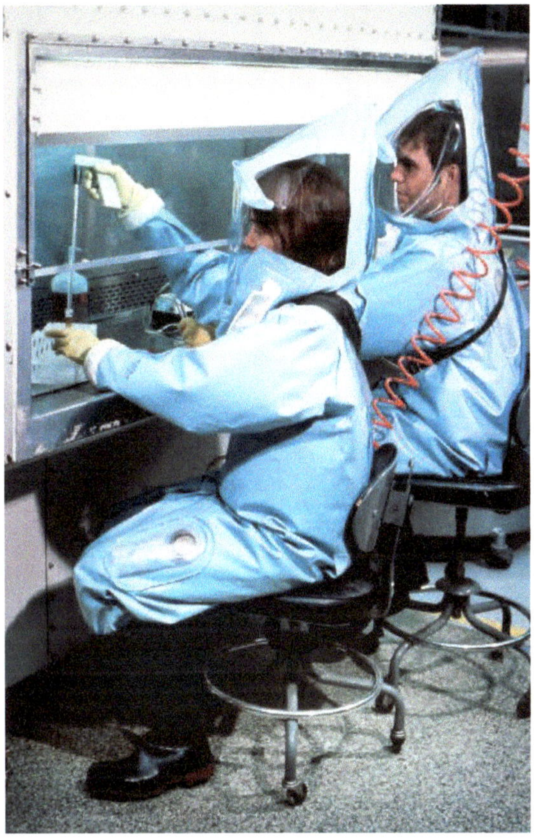

Figure 16-9
Laboratory Technicians Working in a BSL-4 Suit Laboratory

(CDC 2014)

Services that penetrate the laboratory walls, floors, and ceiling, such as plumbing and gas services, must provide assurance that no backflow from the laboratory occurs by fitting the penetrations with two backflow prevention devices in series.

Decontamination of the entire laboratory must be performed using a validated gaseous or vapor method when there have been significant changes in laboratory usage, before major renovations or maintenance shutdowns, and in other situations, as determined by a risk assessment. Consideration of the gaseous or vapor decontamination purging is required in HVAC sizing, equipment selection, and operational controls.

BSCs and other primary containment barrier systems must be installed so that fluctuations of the room air supply and exhaust do not interfere with the laboratory pressure stability or proper operation of the BSCs.

A central vacuum system is not recommended, but if provided it must not serve areas outside the BSL-4 laboratory. Two in-line HEPA filters must be placed near each use point. Filters must be installed to allow in-place decontamination and replacement.

The BSL-4 suit laboratory ventilation system must be designed to provide for a dynamic air barrier in addition to the structural static barrier envelope. This system must be dedicated and nonrecirculating. Only laboratories with the same HVAC requirements (i.e., other BSL-4 and ABSL-4 laboratories) may share ventilation systems, and only if each individual laboratory system is isolated by gastight dampers and HEPA filters. Separate supply and exhaust components of the ventilation system must be designed to maintain the laboratory at negative pressure in relation to surrounding areas and to provide differential pressure/directional airflow to ensure airflow toward areas of the highest potential risk within the laboratory. Differential pressure conditions are typically defined across internal areas.

Figure 16-10
Typical Chemical Shower

(Courtesy of ccrd Partners, Consulting Engineers)

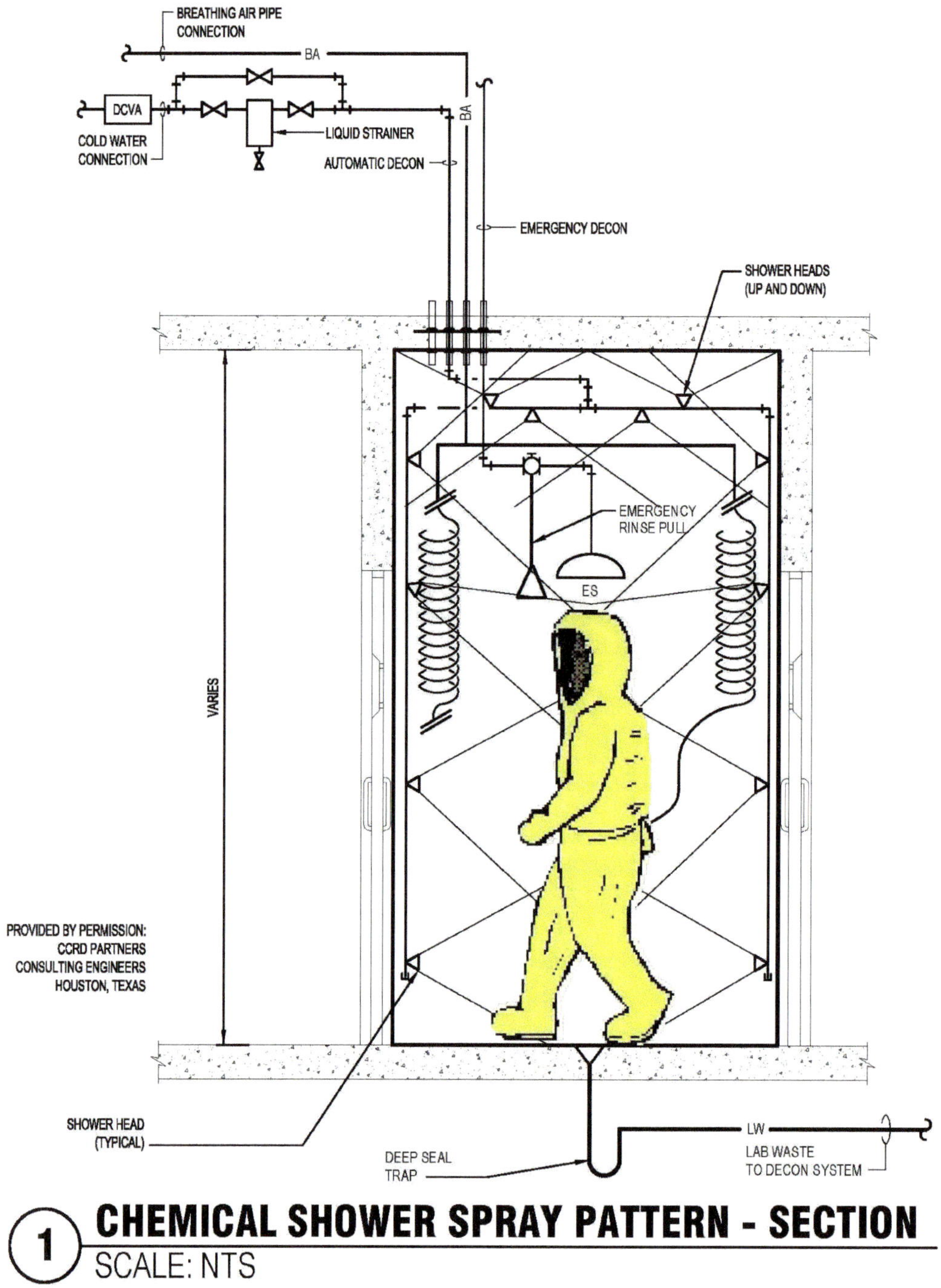

For suit laboratories, *N*+1 redundant supply fans are recommended and *N*+1 redundant exhaust fans are required. Supply and exhaust fans must be interlocked to prevent positive pressurization of the laboratory due to exhaust fan failure. The ventilation system must be monitored and alarmed to indicate malfunction or deviation from design parameters. Visual monitoring devices incorporating a magnetically coupled indicating mechanism must be installed near the clean change room and doors so that proper differential pressures within the laboratory may be verified.

All supply air to the laboratory and the decontamination shower must pass through a single HEPA filter. All exhaust air from the suit laboratory, decontamination shower, and

Figure 16-11
Double HEPA Laboratory Exhaust Filtration System

(Reprinted with permission from Applied Biosafety *[de Kok-Mercado, et al. 2011])*

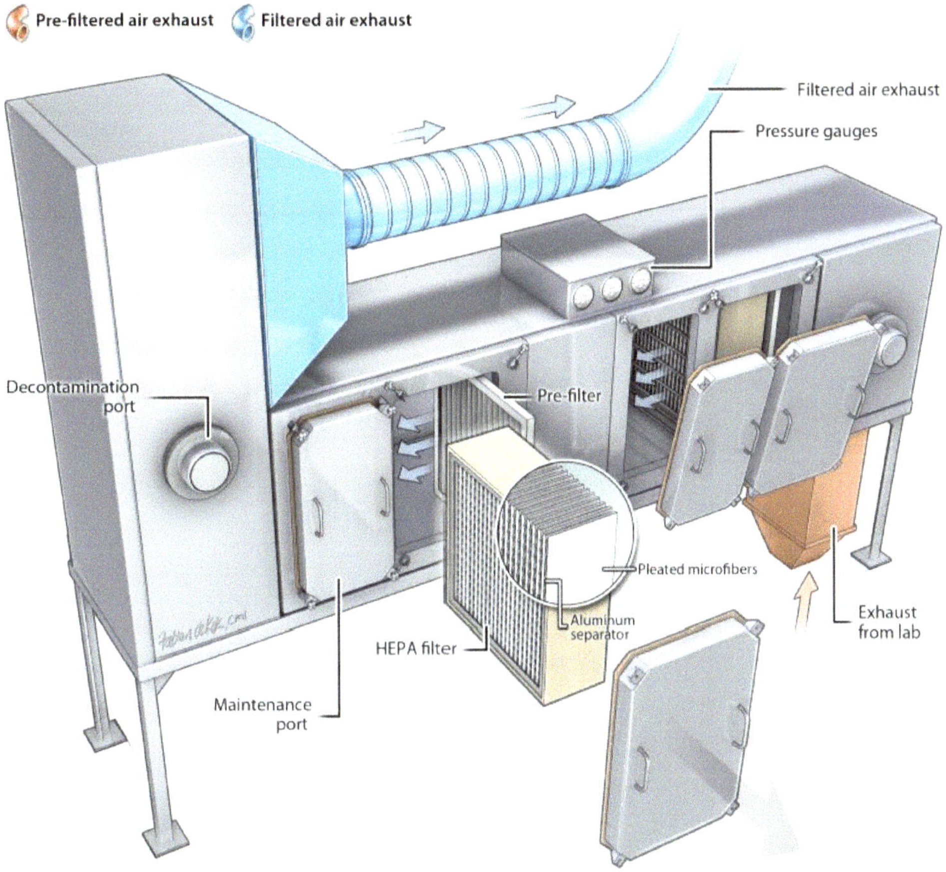

fumigation or decontamination chambers must pass through two HEPA filters, in series, before discharge to the outdoors (see Figure 16-11). Prefilters of the same standard for both supply and exhaust HEPA filters are recommended to prolong the life of the main filter set. All exhaust air discharge must be located away from occupied spaces and outdoor air intakes. All HEPA filters must be located as near as practicable to the laboratory in order to minimize the length of potentially contaminated ductwork. All HEPA filters must be tested and certified annually. Provisions should be considered for decontamination of lengths of potentially contaminated duct. The HEPA filter housings should be designed to allow for in situ decontamination and validation of the filter efficacy prior to removal. The design of the HEPA filter housing must have gastight isolation dampers, decontamination ports, and the ability to scan each filter assembly for leaks. Structural barrier air leakage testing is to be undertaken to identify leakage rates and source areas. Integrity testing can be accomplished using a pressure decay process in an equilibrium pressure/flow test.

Class II BSCs can be connected to the laboratory exhaust system by either a thimble (canopy) connection or a direct (rigid duct) connection. Provisions for proper cabinet performance and air system operation must be verified and coordinated with the overall BSL-4 laboratory design and functional analysis.

Pass-through dunk tanks, fumigation chambers, or equivalent decontamination methods must be provided so that materials and equipment that cannot be decontaminated in the autoclave can be safely removed from the BSL-4 suit laboratory.

Liquid effluents from chemical showers, sinks, floor drains, autoclave chambers, and other sources within the laboratory must be decontaminated by a proven method, preferably heat treatment, before being discharged to the sanitary sewer system. Pressure effects of the ambient pressure and HVAC system pressure controls must be considered in the design of the room pressurization system and impact to liquid level maintenance for P-trap depths. Deep-seal P-traps are required in all instances due to the relative pressures encountered.

A double-door pass-through autoclave must be provided for decontaminating materials passing out of the laboratory. Autoclaves that open outside of the laboratory must be sealed to the primary wall and provided with a sealed control panel and frame bioseal that must be durable and airtight. The autoclave doors must be interlocked so that only one can be opened at any time and must be automatically controlled so that the outside door to the autoclave can only be opened after the decontamination cycle has been completed. Gas and liquid discharge from the autoclave chamber must be decontaminated via a closed system. When feasible, autoclave decontamination processes should be designed so that overpressurization cannot release unfiltered air or steam exposed to infectious material to the environment.

CONTAINABLE SPACES

The USDA and NIH classify some spaces as *containable spaces* for laboratories working with high-consequence agricultural pathogens. According to USDA Manual 242.1, *ARS Facilities Design Standards*:

> In certain facilities, it may be desirable for some spaces surrounding the containment area to act as tertiary barriers. Examples could be: mechanical and utility spaces; interstitial spaces housing ventilation ductwork and utility piping; and attics and double-walled construction surrounding the primary containment zone. No research work or housing of animals takes place in these areas, so they would not be expected to be contaminated. These areas are not considered containment spaces but, if ventilated, are referred to as "containable" spaces. These areas are kept under negative pressure and their exhaust systems are equipped with HEPA filters. Penetrations into these areas were sealed at the time of construction to allow decontamination, but these areas are not required to pass a pressure decay test. Persons leaving these areas are not usually required to shower before leaving the facility. (USDA 2012, p. 225)

The spaces that should be made containable and their required features should be coordinated with the appropriate biosafety office and can include sealing of all penetrations to allow for gaseous decontamination, negative pressurization, exhaust HEPA filters, showers, and change rooms. It may be appropriate to include some of these features for certain spaces that do not need to be classified as containable.

USERS' PROGRAM REQUIREMENTS

To provide the proper HVAC systems for a laboratory, it is critical to understand the use of the building and the needs of the users. Extensive discussions with research or diagnostic program and biosafety personnel are needed to identify their processes, program requirements, the nature of the research, and protocols. Some of the topics to discuss are the following:

- What are the flow diagrams for materials, personnel, live animals, carcasses, animal bedding, samples, laundry, feed, waste, etc.? (This is important because the designer matches the HVAC and plumbing requirements to the biocontainment needs. An example flow diagram is shown in Figure 16-12.)

Figure 16-12
Animal Facility Diagrammatic Model Showing Basic Flows and Relationships

(Reprinted with permission from National Institute of Building Sciences, Whole Building Design Guide *[Stark et al. 2010])*

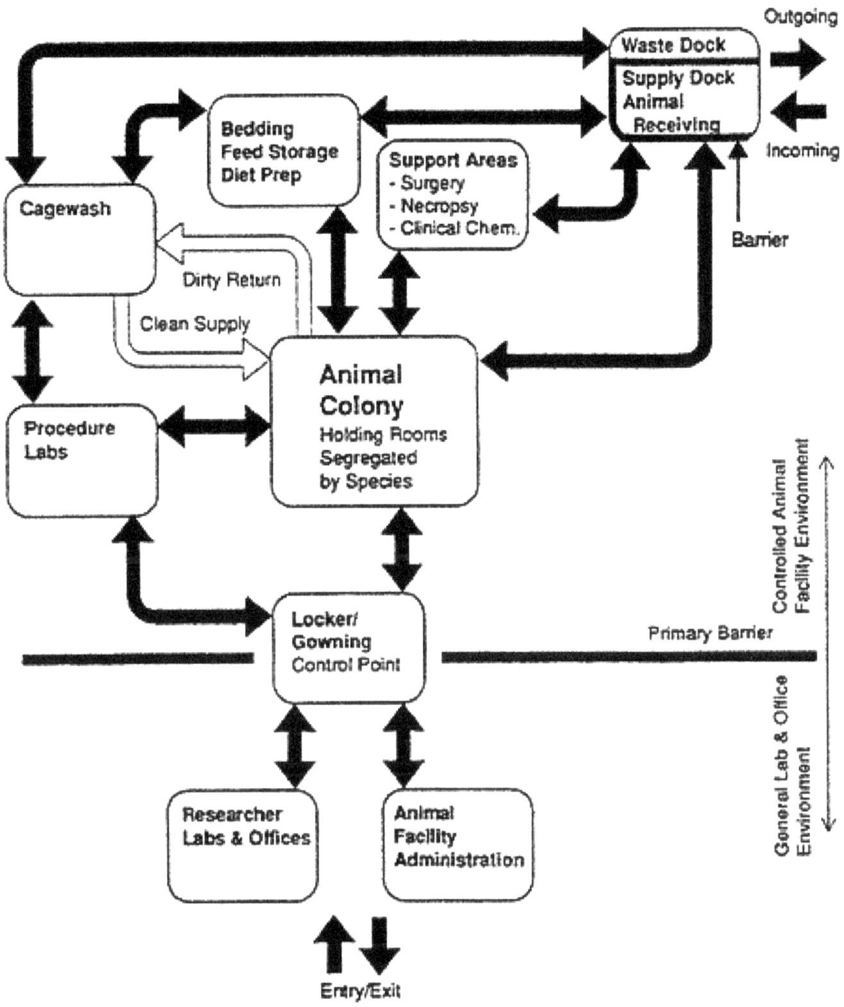

- What are the pressure relationships between the rooms? Are there any relationships that are especially critical and require air locks or other features?
- What organisms are involved and what is the nature of containment? What is the nature of potential accidents so that plans can be put in place to help protect against accidental release of contaminants?
- Are there cross-contamination concerns with animals in adjacent rooms?
- What is the method of decontaminating the rooms? Are there any concerns with compatibility of materials or equipment? Are decontamination ports into the room or ductwork needed? Will the building exhaust be used to evacuate the room after the decontamination process?
- What certification requirements will the occupants need to comply with? What standards or regulations do they need to conform to?
- What is the level of redundancy and emergency power required?
- What is the length of process or experiment in the room? Can the system be shut down for replacing filters between experiments or do redundant filters need to be provided?

The answers to the above questions provide the background to understanding the users' requirements. Room data sheets are also used to gather the technical requirements of each space. Sample room data sheets for various types of laboratories are available with this Guide at www.ashrae.org/LabDG; these sheets are from *NIH Design Policy and Guidelines* (NIH 1999), and updated sheets are anticipated to be available with the forthcoming *Design Requirements Manual* from NIH (n.d.).

ENGINEERING CONSIDERATIONS

SYSTEMS OVERVIEW

The primary functions of the HVAC systems in laboratories are to provide protection to the people working within the laboratory and the surrounding community and to prevent cross-contamination. To support this effort, the HVAC systems supply clean ventilation air, remove thermal loads, control the humidity, and remove odors and contaminants. The engineering design should be reviewed with the facility O&M personnel, as well as the biosafety officer, to verify that it will meet the operational objectives and comply with the institution's standard operating procedures (SOPs).

VENTILATION RATES

The minimum laboratory ventilation rate depends on the specific laboratory requirements. Chapter 16 of *ASHRAE Handbook—HVAC Applications* states that "fixed minimum airflow rates of 4 to 12 air changes per hour (ach) when the space is occupied have been used in the past," but "minimum ventilation rates at the lower end of the 4 to 12 ach range may not be appropriate for all laboratories." The chapter also states that "minimum ventilation rates should be established on a room-by-room basis considering the hazard level of materials expected to be used in the room and the operation and procedures to be performed" (ASHRAE 2015, p. 16.8). Ventilation rates for animal rooms commonly range from 10 to 15 ach according to *Guide for the Care and Use of Laboratory Animals* (ILAR 2011), which provides additional guidance on animal-room ventilation.

During the review of the minimum ventilation rates for a specific project, all stakeholders should understand the limited intent of the general room ventilation as opposed to the primary containment devices such as fume hoods and BSCs. *Prudent Practices in the Laboratory: Handling and Management of Chemical Hazards* (NRC 2011) provides a clear intent for the minimum room ventilation rate:

> General laboratory ventilation. This system should: Provide a source of air for breathing and for input to local ventilation devices**; it should not be relied on for protection from toxic substances released into the laboratory**; ensure that laboratory air is continually replaced, preventing increase of air concentrations of toxic substances during the working day; direct air flow into the laboratory from non-laboratory areas and out to the exterior of the building. (NRC 2011, p. 297, emphasis added)

The actual ventilation rate will therefore need to be carefully evaluated, taking into account the energy impact of higher ventilation rates. The final selected ventilation rate and total supply air provided will be dependent on several factors, including the following:

- Local facility requirements or a risk assessment for minimum airflows.
- Heating and cooling loads including heat gain from laboratory equipment. Care should be taken here to allow for future scientific programmatic changes within a facility.
- Animal species and their populations, both from current design and future flexibility standpoints.
- Recommended or required ambient temperature and humidity levels.
- Animal microenvironment (individually ventilated cages versus static cages).

- Airborne contaminants.
- Total exhaust requirements—from equipment, such as BSCs and fume hoods, and general laboratory exhaust required to maintain containment.

Though some existing research facilities may still incorporate recirculation air with proper dilution and filtration, the National Research Council handbook *Biosafety in the Laboratory: Prudent Practices for Handling and Disposal of Infectious Materials* (NRC 1989) indicates that ventilation systems that serve "laboratory activities that pose the risk of infection via airborne aerosols or droplets" should "provide directional airflow from 'clean' to 'contaminated' areas, and the air should not be recirculated" (p. 20). In some instances it may be possible to use recirculated air if approved by the Authority Having Jurisdiction and the appropriate biosafety professional.

Separation of human occupancies and animal areas is required for human comfort, health, and safety. Physical separation, in addition to separation of ventilation and sanitation services, is required to avoid contamination of the human-occupied areas. In addition, certain animals themselves might have to be separated from other animals. Where select agents are involved (FSAP 2014a, 2014b), there are also requirements to provide separation for security and access control purposes.

Physical separation can be achieved by housing the laboratories and/or animal spaces in a separate building, wing, floor, or room. Good design should ensure that animal housing areas are located close to the laboratories but are separated by a barrier. The separation of animals from laboratories should be reinforced with proper differential pressure or directional airflow control. Depending on the animal species and agent used, the animal holding areas may need to be maintained at a lower pressure than the laboratory.

SPACE PRESSURIZATION AND AIRFLOW

Space pressurization in biocontainment laboratory design is a method of reducing infiltration and preventing the flow of air from less clean and contaminated areas to cleaner areas. The direction of the airflow can be monitored by pressure sensors networked to local or centralized alarms. This enables fast response to any loss of negative pressure. Space pressurization in low-hazard laboratories is usually accomplished by exhausting a different volume of air than is supplied (volumetric airflow tracking). If more air is exhausted than supplied, the space is negatively pressurized; if less air is exhausted than supplied, the space is positively pressurized. Another control method is the use of a variable-air-volume (VAV) system with a room static pressure sensor to maintain a constant pressure differential in the room/area relative to its surroundings (differential pressure sensing). Cascaded pressure control combines the two control methods, using airflow tracking to ensure fast and stable control response and room differential pressure sensing to adjust the airflow tracking offset to maintain the room differential pressure at the desired setpoint.

Room envelope integrity is a critical factor in controlling indoor/outdoor pressure differential and in controlling the sources of airflow into and out of the laboratory. Air should enter and leave through the designated HVAC systems rather than leaks in the envelope. Current architectural envelope designs and commissioning practices should be able to achieve a minimum safe pressure differential level of 0.02 up to 0.20 in. w.g. (5 up to 50 Pa) for an added margin of safety.

Airflow offset between supply and exhaust is needed to maintain space negative pressures. The quantity of air needed to create the pressure difference is directly related to the leakage area of the space. In spaces that are tightly constructed, such as BSL-3 and BSL-4 secondary containment barriers, this leakage area is extremely small or negligible. Because the leakage paths in such spaces are very limited, a small offset in airflow can cause significant pressure differentials. Volumetric airflow tracking is therefore ineffective as a space pressurization method in low-leakage containment spaces. Similarly, differential pressure control and cascade pressure control are unsuitable for these tightly sealed environments where momentary upset conditions can quickly affect the targeted pressure relationship. To maintain a setpoint pressure or pressure differential within the laboratory, it may be necessary implement a pressure control strategy that uses incremental damper and/or feedback controls. The terminal units that serve the supply and exhaust system need to have a suitable response time to maintain the required pressurization control. Where spaces are extremely tight, engineers may also need to consider application of trim volume control devices (i.e., one larger valve operates as constant flow while a trim valve modulates to maintain pressure difference). An alternative is to consider design of engineered transfers that allow larger differential between the supply airflow and exhaust airflow. These engineered transfers require HEPA filtration on BSL-3Ag and BSL-4 spaces where the room envelope is acting as the primary containment barrier.

To measure and control pressurization relationships, a common reference point is needed. The outdoor-air building pressure is commonly used to provide this common pressure reference point. One method of obtaining the building pressure is to use one or more wind-shielded sensors on top of a 20 ft (6 m) pole located on a building's roof. Another is to provide reference pressure ports manifolded to a common header to dampen the effect of wind on any one face of the building. Another method is to place a sensor at each floor level on each exterior-wall exposure (north, south, east, and west orientations). High-speed direct digital controllers enable the sensors to make real-time measurements.

SOLID WASTE DECONTAMINATION METHODS

The following solid waste decontamination procedures are among the methods used in laboratory and animal facilities. Discussion with the appropriate laboratory and biosafety personnel should be conducted to identify the appropriate method for a specific application.

- *Steam sterilization (moist heat).* Autoclaving (a procedure where pressurized steam is applied in a heated vessel for 30 min or longer) is used in many laboratories. Alternative methods are boiling a vessel for 30 min in water containing sodium bicarbonate or using a pressure cooker.
- *Dry heat.* Dry heat forced-air convection sterilizers are also now available that use hot air rather than steam. The sterilization process involves three steps: warm-up, soaking, and cooldown. This process significantly reduces the time required for sterilization and cooldown and eliminates hot steam release.
- *Dunk tanks and air locks/pass boxes.* Materials that cannot be autoclaved or exposed to high heat may be able to be decontaminated chemically in a dunk tank or gaseously in an air lock or pass box with hydrogen peroxide or other gaseous decontamination methods.
- *Incineration.* Waste can be burned instead of being sterilized. However, the waste must be contained in a leakproof container if incineration takes place outside the containment zone.
- *Alkaline hydrolysis.* This method is one method used for the disposal of animal carcasses and uses a pressurized vessel, heat, and sodium hydroxide or potassium hydroxide to sterilize and break down the carcasses.

Waste should be separated into containers for each category of waste: noncontaminated waste, sharps, contaminated material for autoclaving and reuse, contaminated material for disposal, and others. Ventilation needs in areas such as those for red bag storage, incineration, rendering, and liquid effluent decontamination should be considered, as these areas may be exposed to contaminants in the event of an incident. These areas may require washdown provisions, closure and isolation provisions, etc. In addition, ventilation rates should be sufficient to ensure adequate purge of contaminants and odors.

SYSTEM REDUNDANCY MEASURES

System redundancy measures used in microbiological and biomedical laboratories usually depend on how critical the continuous operation of the HVAC system is for the laboratory to function safely. They also depend on how extensive the economic losses will be if the HVAC system or one of its components fails or is shut down for maintenance. Although the implementation of these measures, which includes backup components and emergency power, will decrease the chance of system failure, the possibility cannot be completely eliminated. Because biocontainment facilities rely on directional airflow as one of the primary engineering controls for maintaining containment, exhaust system failure will cause significant negative effects on containment. This may result in contaminants flowing from negative-pressure containment areas to cleaner adjacent areas as well as in allowing contamination within a space to build up beyond permissible thresholds to potentially unsafe levels. Failure of supply air systems, cooling equipment, and heating equipment can be critical if it lasts for an extended period of time. This can result in unacceptably high temperatures and humidity. For example, in animal facilities, loss of supply air systems and temperature controls can result in loss of animals and research. Such catastrophic events should be evaluated in the risk assessment process.

Emergency Power

Emergency power equipment is designed to keep the laboratory systems from shutting down when the normal electrical power supply fails. The emergency power system can be designed to supply enough energy to keep all systems running or to supply only enough power to keep vital systems and components running. The highest priority should be given to life safety systems and the continued operation of the ventilation system to maintain directional airflow and containment. Emergency power must be provided to all high-hazard areas to maintain continuous operation of exhaust fans and life-support equipment, including breathing-air compressors, monitoring and alarm systems, lighting, entry and exit controls, communication systems, and BSCs. It should be noted that systems will shut down briefly then restart on emergency power. Provision of UPSs for all controls and control devices has proved extremely useful in minimizing the restart time of mechanical systems after a power outage.

Multiple or Redundant Fan Systems

Failure or shutdown of the exhaust system in dirty and contaminated areas will likely cause the most problems or health hazards. Without the exhaust fans, the negative pressure in the dirty and contaminated areas will dissipate. This can cause odors and pathogens to infiltrate the cleaner areas and thereby create potentially unsafe environments. Infiltration from the contaminated areas will begin immediately after the exhaust system fails or is shut down. Installation of backup fans and appropriate controls in addition to emergency power generation equipment and on-site fuel storage will provide continuous operation of the exhaust system in most circumstances.

Some national codes such as *2015 International Mechanical Code®* (*IMC®*) (ICC 2015) may require the installation of redundant exhaust fans where subducts replace fail-closed fire/smoke dampers. Elimination of fail-closed dampers is recommended to remove the potential for loss of airflow from a laboratory due to a damper accidentally closing.

AIR DISTRIBUTION

The primary functions of the air distribution system are to distribute the supply air to all areas of the room without causing draft or temperature discomfort and to aid in the adequate capture and purging of light aromatic components, such as ammonia, resulting from urine and of heavy components resulting from solids. For rooms with open cages, conventional air distribution devices will not distribute the air equally to the cages due to the arrangement of the cages relative to each other.

Some items to consider when determining the air distribution strategy and device selections are as follows:

- Air distribution for animal comfort and removal of contaminants
- Confirming that the air distribution does not interfere with the operation of a nearby BSC or fume hood
- Not interfering with the directional airflow, such as blowing air toward a door

For more information on air distribution refer to Chapters 6 and 17.

OPERATIONAL CONSIDERATIONS

Proper O&M of the laboratory systems facilitate a longer service life, reduce the number of emergency situations, and enable the systems to work according to the owner's requirements. Topics of special importance to the O&M of HVAC systems serving laboratory areas include monitoring and alarm requirements, component services, penetration seals, control systems, and documentation.

Monitoring and Alarm Requirements

A system failure can cause a health hazard for the occupants and/or loss of containment. Therefore, there is a need for monitoring and alarming of critical equipment such as fans, and there should be a plan for how to handle the situation if an emergency occurs.

Component Services

Where possible, install devices on the clean side of the HEPA filters. All components that are likely to be contaminated should be installed for easy access and safe handling. Proper clearance and ease of access to components requiring servicing is essential.

Filters likely to be contaminated should be installed in such a way that they can be removed without exposure to laboratory or maintenance personnel. Consult with the facility owner regarding whether he/she prefers the use of bag-in/bag-out filters or shutting down the system and decontaminating the filters and housing before removal. In the case of prefilters, it may be advantageous to install the prefilters within containment areas so that the HVAC system does not need to be shut down for frequent removal and replacement.

Penetration Seals

Ductwork and HVAC components that will be contaminated should be properly sealed to avoid leakage and contamination of areas through which the ductwork passes and contamination of the supply air. The contaminated parts of the HVAC system should be kept under negative pressure, and the clean parts should be kept under positive pressure. After cleaning and maintenance, it is important to make sure the access panels and other openings or connections are properly sealed. All electrical conduits and plumbing penetrations should be sealed to prevent room-to-room cross-contamination.

Control Systems

Control systems play a very crucial part in the maintenance of proper equipment operation, containment, monitoring, and alarms. Control components should be robust, installed in accessible locations, and checked for calibration and operation periodically. Control system design intent should be documented and confirmed during commissioning. The control system documentation should include sequences for normal and abnormal modes of operations. Each failure mode should be documented for performance impact to system stability.

In addition to installing equipment that can handle abnormal operating conditions, laboratory control systems must be able to handle all potential conditions that are likely to occur. The systems must be tested in all modes of operation to ensure that the desired capabilities function properly. See Chapter 11 for details on laboratory control systems.

Documentation

Proper documentation for the design, construction, and commissioning of the laboratory HVAC system should be maintained for the owner and operating personnel. The documentation includes how the systems are intended to function as well as the design parameters. Proper updating of the construction documents is important so that the facility operators have accurate as-built drawings. The completed commissioning report and testing and balancing report are very helpful to confirm that the system was operating correctly when the project was completed, and they also serve as a baseline for future performance comparisons.

SPECIAL ANIMAL CONSIDERATIONS

OPERATIONAL AND SAFETY PROTOCOLS

The spread of pathogens and zoonotic agents must be controlled to avoid infecting disease-free animals and humans. Depending on the pathogen, disease can spread in different ways, such as through the air, direct contact, indirect contact, intermediate hosts, placental transmission, or a combination of these methods. The sensitivities of a pathogen to temperature, pressure, pH, and chemicals must be determined for inactivating the pathogen in rooms and on equipment that have been contaminated. Pathogens are less sensitive to inactivating methods as their associated risks of spread increase. This can result in an increased severity of the disease and thereby increased stringency of precautions. Special stringent precautions must be taken if a pathogen can cause health risks for humans. These are primarily accomplished through the use of operational and safety protocols and need to be coordinated with the design intent.

Understanding these protocols is very important to providing the proper infrastructure. Some examples of aspects to consider include the following:

- Decontamination of the contaminated materials (cages, carcasses, clothing, trash, samples, equipment, electronics) and their removal from the containment area (dunk tanks, pass-through boxes, autoclaves, interlocking doors with interlocking air supply/exhaust, etc.)
- Entry of laboratory personnel, animals, and materials into the space and how this affects pressurization relationships

- Liquid waste treatment
- Provisions for room decontamination (the method of decontamination and the required features, such as ports located outside the containment area or in the ductwork)
- Required pressure relationships between the spaces and the potential need for air locks or pressurized vestibules
- Washdown requirements for animal holding areas
- Showering requirements
- Maintenance provisions for critical equipment (access from within or outside the containment area)
- Type(s) of BSCs required

VENTILATED CAGE RACK SYSTEMS

For small animals, filtered cage systems may be used for housing. The systems may use a forced ventilation system to provide sufficient air to maintain contamination thresholds and maintain temperature control. Typically, ventilated racks are provided with multiple closed cages that are provided with supply and exhaust air connections from manifolds that are integral to the rack.

There are four common methods used for cage rack ventilation (see Figure 16-13):

- Circulating air with rack-mounted supply and exhaust fans. Supply air is prefiltered and passes through final HEPA filters. Exhaust air is also HEPA filtered to remove contaminants prior to discharging clean air to the animal holding room.
- Circulating air with rack-mounted supply and exhaust fans, but rather than discharging to the room, the exhaust is connected to the building exhaust system via a thimble connection or capture hood.
- Supplying conditioned air from the room by rack-mounted supply fans with contaminated air discharged from the racks by rack exhaust fans connected directly to the building exhaust system.
- Rack-mounted supply and exhaust fans connected directly to the building system.

Figure 16-13 A Typical Cage Rack System for Mice

(Wilkins and Waters 2004)

The diffusers and exhaust openings used with these systems should be located and constructed so that they prevent the creation drafts but provide for a uniform low-velocity ventilation throughout the occupied spaces. In addition, they should be easy to clean. Care should be taken to adequately screen the room and outside openings of the ventilation system to prevent the escape of animals as well as the entry of vermin. The screens should be easily removable and cleaned periodically to avoid restricting airflow. Screen sieve size, location, and access for service depend on room configuration, air distribution, exhaust layout, and animal species. For instance, an insect space requires a different type of screen size, attachment, and placement than, say, the screens serving small-animal holding-room ventilation systems.

It is important to understand that increasing the air circulation in an animal room does not guarantee that the animal cages receive more air. However, animal cages that have only the top open usually have an increased amount of air circulation when there is an increased amount of air supplied to the room.

Cage racks have grown in acceptance since they were first introduced in the 1970s because they allow higher population densities to be housed in animal holding rooms. As animal populations in laboratories have grown, so has the need to provide a safe and comfortable environment for both the laboratory workers and the animals. Ventilation systems that deliver filtered, conditioned air to the cages and remove cage exhaust air from the holding rooms at a low noise level are critical in maintaining safe and comfortable environments in laboratories.

The following requirements should be considered when designing a ventilation system for cage racks:

- Supply and exhaust airflow for each rack must be supplied at a constant volume and precisely controlled.
- Supply air to and exhaust air form each rack must be HEPA filtered.
- Supply airflow should be on the order of 50 ach to each cage. Air changes greater than this may cause excessive noise that can affect the animals.

In an animal holding room, the room air is typically close to the temperature and humidity levels required for cages. Using the laboratory's central air-handling system as the source of supply air to the ventilated racks requires additional redundancy and environmental controls, which can increase the installed costs. High-performance, pressure-independent airflow controls are typically more effective than dampers and flow- or pressure-measuring devices for connecting racks to the building system.

ENVIRONMENTAL CONDITIONS

Maintenance of operating systems for animal housing can be of significant operating cost; hence, prudent engineering design should address energy efficiency either by consideration of energy avoidance or by way of energy recovery. Regardless of the method chosen, ethical treatment of the research animals is of paramount concern, and less-than-desirable conditions should be avoided and do not represent best practices.

Each species of animal has a unique range of preferred environmental conditions that must be present for the animal to survive and reproduce. Even within a specific species, there is variability among subclasses. For example, a hairless mouse requires a warmer temperature than a mouse with hair for optimal conditions. Some animals, such as squirrels and cottontail rabbits, can survive in captivity but will not reproduce.

Minimum space requirements are generally dependent on the size of the animal. The HVAC system is responsible for maintaining the temperature and humidity in the animal room (macroenvironment) and animal cage (microenvironment). The specific temperature and humidity to be maintained are dependent upon the animal's metabolism and insulation level. Changes in temperature and humidity affect the amount of food consumed and thus the weight of the animal. For these reasons, holding rooms in animal facilities should be designed for flexibility to accommodate a range of species; they should have the ability to maintain different temperatures and relative humidities to support the animal species anticipated to be housed. There are associated capital costs, operational costs, and space requirements associated with designing a flexible system; these requirements should be coordinated with each individual facility owner and laboratory personnel, and the limitations should be clearly defined.

Noise and vibration resulting from an HVAC system can have a direct impact on an animal's response to an experiment. This is because while some animals are more visually dependent, others animals rely heavily on sound for courtship, mating behavior, and predator detection. Feeding and cleaning operations, heating and ventilation equipment, and the animals themselves are sources of noise in an animal laboratory. Extremely high noise levels can cause temporary and permanent hearing loss and stress reactions for both the animals and their handlers. Noise levels should be verified for specific species prior to design of the animal laboratory.

Lighting levels and diurnal or daily cycles preferred by an animal species depend on their visual capability and circadian rhythm (*circadian* describes a periodicity that is close to one day). The level of lighting should be kept similar to that of the animal's natural environment. Inappropriate light levels can result in damage to the animals' vision and cause them undue stress. Cycling the lighting on and off influences when the animal is active. Many animals are active and reproduce at night.

In addition to the environmental conditions that are controlled by HVAC and electrical systems (temperature, humidity, acoustics, and light), there are other conditions such as space requirements, diet, social interaction, and reproduction that can be related to the environmental conditions and must be accounted for. Chapter 3 of *Guide for the Care and Use of Laboratory Animals* (ILAR 2011) provides detailed guidance on the appropriate environment, housing, and management for typical laboratory animals used or produced for research, testing, and teaching. In general, an animal does better when the environmental conditions in the laboratory approximate its natural habitat.

SANITATION AND CLEANABILITY

Sanitation and the cleanability of animal areas are the most significant factors that determine the local environmental conditions for the animals. Poor sanitary conditions can cause toxic buildup and unacceptable odor levels that result in both animal and human health problems. The laboratory ventilation system cannot compensate for poor sanitation procedures.

Design issues that are critical to maintaining clean animal laboratory and animal holding areas include the following:

- Adequate drainage (if required)
- Floors that are moisture-resistant, nonabsorbent, impact-resistant, chemically resistant, and relatively smooth (unless a textured surface is required)
- Washable paint on ceilings and walls
- Walls that can be cleaned with chemicals and water under high pressure (typically for large-animal holding areas)
- Walls, ceilings, floors, doors, windows, and the joints between these that are free of cracks and crevices

Design features that are critical for maintaining clean animal laboratory and animal holding area HVAC components include the following:

- All penetrations (including edges of diffusers and grilles) must be sealed.
- Aluminum or stainless steel air devices should be used.
- Piping, wires, and ductwork should not be exposed unless they are easily cleanable and are provided with standoff mounting that affords easy cleaning completely around the device or assembly.
- All materials used should have smooth, nonporous surfaces that are readily cleanable.

For some animal rooms, the exhaust ducts of the HVAC system may become excessively dirty or contaminated by pathogens and, consequently, will require periodic internal cleaning. The ducts can be designed to accommodate cleaning by installing access panels or doors so that every section of the internal ductwork can be easily reached. For BSL-3Ag and ABSL-4 facilities, contaminated exhaust ductwork between the room and exhaust HEPA filter must be gastight. No access panels are allowed for BSL-3Ag and ABSL-4 laboratories, and all air balance ports should be closed in a positive manner to eliminate inadvertent access or opening.

Provisions can be made to decontaminate or clean closed duct systems with decontamination ports and condensate drainage outlets. Where access to the exhaust ducts is needed, the distance between the room grilles and the HEPA filter bioseal damper should be as short as possible and without excessive offsets. In ABSL-3 enhanced laboratories, where there is a need for extensive horizontal ductwork that has to be decontaminated, internal swabbing is more involved because of the disassembly of the duct and cut-in needs. Consideration should be given to identifying access points and possible use of flanged duct sections so that sections of the duct can be removed to facilitate access for cleaning.

REFERENCES

ABSA. n.d. Risk group classification for infectious agents. Mundelein, IL: American Biological Safety Association. www.absa.org/riskgroups.

ACME. 2003. Arthropod containment levels (ACLs). *Vector-Borne and Zoonotic Diseases*. American Committee of Medical Entomology, American Society of Tropical Medicine and Hygiene. http://online.liebertpub.com/doi/pdf/10.1089/153036603322163475.

AIHA. 2012. ANSI/AIHA/ASSE Z9.5-2012, *Laboratory Ventilation.* Falls Church, VA: American Industrial Hygiene Association.

ASHRAE. 2015. Chapter 16, Laboratories. In *ASHRAE handbook—HVAC applications.* Atlanta: ASHRAE.

ASSE. 2014. ANSI/ASSE Z9.14-2014, *Testing and performance-verification methodologies for ventilation systems for biosafety level 3 (BSL-3) and animal biosafety level 3 (ABSL-3) facilities.* Park Ridge, IL: American Society of Safety Engineers.

CDC. n.d. 4 Biosafety Lab Levels. Atlanta: Centers for Disease Control and Prevention. www.cdc.gov/phpr/documents/BSL_infographic_final.pdf.

CDC. n.d. Quick learn lesson: Recognizing the biosafety levels, p. 4. Atlanta: Centers for Disease Control and Prevention. www.cdc.gov/training/QuickLearns/biosafety.

CDC. 2009a. *Biosafety in Microbiological and Biomedical Laboratories* (*BMBL*), 5th Edition. DHHS Publication No. (CDC) 21-1112. Atlanta: Centers for Disease Control and Prevention, U.S. Department of Health and Human Services. www.cdc.gov/biosafety/publications/bmbl5.

CDC. 2009b. Primary containment for biohazards: Selection, installation and use of biological safety cabinets, Appendix A of *Biosafety in Microbiological and Biomedical Laboratories* (*BMBL*). Atlanta: Centers for Disease Control and Prevention. www.cdc.gov/biosafety/publications/bmbl5/BMBL5_appendixA.pdf.

CDC. 2014. Newsroom Image Library - CDC Leaders & Scientists. Atlanta: Centers for Disease Control and Prevention. www.cdc.gov/media/subtopic/library/scientists.htm.

de Kok-Mercado, F., F.M. Kutlak, and P.B. Jahrling. 2011. The NIAID integrated research facility at Fort Detrick. *Applied Biosafety: Journal of the American Biological Safety Association* 16(2). www.absa.org/abj/abj/111602Jahrling.pdf.

EU. 2000. Directive 2000/54/EC of the European Parliament and of the Council of 18 September 2000 on the protection of workers from risks related to exposure to biological agents at work. (Seventh individual directive within the meaning of Article 16(1) of Directive 89/391/EC.) *Official Journal of the European Communities* L262/21. European Union.

FSAP. 2014a. Regulations 7 CFR Part 331, 9 CFR Part 121, and 42 CFR Part 73. Washington, DC: Federal Select Agents Program, Centers for Disease Control and Prevention and USDA Animal and Plant Health Inspection Services. www.selectagents.gov/Regulations.html.

FSAP. 2014b. Select agents and toxins list. Washington, DC: Federal Select Agents Program, Centers for Disease Control and Prevention and USDA Animal and Plant Health Inspection Services. www.selectagents.gov/SelectAgentsandToxinsList.html.

Government of Canada. 2013. Chapter 4, Risk groups, containment levels, and risk assessments. In *Canadian biosafety standards and guidelines (CBSG).* Ottawa, Canada: Government of Canada. http://canadianbiosafetystandards.collaboration.gc.ca/cbsg-nldcb/index-eng.php?page=12.

IBC. 2010. NIH risk group classifications. Troy, NY: Institutional Biosafety Committee, Rensselaer Polytechnic Institute. https://www.rpi.edu/research/office/ibc/riskgroup classifications.html.

ICC. 2015. *2015 International mechanical code® (IMC®)*. Washington, DC: International Code Council.

ILAR. 2011. *Guide for the Care and Use of Laboratory Animals*, 8th ed. Washington, DC: Institute for Laboratory Animal Research, Division on Earth and Life Studies, National Research Council. http://grants.nih.gov/grants/olaw/Guide-for-the-care-and-Use-of-Laboratory-Animals.pdf

NIAID. n.d. Untitled image of BSL-4 suit laboratory room sequence. Bethesda, MD: National Institute of Allergy and Infectious Diseases, National Institutes of Health. www.niaid.nih.gov/SiteCollectionImages/about/organization/dir/BSL4labBig.jpg.

NIH. n.d. *Design Requirements Manual* (*DRM*). Bethesda, MD: National Institutes of Health, Division of Technical Resources. Forthcoming. For more information, visit www.orf.od.nih.gov/PoliciesAndGuidelines/BiomedicalandAnimalResearchFacilities DesignPoliciesandGuidelines/Pages/default.aspx.

NIH. 1999. *NIH Design Policy and Guidelines*. Bethesda, MD: National Institutes of Health, Division of Technical Resources.

NIH. 2013a. *NIH guidelines for research involving recombinant or synthetic nucleic acid molecules*. Bethesda, MD: National Institutes of Health, Office of Biotechnology Activities. http://osp.od.nih.gov/sites/default/files/NIH_Guidelines_0.pdf.

NIH. 2013b. *Design Requirements Manual*. Bethesda, MD: National Institutes of Health, Division of Technical Resources. www.orf.od.nih.gov/PoliciesAndGuidelines/

BiomedicalandAnimalResearchFacilitiesDesignPoliciesandGuidelines/Pages/DesignRequirementsManualPDF.aspx.

NRC. 1989. *Biosafety in the Laboratory: Prudent Practices for Handling and Disposal of Infectious Materials*. 1989. Washington, DC: National Research Council, National Academies Press. www.nap.edu/catalog/1197.html.

NRC. 2011. *Prudent Practices in the Laboratory: Handling and Management of Chemical Hazards*. 2011. Washington, DC: National Research Council. www.nap.edu/catalog.php?record_id=12654.

Olsen, C.W. 2000. Zoonotic diseases tutorial. Madison, WI: School of Veterinary Medicine, University of Wisconsin-Madison. www.vetmed.wisc.edu/pbs/zoonoses.

SA/SNZ. 2002. AS/NZS 2243.3:2002, *Australian/New Zealand Standard™: Safety in laboratories, Part 3: Microbiological aspects and containment facilities*. Sydney, Australia: Standards Australia International Ltd.

Stark, S., J. Petitto, and S. Darr. 2010. Animal research facility. *Whole Building Design Guide*. www.wbdg.org/design/animal_research.php.

U.S. Army. 2013. *Safety Standards for Microbiological and Biomedical Laboratories*, Army Pamphlet 385-69. Department of the Army. Washington, DC: Government Printing Office. www.apd.army.mil/pdffiles/p385_69.pdf.

USDA. 2012. Manual 242.1, *ARS Facilities Design Standards*. Washington, DC: United States Department of Agriculture. www.afm.ars.usda.gov/ppweb/pdf/242-01m.pdf.

WHO. 2004. *Laboratory biosafety manual*, 3d ed. Geneva: World Health Organization. www.who.int/csr/resources/publications/biosafety/en/Biosafety7.pdf. Available in languages other than English at www.who.int/csr/resources/publications/biosafety/WHO_CDS_CSR_LYO_2004_11/en/index.html.

Wilkins, C., and B. Waters. 2004. HVAC design in animal facilities. *ASHRAE Journal* 46(9):38.

BIBLIOGRAPHY

FASS. 2010. *Guide for the care and use of agricultural animals in research and teaching*, 3d ed. Champaign, IL: Federation of Animal Science Societies. www.fass.org/docs/agguide3rd/Ag_Guide_3rd_ed.pdf.

PHAC. 2015. Pathogen safety data sheets and risk assessment. Ottawa, Ontario: Public Health Agency of Canada. www.phac-aspc.gc.ca/lab-bio/res/psds-ftss/index-eng.php.

CFD Modeling of Laboratory Ventilation

17

OVERVIEW

The laboratory ventilation system is the primary means for the removal of airborne contaminants generated within a laboratory. The system's ability to remove the contaminants from the laboratory space is defined as the *ventilation effectiveness*. As designers consider reducing ventilation rates to make the laboratory more energy efficient, modeling the ventilation effectiveness becomes a critical design issue.

Previous chapters in this Guide provide information on best-practice approaches to laboratory ventilation. Specifically, Chapter 4 discusses minimum air change requirements and Chapter 12 describes airflow patterns within the laboratory. This chapter describes methods to model airflow patterns within a laboratory using computational fluid dynamics (CFD), which is relatively new to laboratory design but has the potential for refining exhaust and supply system design to improve safety and reduce energy costs. There are some guidelines on using CFD for laboratory ventilation given the current state of CFD development. The following concepts and applications are presented in this chapter from a designer's point of view:

- Uses of CFD in laboratories
- Introduction to CFD modeling
- Types of CFD models
- Typical stages in CFD computations
- Recommendations for conducting CFD modeling
- Interpreting CFD results

There are also several alternatives to CFD modeling. For an existing laboratory or a mock-up, ventilation effectiveness can be measured with tracer gas testing using ASHRAE Standard 129, *Measuring Air-Change Effectiveness* (ASHRAE 2002), instead of modeling. A modeling alternative is the multizone airflow model. Multizone models calculate airflow and contamination transport through a collection of rooms, hallways, and ducts. The airflow between zones is calculated from the pressure-flow characteristics of the path model and pressure differences across the paths. The multizone model considers mean airflow, temperature gradients, mechanical devices, and pressure losses. When additional detail within the zone is necessary or when contaminant generation is dependent on airflow, then a microscopic approach should be considered such as that provided by CFD.

USES OF CFD IN LABORATORIES

CFD can be used to predict several aspects of laboratory ventilation. But please note that not all laboratory rooms require CFD modeling.

DESIGN OF SUPPLY DIFFUSERS

Large exhaust flows from fume hoods may be concentrated into one part of a room. The supply air traveling toward the exhausts may create disruptive room currents with higher air speed than desired around the fume hoods. One modeling strategy in selecting a diffuser layout is to compare a proposed design to the "ideal" case of a uniform low-velocity supply that covers the entire exposed ceiling. An example is shown in Figures 17-1 and 17-2. Figure 17-1 shows the room layout. Figure 17-2 compares an ideal case to two diffuser configurations, illustrating the modeling results within the central portion of the room. The results shown, in plan view, are the velocity patterns at 4 ft (1.2 m) above the floor level. The ideal case (Figure 17-2a) shows relatively low velocities near all fume hood entrances at the top and bottom edges of the diagram and at the corridor near the top left corner. Figures 17-2 and 17-2b show the increased room velocities when real diffusers are used.

THERMAL COMFORT AND COOLING REQUIREMENTS

Thermal comfort is important for personnel productivity. Thermal comfort is a complicated function of air temperature, air velocities, radiant energy (from windows, walls,

Figure 17-1
Example Room Layout
(Courtesy of Rowan Williams Davies & Irwin Inc.)

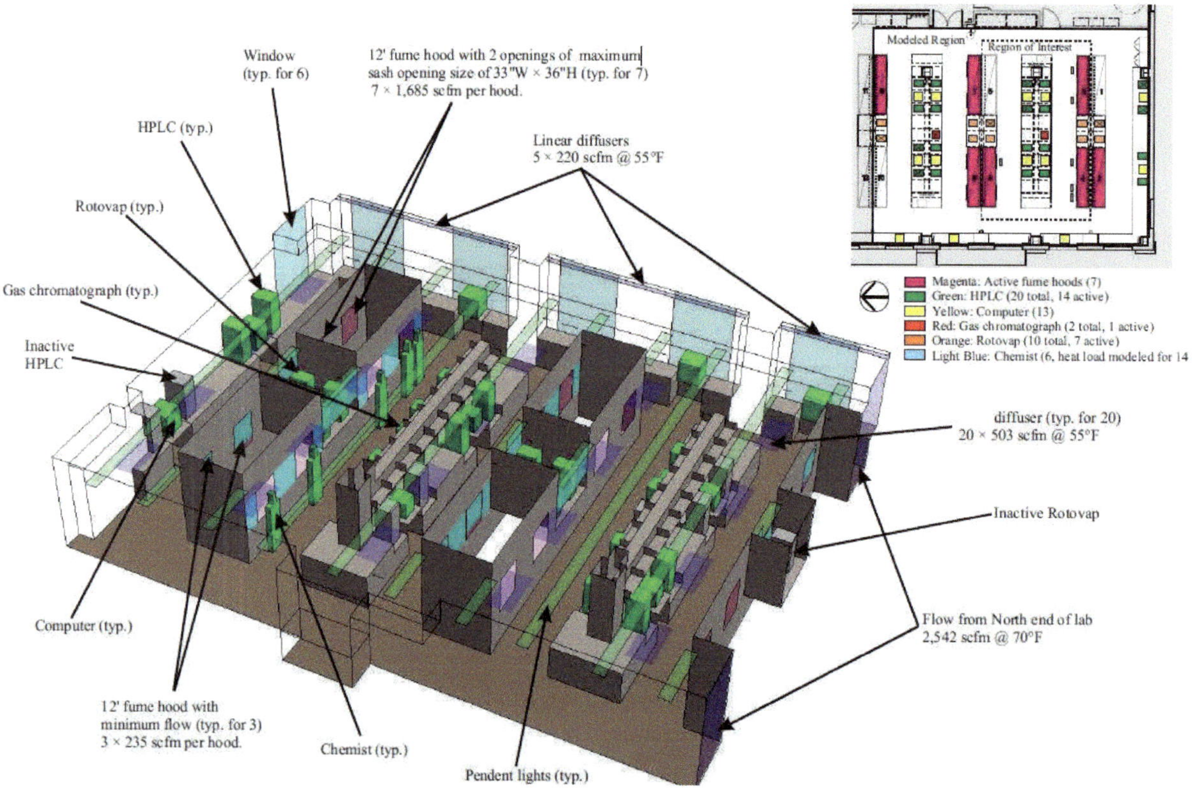

and equipment), personnel activity levels, and clothing levels. Equipment also requires cooling for proper operation. For personnel productivity, CFD modeling can provide temperature and air velocities to help determine human comfort levels (i.e., if adequate cooling is occurring or if there is too much cooling). For equipment, CFD modeling can establish temperatures in the room where the equipment is located to help determine if adequate cooling is occurring.

OTHER SAFETY AND COMFORT ISSUES

Other safety and comfort issues may be addressed by CFD predictions of velocities and chemical concentrations within the laboratory air space. Examples include the following:

- Excessive room air velocities at workbenches and other sensitive areas.
- The containment capability of fume hoods.
- Where contaminants from releases or spills may be transported within the room.

Figure 17-2
Comparison of Two Diffuser Configurations to Ideal Uniform Supply

(Courtesy of Rowan Williams Davies & Irwin Inc.)

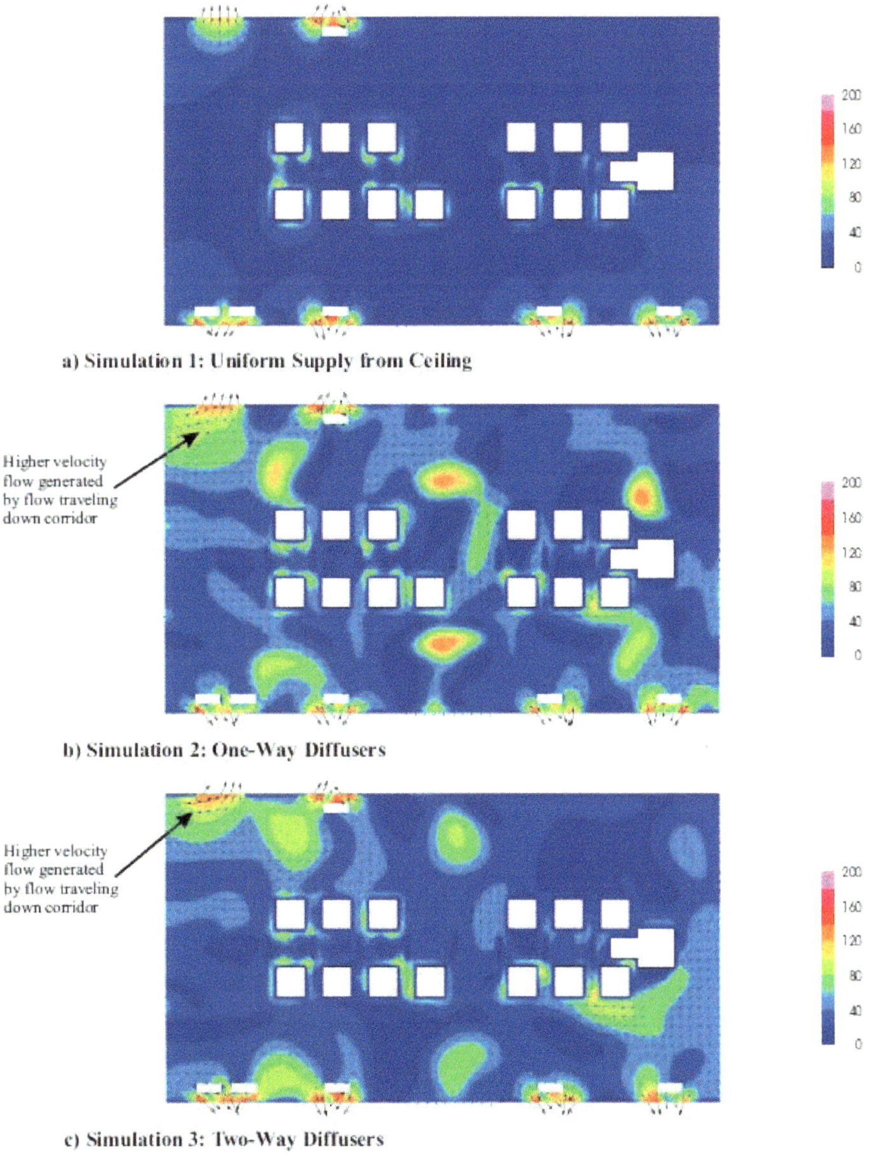

- Where lower explosion limits (LELs) may be exceeded within a room for various releases.
- How specialty hoods with unusual characteristics, including those with multiple openings, behave.

INTRODUCTION TO CFD MODELING

CFD is an intensive computer-based method for calculating detailed airflow patterns and contaminant concentrations within a room space. Room temperature, air speed and direction, humidity, static pressure, and contaminant concentration can be calculated over millions of locations within a room and possibly within associated supply and exhaust devices. The results can then be presented graphically in many ways. Figure 17-3 illustrates a cross-section view of velocity near two diffuser outlets. The figure shows only a

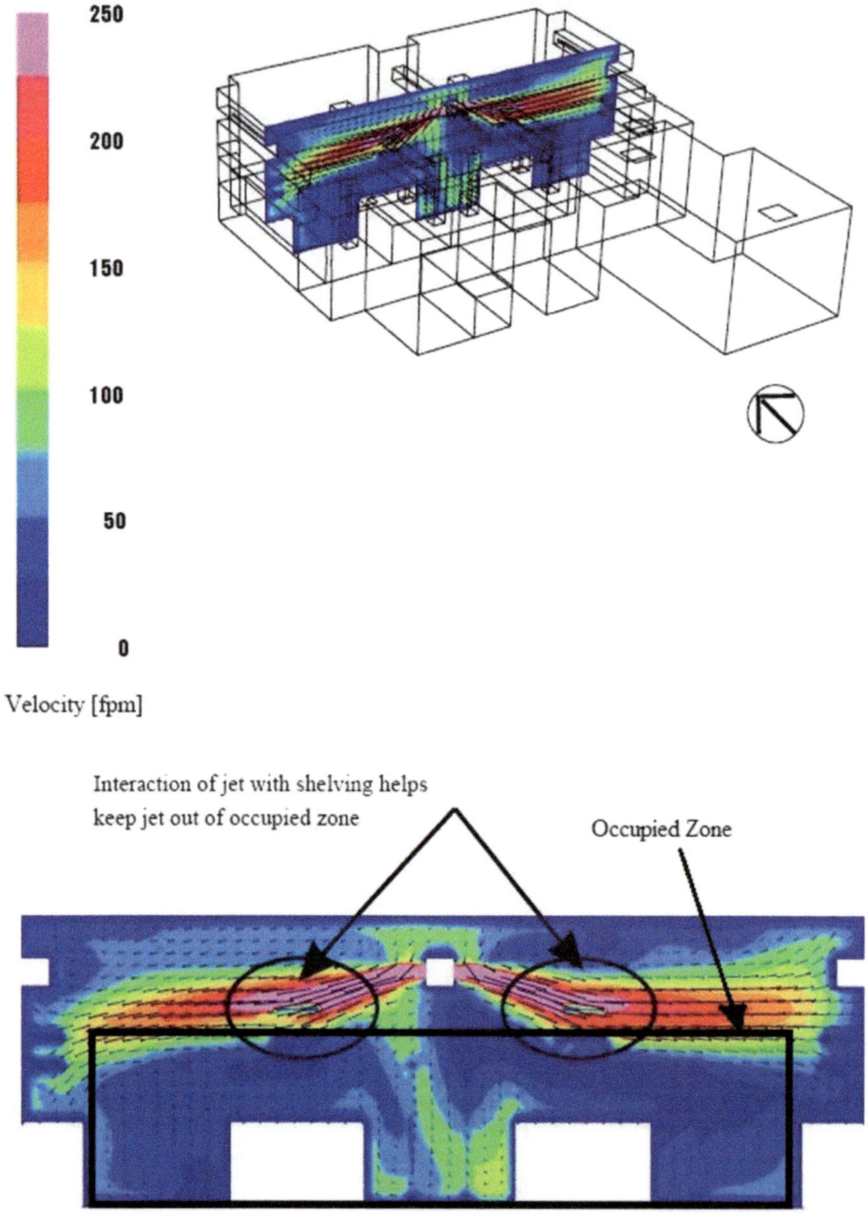

Figure 17-3 Example CFD Output—Single Cross-Section of Velocities

(Courtesy of Rowan Williams Davies & Irwin Inc.)

small portion of the calculations that are actually required to be performed for the entire room.

CFD modeling can be used for proposed or existing laboratories, and it can be useful for laboratory ventilation system designs. The detailed flow fields produced by CFD can, in principle, help with such issues as supply louver design and placement, fume hood design and placement, control of heat or contaminant sources, indoor air quality (IAQ), thermal comfort, and occupational safety.

Although there are many commercial CFD providers, CFD modeling is still an emerging technology that takes effort to master. Simply running a CFD modeling program does not necessarily provide meaningful or useful results.

TYPES OF CFD MODELS

There are two main types of CFD models of interest for laboratory ventilation: turbulent transport Reynolds-averaged Navier Stokes (RANS) models and large-eddy simulation (LES) models.

The turbulent transport RANS models are less detailed in modeling sophistication. RANS models do not explicitly model turbulent air velocity fluctuations because this is well beyond current computational abilities. Instead, turbulent fluctuations are fully parameterized using turbulence closure models. RANS models also typically (but not always) produce only time-averaged values of the computed quantities at each grid point. RANS models are less computer resource intensive and are the more common approach at this time. Within this type there are various turbulence closure models that are available through commercial CFD providers.

Large-eddy simulation (LES) models are a more detailed approach with direct simulation of larger turbulent eddy scales and parameterization of smaller scales. LES models are intrinsically transient and produce a time history of computed results at each grid point in the room. LES models can show important time-dependent fluctuations that indicate periodic disruption of fume hood airflow by room currents. These models require finer grid spacing and time resolution and thus need much more computational power than the RANS models. LES models have recently become economically feasible at the engineering design level. At this time, LES models do take a significant amount of time to run, but they can be worth the cost in some situations, such as those requiring dynamic airflows within a fume hood or transient modeling of chemical spills.

TYPICAL STAGES IN CFD COMPUTATIONS

CFD computations typically involve the following stages:

- Model development
- Grid generation
- Boundary conditions
- Solution
- Postprocessing

Model development is the creation of a three-dimensional (3-D) computer model of the room geometry using a computer-aided design (CAD) package. The model geometry includes the interior of the space being modeled and any obstructions to flow present within the space. Supply and exhaust points are identified within the model and are represented to provide similar flow characteristics to actual supply and exhaust points. During the model geometry creation stage, some simplifications are made, typically based on experience and an understanding of the flow details of interest. For example, furniture and human shape geometries are often simplified to represent approximate massing. Another important simplification is the modeling of supply diffuser geometries. This simplification requires careful attention to produce correct airflow characteristics and is discussed in more detail in the following section.

Grid generation involves the subdivision of the room into often millions of 3-D cells within which the equations of motion will be solved. Then, boundary conditions, or the factors influencing the flow within the space, are assigned to the model. These conditions typically include air supply flow rates and temperatures, exhaust rates, heat flows through surfaces (e.g., walls, roof, floor), radiation and outdoor lighting through windows, and internal heat sources and contaminants (e.g., equipment, humans, animals).

The solution stage consists of a computer solving a complex and coupled set of equations enforcing the conservation of mass, momentum, energy, and various chemical species of interest between each cell within the space. Within each cell is stored the variables of interest (e.g., temperature, velocity, concentration, and turbulence).

Finally, the postprocessing stage allows the engineer to visualize and probe the resulting flow in both space and time, through colorful still pictures or animations. These results provide designers and decision makers with an understanding of flow patterns and possibly the causes for poor circulation that can be corrected.

RECOMMENDATIONS FOR CONDUCTING CFD MODELING

CFD models have a variety of inputs and procedures that may be unfamiliar to HVAC engineers. This section presents some recommendations for implementing CFD models, but note that these guidelines cannot be substituted for direct experience. Commercial CFD developers typically provide guidance and training applicable to their specific products. End users should gain experience first with standard room applications. More detailed guidance on CFD modeling can be obtained from Chapter 13 of *ASHRAE Handbook—Fundamentals* (2013a).

- It is recommended that LES modeling be used for detailed studies of airflow around and within fume hoods rather than RANS models. Leakage from a fume hood is often a transient phenomenon that is better simulated by a detailed model such as the LES model. In this case, the extra costs of the LES model are justified.
- Adequate grid spacing used in the room is important for accurate results. Recommended spacing is 1 in. (25 mm) or less in important areas, such as the face of a fume hood, and other areas where large variations are expected. No more than 6 in. (150 mm) should be used in the general air space far from surfaces and boundaries where small gradients are expected. The grid spacing should be varied in size (for example, by doubling the number of grid points) for some runs to show that the results are independent of grid spacing. As computer speed and memory storage become more economical, the grid points should be increased as much as possible.
- For transient calculations, time steps should be appropriately small to ensure convergence of the models. Time steps decrease as grid spacing decreases, which increases the need for computer resources. Time steps should also be varied to show independence.

- Internal heat sources and sinks should be modeled. These include workers, animals, equipment, lighting, radiant cooling panels, active or passive beams, and outdoor sunlight.
- Boundary influences such as walls, windows, people, furniture, supply diffusers, exhausts, and equipment geometry should also be included.
- Correct modeling of supply diffusers is critical to modeling laboratory ventilation. Diffusers are too small compared to the room to accurately model with normal grid spacing. The direction, temperature, and speed of air can vary dramatically depending on the location of the air within the exit of the diffuser.

  To overcome this problem, several methods have been developed to model the overall effects of the diffuser in terms of momentum and directionality. Chapter 13 of *ASHRAE Handbook—Fundamentals* (2013a) has more information on these methods. The methods should be evaluated using simple room geometries to match the published throw and directions of a diffuser model using manufacturer-supplied data or other measurement results.

  Figures 17-4 and 17-5 illustrate the possible differences that can arise in modeling supply diffusers. Figure 17-4 shows the exhaust with the correct flow rate but an incorrect velocity due to exaggeration of the outlet area. Figure 17-5 has a correction to better model the supply momentum to match empirical data from the manufacturer. However, in some cases the manufacturer data may be suspect. For example, some diffusers tend to dump colder air downwards rather than distribute the exhaust in many directions as predicted.

INTERPRETING CFD RESULTS

CFD results should be interpreted to improve laboratory designs. CFD programs have the ability to produce many types of graphical outputs, which by themselves may have limited usability except for visualization and understanding flow phenomena. Comparing the results to quantitative design criteria could add more useful information. The following subsections discuss suggestions for design criteria for use in interpreting CFD results and evaluating designs.

VENTILATION EFFECTIVENESS AND AGE OF AIR

Two common and related metrics are ventilation effectiveness and the age of air. Ventilation effectiveness at a given room location is defined as the room ventilation time constant divided by the age of the air at a particular location. The ventilation time constant is the room volume divided by the overall room ventilation rate and is the inverse of the room air change rate. The age of air is the time that a parcel of air at a particular location has been in the room since entering from the supply system or being transferred from another room. The age of air is a commonly available result determined from CFD packages by tracking parcels of air as they travel in the room. The longer the age of the air, the more odors and other contaminants are likely to be present at a given location within the room. CFD modeling can compare two configurations to determine which provides the lowest air age.

THERMAL COMFORT

As previously stated, thermal comfort is a complicated physiological and psychological reaction to several environmental factors. CFD can provide the primary variables of temperature and air speed at any location within the air space. However, examining only the temperature and velocity fields from CFD results may not be enough for designers to

Figure 17-4
Incorrect Modeling of Supply Momentum

(Courtesy of Rowan Williams Davies & Irwin Inc.)

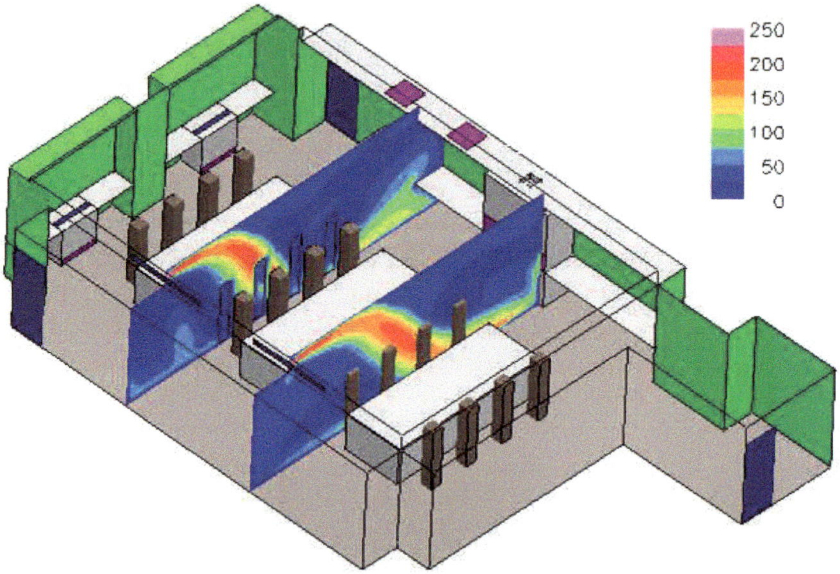

Figure 17-5
Corrected Momentum to Match Diffuser Throw Data

(Courtesy of Rowan Williams Davies & Irwin Inc.)

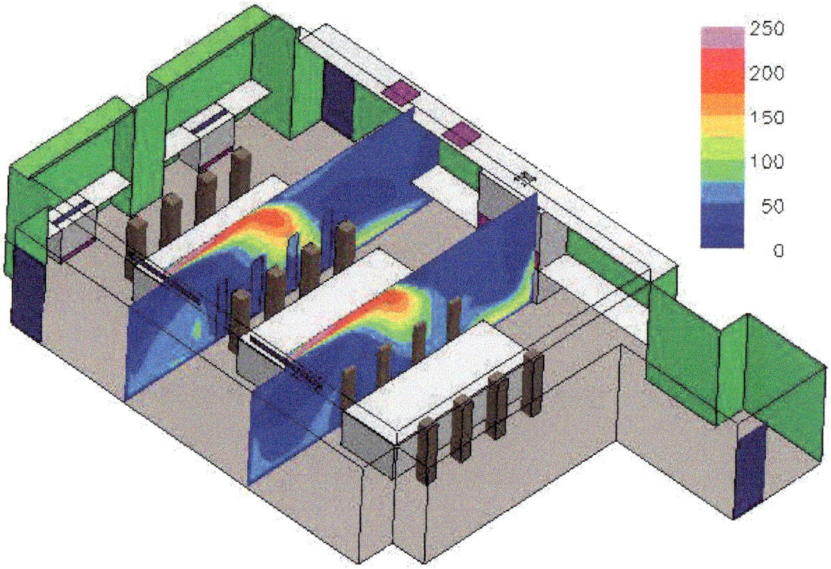

fully assess comfort. To determine the acceptability of the thermal conditions from the temperature and velocity, it is recommended that the latest edition of ANSI/ASHRAE Standard 55, *Thermal Environmental Conditions for Human Occupancy* (ASHRAE 2013b), be used to establish comfort at various fixed points of interest.

FUME HOOD LEAKAGE

Fume hood leakage can be reported in several ways. One way is to simulate a release scenario, much like that specified in ASHRAE Standard 110 for tracer gas testing (ASHRAE 1995), by releasing 8.5 cfm (4 L/min) of tracer gas in the hood and reporting concentrations at the breathing zone. (Note that Standard 110 is a test method and does not specify an acceptable concentration.) Another method includes examining a cross section of concentrations and velocities at the hood face; locations of outward-moving high concentrations into the room indicate undesirable leakage. The reporting could also include a simple plot of concentrations at the hood face, which could then be used to look for leakage areas. Alternatively, CFD postprocessing could calculate the mass emission rate outward (the product of concentration, outward velocity, and face area over all calculation points along the face).

ROOM AIR CURRENTS

For understanding room ventilation patterns, visualization of room velocities and temperatures is useful. There are a number of ways to achieve visualization using CFD results. Traditional cross sections in vertical and horizontal planes are helpful but cannot easily show the entire room. One way to show more of the room is to use a series of cross sections in an animation. Another way is to have a 3-D surface created that shows velocities (or other CFD output variables) above a certain value. For example, a high-velocity jet above, say, 100 fpm (0.5 m/s) can be illustrated as a 3-D "tube" that emerges from the diffuser and extends into the room. A cold supply plume with a temperature below a certain level could also be shown as a 3-D object. Another whole-room approach is an animation showing numerous virtual particles that travel through the room over the calculated velocity field. The particles would tend to travel from the supply diffuser into the exhaust system. They can be color coded with travel time so that areas of the room with stale air (particles with longer travel times) are more visible. This would help indicate where stale air is accumulating and thus where indoor air quality may be diminished.

For evaluating cross drafts, CFD models can report maximum and average room velocities at specific points in front of fume hoods. The cross-draft velocities can be compared to common guidelines, such as cross-draft velocities no larger than 50% of fume hood face velocities.

High velocities may not be desired at workbenches with fragile or sensitive experiments. Velocity predictions at specific points in the room are easily generated from CFD results and could be plotted and compared to a requirement.

CONCENTRATION LEVELS FROM A SPILL OR VAPOR RELEASE

CFD results can easily be used to help designers visualize spills or accidental releases of contaminants. For concentration fields where a contaminant is typically released from a small source, the best visualization is with a 3-D smoke plume that shows concentrations above a fixed level. The plume could be an averaged plume with a steady-state RANS calculation or a growing and moving plume with a transient RANS or LES model. Also, concentrations at specific locations can be reported and graphed, for example, at possible ignition points when evaluating LEL situations or at breathing zones.

REFERENCES

ASHRAE. 1995. ASHRAE Standard 110-1995, *Method of testing performance of laboratory fume hoods*. Atlanta: ASHRAE.

ASHRAE. 2002. ASHRAE Standard 129-1997 (RA 2002), *Measuring air-change effectiveness*. Atlanta: ASHRAE.

ASHRAE. 2013a. Chapter 13, Indoor environmental modeling. In *ASHRAE handbook—Fundamentals*. Atlanta: ASHRAE.

ASHRAE. 2013b. ANSI/ASHRAE Standard 55-2013, *Thermal environmental conditions for human occupancy*. Atlanta: ASHRAE.

Sustainable Design

18

OVERVIEW

Matthew May describes sustainability as "the ability to maintain something at a certain level, indefinitely." He further defines the concept by describing two necessary criteria in sustainable design:

> The first is that to be sustainable, any given asset, no matter what it is, must be kept whole, without making significant trade-offs that undermine the capital used to generate and maintain it. The second follows from the first: sustainability hinges on the ability to see finite resources as the very source of innovation. (May 2009, p. 143)

From this general perspective on sustainability, *ASHRAE GreenGuide: Design, Construction, and Operation of Sustainable Buildings* offers practical application of the concept of sustainable design to mechanical building systems. Specifically, it states six goals for green/sustainable design (ASHRAE 2013a, p. 4):

- Minimizing natural resource consumption through more efficient utilization of nonrenewable energy and other natural resources, land, water, and construction materials, including utilization of renewable energy resources to strive to achieve net zero energy consumption.
- Minimizing emissions that negatively impact our global atmosphere and ultimately the indoor environment, especially those related to indoor air quality (IAQ), greenhouse gases, global warming, particulates, or acid rain.
- Minimizing discharge of solid waste and liquid effluents, including demolition and occupant waste, sewer, and stormwater, and the associated infrastructure required to accommodate removal.
- Minimizing negative impacts on the building site.
- Optimizing the quality of the indoor environment, including air quality, thermal regime, illumination, acoustics/noise, and visual aspects to provide comfortable human physiological and psychological perceptions.
- Optimizing the integration of the new building project within the overall built and urban environment. A truly green/sustainable building should not be thought of or considered in a vacuum, but rather in how it integrates within the overall societal context.

These are broad goals, and all of them require a collaborative effort among the disciplines charged with the task of designing a laboratory facility. A common term for this collaborative effort is *integrated design*. Keeler and Burke (2009, p. 1) assert that green design and integrated building design are "equivalent terms." To capture all of these ideas, as well as several ASHRAE initiatives, this chapter uses the phrase *high-performance building design* to describe this integrated design process.

Laboratory facilities present unique challenges and opportunities to incorporate high-performance building design concepts into the design process. First, the energy use intensity (EUI) of a laboratory is 4 to 100 times greater than that of conventional buildings (Sartor et al. 2000). This is due to the energy-intensive processes located in the laboratory and the corresponding building systems that support them. Second, the safety needs of laboratories typically constrain the engineering design options. For example, ventilation systems in laboratories using hazardous chemicals must provide outdoor air to compensate for air exhausted to the building's exterior. Thus, the design outdoor airflow rate is driven by the local ventilation devices in the laboratory facility and is typically greater than the outdoor air requirements dictated by local building codes. Finally, the laboratory program may constrain the mechanical system design options by creating mandates for flexibility and redundancy that may lead to inefficiencies in the mechanical system design (e.g., large amounts of reheat energy due to high air change rates). These design restrictions may be the reason laboratory design teams have avoided the high-performance building design concept in the past. Some estimates, however, suggest that use of high-performance building design methods can result in a 50% energy consumption reduction in new and existing laboratory facilities (Mills et al. 1996). Therefore, although the inherent restrictions of laboratory facility design pose challenges to minimizing energy consumption, they also present opportunities to mitigate the energy costs associated with them.

This chapter covers the following, which will help engineers design, build, and operate sustainable laboratories:

- High-performance building design process
- Computer modeling
- Green tips for laboratories
- Ongoing commissioning
- Laboratory sustainability checklist

HIGH-PERFORMANCE BUILDING DESIGN PROCESS

NREL (Hayter et al. 2000) suggests the following design process for high-performance buildings:

- Predesign phase
  - Simulate a base-case building model and establish energy-use targets.
  - Complete analysis.
  - Brainstorm solutions with all design team members.
  - Perform energy simulations on base-case variants considering economic criteria.
- Design phase
  - Optimize sequences of operation.
  - Prepare preliminary architectural drawings.
  - Design the HVAC, plumbing, water conservation, and lighting systems.
  - Finalize plans and specifications.

- Construction/occupation phase
  - Rerun energy simulations before making construction design changes.
  - Commission all equipment and controls. Educate building operators to ensure that they operate the building as intended.

ASHRAE (2013b) recommends the following sequence, which complements NREL's procedure, when evaluating a project for energy efficiency:

- Minimize the impact of the building's functional requirements.
- Minimize heating and cooling loads.
- Maximize subsystems' efficiencies.
- Study alternative ways to integrate subsystems.

ASHRAE's recommended steps should be incorporated during the predesign steps in the aforementioned NREL process. The process may be used during any building design project, regardless of whether the concept of sustainability is specifically mentioned as a criterion for the design. Thus, this project methodology should be considered "standard of care" for the building science industry.

The project team members typically include the laboratory building owner, developer, facility operator, architect, construction manager, laboratory planner, mechanical/plumbing/fire protection engineers, energy simulation engineer, electrical/lighting engineer, structural engineer, civil engineer, commissioning authority, and cost estimator. As a result, the high-performance building design process encompasses design, construction, commissioning, operation, maintenance, occupancy, environmental impacts, and end-of-life recycling.

ASHRAE GreenGuide (2013a), the Leadership in Energy and Environmental Design® (LEED®) Green Building Rating System process promulgated by U.S. Green Building Council (USGBC), International Institute of Sustainable Laboratories (I^2SL) (the continuity of Laboratories for the 21st Century [Labs21], which was a joint initiative between the U.S. Department of Energy and U.S. Environmental Protection Agency), and National Institute of Building Sciences (NIBS) through their *Whole Building Design Guide* web portal are excellent guides to this process. Finally, *International Green Construction Code®* powered by ANSI/ASHRAE/ICC/USGBC/IES Standard 189.1 (IgCC/189.1) (ICC 2018) offers a systematic approach to high-performance building design and should be consulted by the practitioners of this design process.

COMPUTER MODELING

By comparing design options prior to construction, computer modeling optimizes performance and thus plays a critical role in the high-performance building design process. Modeling must begin during the predesign phase of a project and be modified, as needed, throughout the life of the project. Reviewed in this Guide are the four most commonly used modeling tools: energy modeling, exergy analysis, life-cycle cost analysis (LCCA), and building information modeling (BIM). These tools are not competing; they all model important aspects of the building and aid in decision making.

ENERGY MODELING

Upon completion of the Basis of Design (BOD), an energy model of the project should be generated. The method of evaluating energy use should conform to ANSI/ASHRAE Standard 140, *Standard Method of Test for the Evaluation of Building Energy*

Figure 18-1
Baseline Building Energy Use

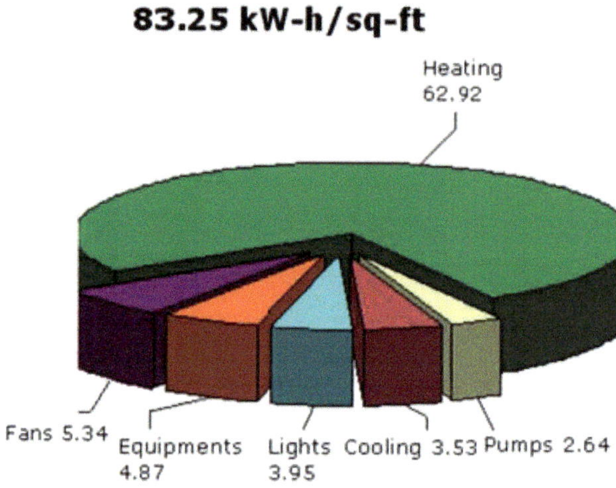

Figure 18-2
Proposed Building Energy Use

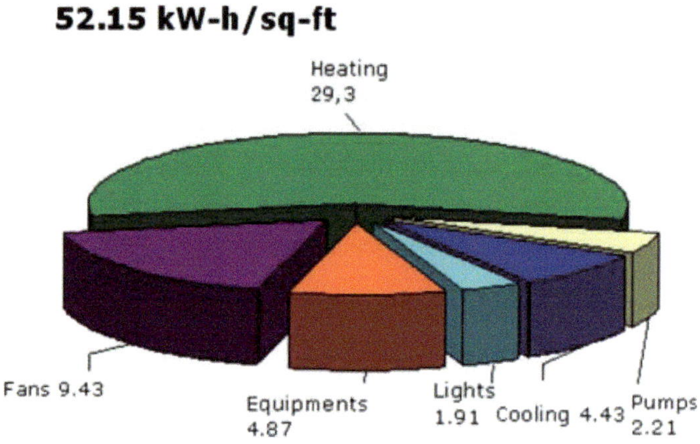

Analysis Computer Programs (ASHRAE 2014), as required to achieve LEED certification.

Energy modeling consists of creating two models to compare (ASHRAE 2013c). The first model consists of the baseline building, and the second is the proposed design for the project. The proposed building model is evaluated against the baseline and reevaluated at each step in the development of the project to review the impact of design decisions on energy costs. The comparison of Figures 18-1 and 18-2 shows energy-use results of a baseline model (Figure 18-1) and a proposed project model (Figure 18-2) and provides an example of the energy reduction that can result from the high-performance building design process.

Benchmarking the proposed model against the baseline model may provide insight about the relative efficiency of the building. One metric used in benchmarking is the annual energy use intensity (EUI) (EPA n.d.), in units of $kWh/ft^2 \cdot yr$ ($kWh/m^2 \cdot yr$). This value should be correlated with the processes in the laboratory. For example, the density of fume hoods in a facility, which is normally expressed as the quantity of fume hoods per 5000 ft^2 (465 m^2), substantially affects EUI. Generally, this value varies between 0.5 and 10 depending on the laboratory's functions. Another interesting benchmark is the number of watts of power per total airflow rate supplied, returned, and exhausted. This value normally varies between 0.3 and 2.0 W/cfm (0.6 and 4.3 W·L/s). Finally, the population density of a laboratory, normally given in ft^2/person (±500 to 600 ft^2 [±45 to 55 m^2] is typical) is an important correlation. A summary graph of a benchmarking study is shown

Figure 18-3 Energy Consumption Benchmarking (kW·h/ft²/yr)

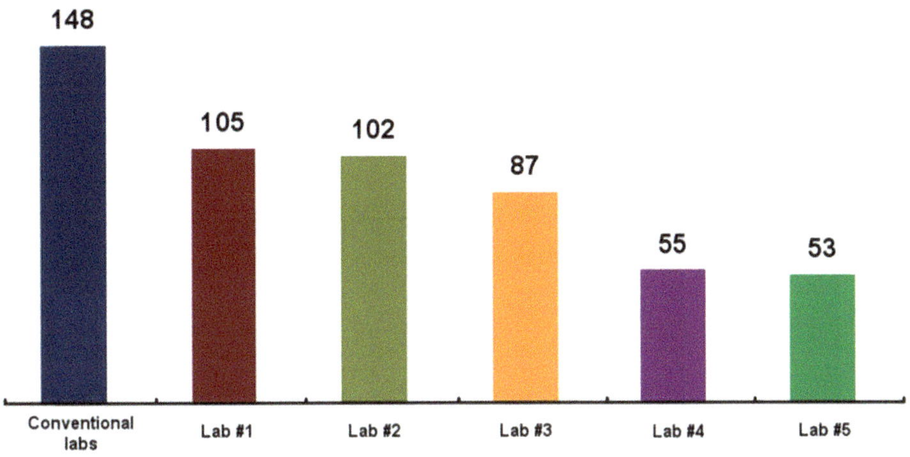

in Figure 18-3. The Laboratory Energy Efficiency Profiler (LEEP) Tool (I^2SL n.d.) should be consulted to obtain additional benchmarking data.

EXERGY ANALYSIS

Exergy is defined as the maximum theoretical work obtainable when two systems at different states are brought into equilibrium with each other (Moran and Shapiro 2000). Exergy analysis, sometimes referred to as *availability analysis*, is a form of energy modeling that uses the first and second laws of thermodynamics to evaluate the appropriateness of the energy source to the application. Its goal is to match the energy quality of the source energy with the final energy use.

Although exergy analysis is a method of evaluating the efficiency of building systems, it is typically not required by building code authorities or nongovernmental organizations such as USGBC. However, exergy analysis allows components of the system to be efficiently matched to energy sources. Case studies demonstrate the effectiveness of exergy analysis in improving the energy performance of building systems (Itard 2005).

LIFE-CYCLE COST ANALYSIS

In addition to the energy models, a life-cycle cost analysis (LCCA) should be performed to evaluate the cost-effectiveness of the building and system options. National Institute of Standards and Technology (NIST) provides excellent guidance for performing a LCCA (Fuller and Petersen 1995; BLCC n.d.).

BUILDING INFORMATION MODELING (BIM)

ASHRAE defines BIM as follows:

> Building information modeling (BIM) is a building design and documentation methodology that relies on the creation and collection of interrelated computable information about a building project so that reliable, coordinated, and internally consistent digital representations of the building are available for design decision making, production of high-quality construction documents, construction planning, and predicting performance in various ways. (ASHRAE 2015, p. 40.15)

BIM is seeing increasing usage among the design community and owners who desire a tool to facilitate rapid and accurate depictions of their facilities and that can easily be updated. This process may involve several software tools linked to automated graphics software. A number of organizations offer guidance on such tools, including U.S. General Services Administration (GSA) through their website (GSA 2015), NIBS through their *Whole Building Design Guide* web portal (Smith and Edgar 2015), and various software providers as included on the CAD/BIM Technology Center website (https:cad bim.usace.army.mil).

GREEN TIPS FOR LABORATORIES

Considerations when designing a high-performance laboratory building include completing a systematic review to minimize the use of natural resources, protecting the exterior and interior laboratory environments, minimizing energy use by design, recovering as much energy as possible, fine-tuning controls for energy efficiency, and choosing the best source of energy from off site or on site. Each of these elements is discussed in detail in the following subsections:

- Minimizing natural resources consumption
- Protecting the environment
- Energy efficiency—Design
- Energy efficiency—Recovery
- Energy efficiency—Source
- Energy efficiency—Controls
- On-site energy production

MINIMIZING NATURAL RESOURCES CONSUMPTION

Minimizing natural resources consumption is a goal of all high-performing buildings, including laboratories. Water conservation is very important; opportunities for achieving this are discussed in this section, as are strategies for downsizing the needs for materials and equipment.

Water Conservation

Many opportunities are available for conserving water in laboratories. For example, rainwater can be recovered in a cistern, filtered, and treated for use in toilets and urinals and as makeup water in cooling towers (see Figure 18-4). Other water conservation measures to consider include the following:

- Using low-flow plumbing fixtures.
- Using waterless urinals with compatible drain pipe materials.
- Using a closed-loop system instead of city water for process cooling. The energy of such a cooling loop can be recovered with a heat pump and used to heat the building.
- Using vacuum pumps (local or centralized) to avoid using tap water to generate a vacuum.
- Recovering gray water from showers, lavatories, research animal cage and rack washers, and water rejected from reverse osmosis and boiler blowdown and storage in a cistern for use in flushing toilets and urinals, landscape irrigation, or trench and drain flushing in animal facilities.
- Recovering cooling-coil condensate for cooling tower water makeup.
- Installing drift eliminators on cooling towers to minimize water loss.
- Using electronic faucets and flush valve controls.
- Using sustainably designed research animal cage and rack washers, animal watering systems, autoclaves, and aquatic systems for laboratory research involving amphibians and/or fish.

Figure 18-4
Rainwater Recovery

(Courtesy of Pageau Morel)

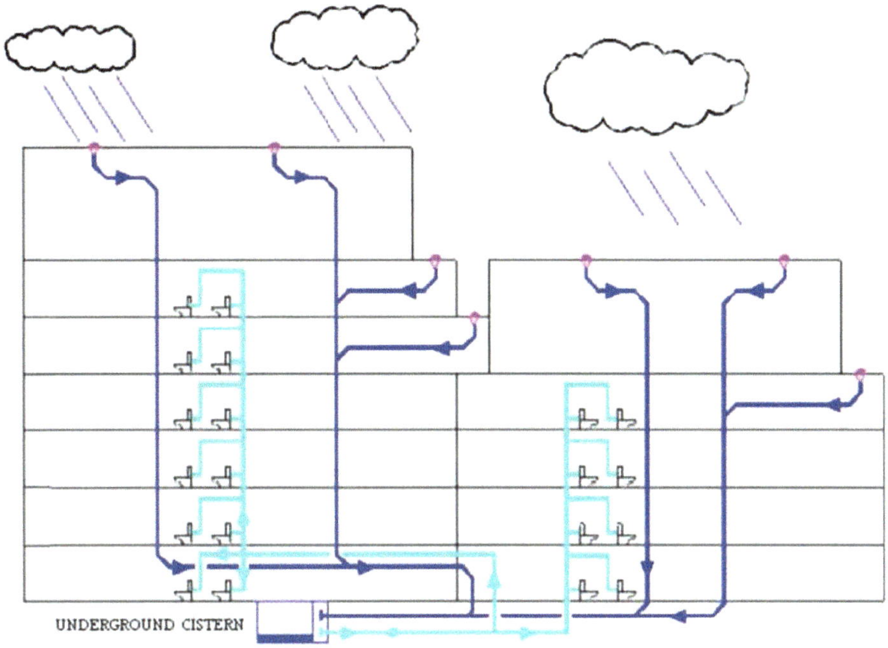

Other Strategies

Other potential opportunities for minimizing natural resources consumption include the following:

- Identify synergies between functions and the potential to share services and equipment to reduce and optimize the construction area of the building.
- Rightsize HVAC systems (Labs21 2005a). For example, manifold exhaust ducts and use diversity to size equipment and ductwork to minimize resource consumption.
- Locate laboratories with intensive needs, such as those with many fume hoods and bench services, near the mechanical room to reduce ductwork and piping.
- Create flexibility by providing easy access for the addition of required services rather than providing all possible services on day one.
- Install multiple small systems instead of one large system (for example, install two units at 50% total required capacity instead of one at 100%) to facilitate maintenance and allow for continuity of certain services.
- Zone mechanical services to permit local shutdown while maintaining services to other areas.
- Use energy-efficient laboratory fume hoods.
- Use demand-based control of laboratory air change rates.
- Specify premium high-efficiency equipment.
- Select equipment with part-load operation and variable conditions in mind.
- Specify low-pressure-drop design.
- Use energy recovery from exhaust air or process cooling water when this is allowed.
- Consider on-site power generation.
- Consider using renewable energy.
- Consider installing active or passive beams in laboratories (Labs21 2009).

PROTECTING THE ENVIRONMENT

Laboratory systems are designed to protect their occupants by rejecting contaminated air to the exterior. Depending on the contaminants, different technologies exist to clean the air before exhausting it outdoors. Chemical adsorption filters and wet scrubbers are often used to remove contaminants. Laboratories should also have a management plan for used chemicals to avoid rejecting them into the drainage system. In addition to these measures, the following are guidelines to protecting the environment from the chemicals used in a laboratory:

- Minimize emissions that negatively impact the indoor environment and the atmosphere of our planet.
- Minimize discharge of solid waste and liquid effluents.
- Minimize negative impacts on site ecosystems.
- Maximize the air quality of the indoor environment.

ENERGY EFFICIENCY—DESIGN

The following measures will all help reduce the energy consumption of a laboratory.

Building Thermal Envelope

Energy models and simulation software should be used to select the best thermal performance values of building components and optimize window solar heat gain coefficients (SHGCs). Such analysis must be done considering the life-cycle costs of the building envelope components.

Exhaust Fans

Sizing and selection of exhaust fans should be based both on the amount of air to be exhausted at different times of the day and on the maintenance of a minimum discharge velocity within the exhaust stack(s). All efforts should be made to minimize the amount of exhausted air while preventing reentrainment of contaminated air into outdoor air intakes. A wind wake analysis is typically done to verify reentrainment is not a problem.

Air Change Rate—Outdoor Air

Because all air supplied in a laboratory where fume hoods are located must be exhausted, it should be kept to a minimum without compromising the safety of the users. Minimum flow rates should be established for the following four conditions: day (occupied), day (unoccupied), nights and weekends, and emergency purge.

Air Change Rate—Local Cooling

Ventilation rates driven by the cooling load results in more air than that required by fume hoods needing to be exhausted. Local cooling with fan-coil units, active or passive beams (sensible cooling only), or fan-powered induction units reduces the amount of fresh air exhausted from the laboratory to the minimum amounts required for user safety.

Air Change Rate—VAV System

If the ventilation rate is driven by fume hood exhaust, variable airflow devices on supply and exhaust ducts will keep exhausted airflow to a minimum. It should also be noted, as shown in Table 18-1, that reducing the airflow rates also tremendously reduces the energy consumption of the fans.

Air Change Rate—IAQ Sensors, Demand-Based Control

With this approach, a laboratory's minimum airflow rate is varied based on measurements of the laboratory's actual air quality level or "air cleanliness." Sensors used to

Table 18-1 Effect of Reducing Air Changes

Air Changes per Hour, ach	Supply, cfm (L/s)	Heating/Cooling Ratio	Fan Power Ratio
10	675 (320)	1	1
6	405 (190)	0.6	0.22
4	270 (125)	0.4	0.06

determine air quality should be evaluated based on their ability to detect the chemicals used in the space. When air contaminants are measured above a given threshold, the minimum air change rate is proportionally increased to a level sufficient to purge the room. When contaminants are below the determined threshold, lower minimum airflow rates (as low as 2 ach depending on the application) may be appropriate when the fume hood exhaust or room cooling load requirements do not require higher airflow rates.

Air Change Rate—Fume Hood Selection

As a general approach, the design engineer should analyze the laboratory type in conjunction with the minimum air change rate required. If there is only one 4 ft (1.2 m) wide fume hood in a large laboratory and the maximum exhaust of the hood is smaller than the minimum air change rate required in the laboratory, the fume hood could be either a constant-volume exhaust type or a constant-volume low-flow high-performance type.

When the air change rate is driven by the fume hood's requirements, all efforts should be taken to reduce the exhaust flow according to one or more of the following control strategies: variable-air-volume (VAV) exhaust and supply, zone presence sensors in front of the hood's sash, or laboratory occupancy sensors to reset face velocity at the sash opening.

Using horizontal sashes on fume hoods reduces the amount of exhausted air by 50% or more. This option must be discussed with users to ensure compatibility with their laboratory usage.

In teaching laboratories, it might be feasible to dedicate a single hood for chemical holding. In such a case, hoods used by students could be emptied outside of teaching hours, thus allowing these hoods to be completely shut off for exhausting.

Air Change Rate—Animal Holding Rooms

Use ventilated cage racks and demand-based control to reduce the air change rate of animal holding rooms.

Air Change Rate—Users

As a general rule, a fume hood's sash remains closed when the hood not in use. When the air change rate of a laboratory is fume-hood driven, it is possible, with programming, to activate an audio signal (a beep) from the hood if the sash stays open for more than the allotted time (15 min, for example), alerting users to close the sash.

The opening of a vertical sash can be mechanically limited to 14 or 18 in. (356 or 457 mm) with a sash stop to reduce the amount of exhausted air when in use.

Different types of system arrangements are presented in Figures 18-5 to 18-8. The objective of these designs is to reduce the fresh air intake while maintaining a safe laboratory environment.

The conventional system in Figure 18-5 shows the use of a runaround energy recovery loop applied to both fume hood and general laboratory exhausts. The figure also shows the use of a system dedicated to office areas.

The modified conventional system in Figure 18-6 shows air supplied to both offices and laboratories from the same system. The total amount of air treated is less than in a conventional system (20,000 cfm [9440 L/s] in this example) since return air from the offices essentially serves as makeup air for the laboratories.

In Figure 18-7, dedicated fan-coil units cool and recirculate air inside each laboratory, reducing the quantity of outdoor air required. Also, general laboratory exhaust is separated from fume hood exhaust so that an enthalpy wheel can be used to recover energy from the general exhaust. Note that an enthalpy wheel is much more efficient than a run-around loop but, unlike the loop, it cannot prevent cross-contamination of the airstreams. Enthalpy wheels should use purge to minimize the potential for cross-contamination.

Figure 18-8 shows active-beam technology used as an alternative to fan-coil units to provide the additional cooling required.

Reduction in air changes per hour (ACH) or reduction in sash opening does not necessarily translate into a proportional reduction of the exhaust airflow rate. Table 18-2 shows the various reductions attained in the laboratory exhaust of an existing laboratory building based on the relation of ACH to sash opening percentages.

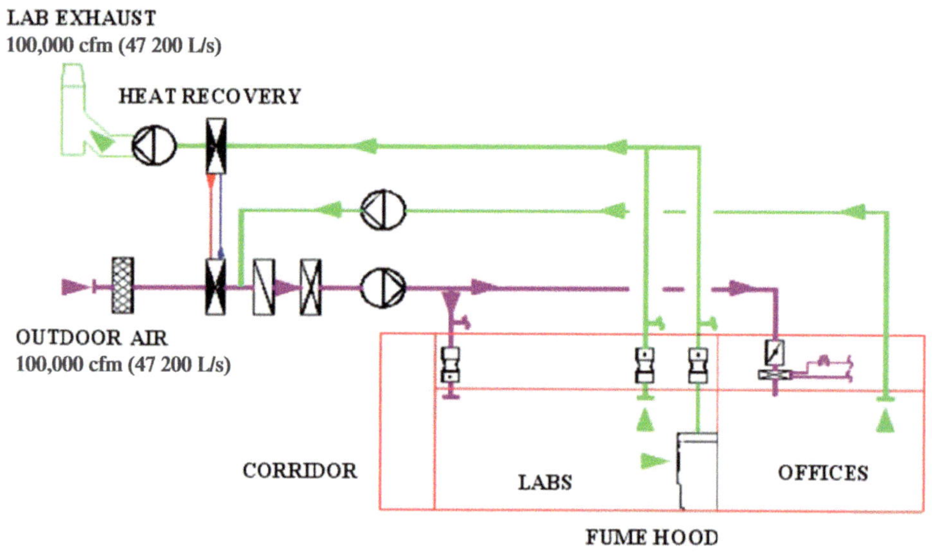

Figure 18-5 Conventional System

(Courtesy of Pageau Morel)

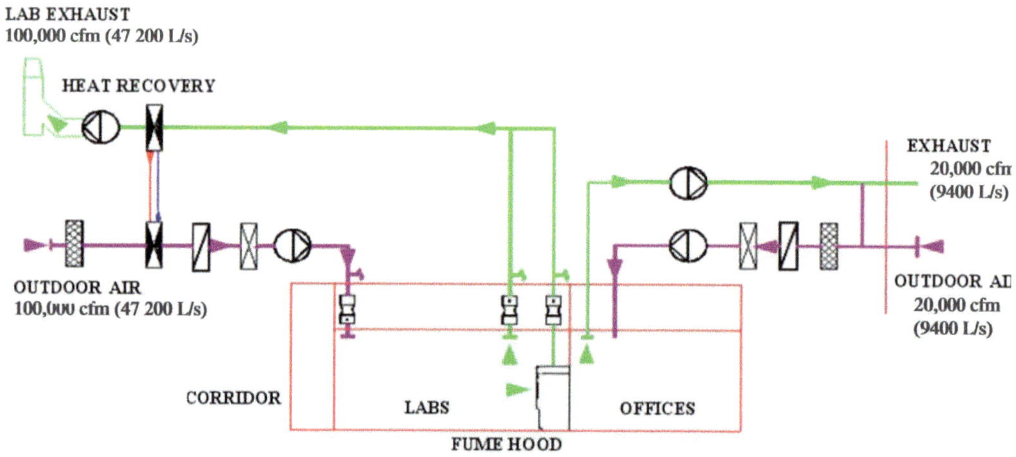

Figure 18-6 Modified Conventional System

(Courtesy of Pageau Morel)

Figure 18-7
Fan-Coil and Separate General Laboratory Exhaust
(Courtesy of Pageau Morel)

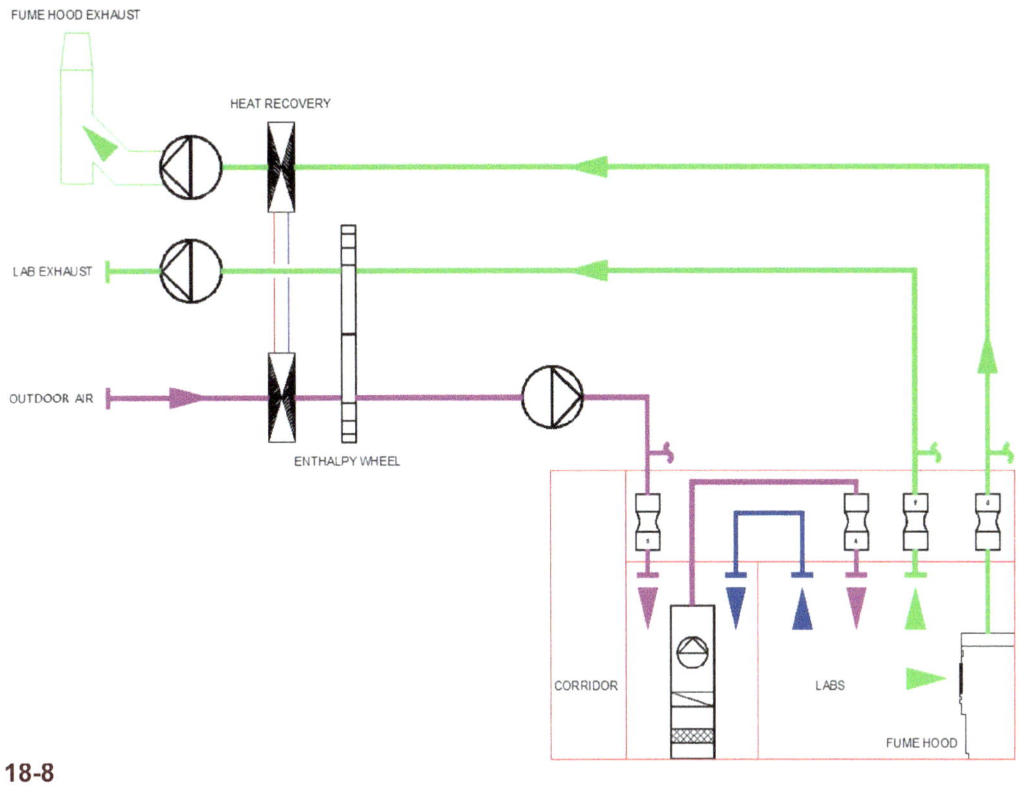

Figure 18-8
Active Beams and Separate General Laboratory Exhaust
(Courtesy of Pageau Morel)

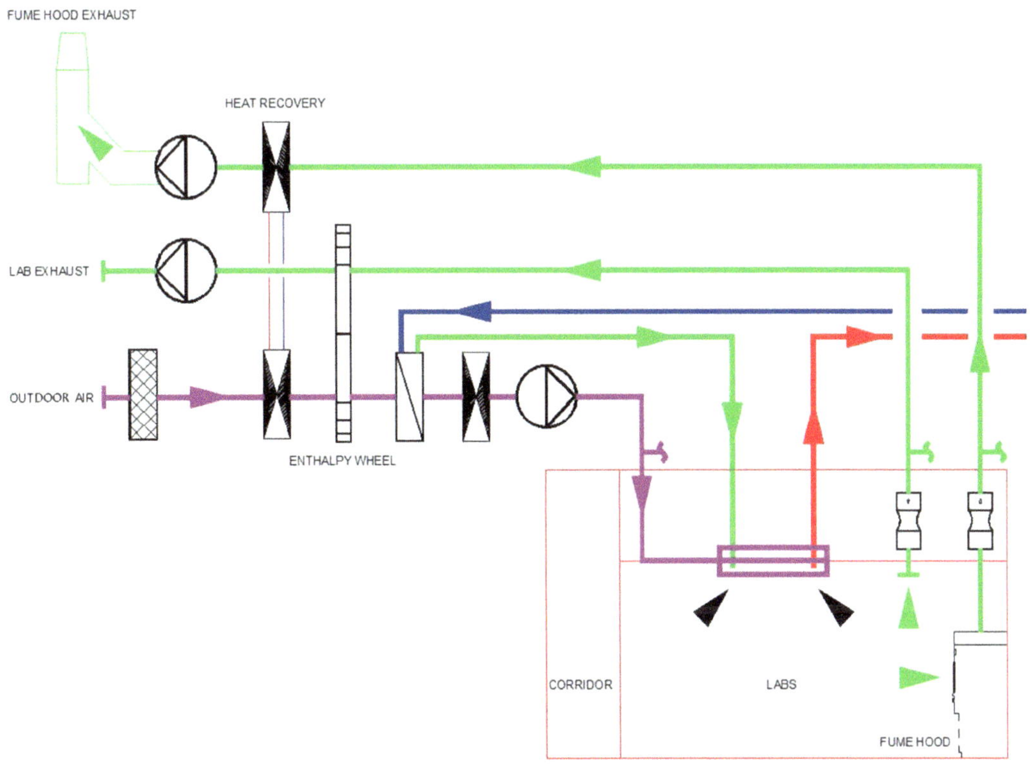

Table 18-2 Example of Real Laboratory Building's Airflow

		Total Laboratory Exhaust Including Fume Hoods, cfm (L/s)				
Sash Open Percentage:		100%	75%	50%	25%	Closed
	10 ach	235,400 (111 085)	207,700 (98 015)	182,900 (86 310)	165,200 (77 960)	141,650 (66 845)
Total	6 ach	212,750 (100 400)	183,850 (86 760)	156,700 (73 950)	133,200 (62 860)	114,050 (53 820)
	4 ach	207,800 (98 060)	177,800 (55 590)	148,300 (69 980)	120,350 (56 795)	97,400 (45 960)

Canopy Hoods

Canopy hoods are normally used to exhaust heat and humidity. Blocking the center of the canopy hood can improve efficiency and reduce the amount of exhausted air. A damper controlled by a timer should be used when applicable.

Ductwork Sealing

Supply, return, and exhaust ducts should be tightly sealed to reduce air leakage and associated energy losses.

Plug-Load Reduction

In many laboratory projects, laboratory equipment is purchased by people outside the construction team. Energy efficiency criteria such as those developed for the ENERGY STAR program (EPA 2015) should be used to evaluate and select the equipment. Engineers should collaborate to define or establish energy performance criteria for equipment not covered in the ENERGY STAR program. For more information on plug-load reduction, see the Labs21 best practice guide on rightsizing (Labs21 2005a).

Zone Presence Sensors

Zone presence sensors can be used to control lighting and fume hood airflow in laboratories. Sensors designed to monitor the region facing a fume hood to sense the presence of an operator should preferably be dual technology (infrared and ultrasonic). Normally the "occupied" status is maintained ±15 min (adjustable from the energy management and control system [EMCS]) after occupants leave the laboratory.

Care should be taken when using these sensors to reset the minimum air change rates. If excess contaminant levels are detected during airflow rate levels set for unoccupied mode, occupied-mode airflow rates must be reestablished to purge the laboratory of any excess contaminants.

Low-Velocity HVAC Units

Selecting air-handling systems' components at 300 fpm (1.5 m/s) instead of the standard 500 fpm (2.5 m/s) face velocity can reduce power consumption by 78%. This also reduces the potential for water entrainment at the cooling/dehumidifying coils. On the other hand, it means using a little more building space (floor space or mechanical room volume). The net result is usually positive.

High-Efficiency Duct Fittings (Low Pressure Losses)

Often neglected, particularly in exhaust systems, is that all duct fittings on collectors, branches, reducers, etc. should be designed and built to regain as much static pressure as possible and reduce the energy consumption of the fans.

Figure 18-9 Direct-Drive Coupling

(Courtesy of Pageau Morel)

Direct-Drive Fans

As often as possible, select direct-drive fans to eliminate belt losses and reduce maintenance and dust generation from belt wear and tear. Fans should be selected for the highest efficiency at their most probable flow, not the maximum flow. Direct-drive fans are normally connected to a variable-frequency drive (VFD). Fan selection should be done at 80% of the nominal motor revolutions per minute (RPM) to maintain some capacity to increase the flow of the fan depending on final site conditions. Figure 18-9 shows an example of a direct-drive fan.

Exhaust Network Design

Design networks in different pressure zones with separate fans to avoid operating the whole system at high pressure such as is required by remote laboratories or biological safety cabinets (BSCs).

Exhaust Grouping

Group small independent exhausts with larger exhausts (where separation is not required for safety or for concerns over material compatibility) to take advantage of dilution.

VFDs on Motors

With the prices of drives being more and more affordable, it has become good practice to install VFDs on pumps and fans whenever flow variation is significant, because this tremendously reduces the energy consumption of motors. To operate with a drive, a motor must be classified as "inverter duty" (EERE 2012). Analysis on the importance of keeping a fan running within a group of exhaust fans should be done to determine whether an electric bypass is required on the VFDs.

Bypass Dampers

Bypass dampers located upstream of cooling coils and heating coils reduce static pressure losses whenever these coils are not in use.

Filters

Selection of filters should be done considering allowable pressure losses, replacement costs, disposal costs, and associated energy consumption. Such an analysis should be done on both supply and exhaust networks.

Table 18-3
Summary of Design Practice for Laboratory HVAC Systems
(Adapted from Labs21 [2005b], Table 2)

Component	Standard	Good	Better
Air handler face velocity	500 fpm (2.5 m/s)	400 fpm (2.0 m/s)	300 fpm (1.5 m/s)
Air-handling unit	2.5 in. w.g. (623 Pa)	1.5 in. w.g. (374 Pa)	0.75 in. w.g. (187 Pa)
Energy recovery device pressure drop	1.00 in. w.g. (249 Pa)	0.60 in. w.g. (149 Pa)	0.35 in. w.g. (87. Pa)
VAV control devices pressure drop	N/A	0.3 to 0.6 in. w.g. (75 to 149 Pa)	0.1 in. w.g. (25 Pa)
Zone temperature control coils pressure drop	0.5 in. w.g. (125 Pa)	0.30 in. w.g. (75 Pa)	0 (no coils)
Total supply and exhaust ductwork pressure drop	4.0 in. w.g. (996 Pa)	2.25 in. w.g. (560 Pa)	1.20 in. w.g. (299 Pa)
Exhaust stack pressure drop	0.7 in. w.g. (174 Pa) full design flow through entire exhaust system, constant volume	0.7 in. w.g. (174 Pa) full design flow through fan and stack only, VAV system with bypass	0.75 in. w.g. (187 Pa) averaging half the design flow, VAV system with multiple stacks
Noise control (silencers)	1.0 in. w.g. (249 Pa)	0.25 in. w.g. (62 Pa)	0 (no device)
Total pressure drop	9.7 in. w.g. (2416 Pa)	6.2 in. w.g. (1544 Pa)	3.2 in. w.g. (797 Pa)
Approximate fan power requirement, W/cfm (W·L/s)	1.8 (3.8)	1.2 (2.54)	0.6 (1.27)

Reduction of HVAC System Pressure Losses

Reduction of static pressure losses in HVAC system design will result in lower motor horsepower and energy consumption. Table 18-3 presents some design parameters that can be used when designing HVAC systems.

ENERGY EFFICIENCY—RECOVERY

As presented in Chapter 10, there are many options for heat recovery on air exhausts. If possible, separate fume hood and general laboratory exhausts and low-temperature heating loops should be applied in laboratory buildings, as discussed in the following subsections.

Separate Fume Hood and General Laboratory Exhausts

Because of the risk of cross-contamination, fume hood exhaust recovery is normally done with lower-efficiency devices such as glycol heat recovery runaround loops or air-to-air heat exchangers. Caution should be used with air-to-air heat recovery so that cross-contamination does not occur or that it is below the permissible limits stated in ASHRAE Standard 62.1 (ASHRAE 2013d).

Having a separate general laboratory exhaust provides an opportunity to install a much more efficient energy device such as an enthalpy "total energy" wheel on this portion of the evacuated air.

Low-Temperature Heating Loop

A low-temperature (110°F to 90°F [43°C to 32°C]) heating network provides an opportunity to recover heat from boiler stacks, chiller condensers, heat pumps, or ground-source heat pumps. Condensing boilers can be connected to this loop for high-efficiency heating. This heating loop can then be used for economical heating (air heating), preheat,

Figure 18-10 Low-Temperature Heating Loop

(Courtesy of Pageau Morel)

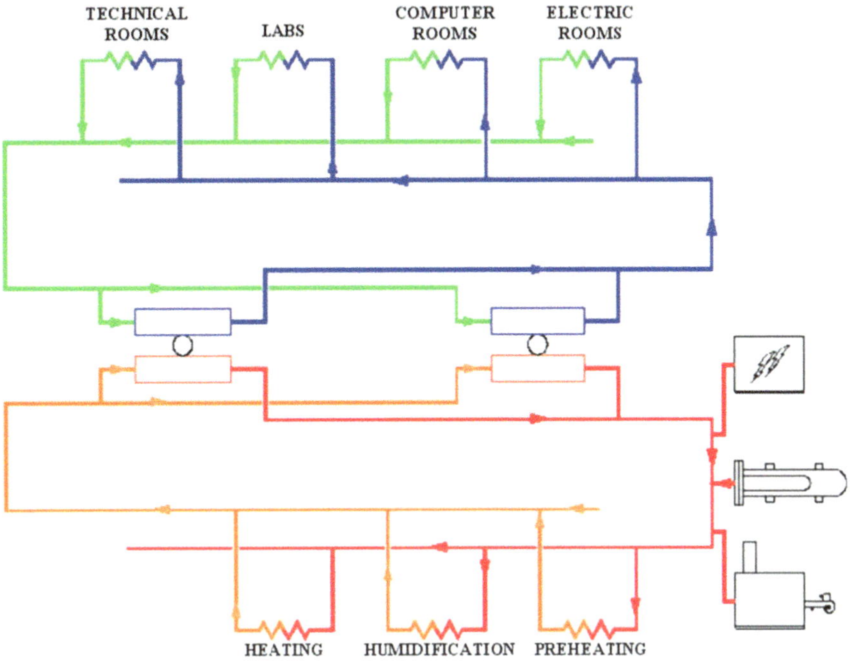

reheat, domestic hot-water preheat, and even adiabatic humidification. Because many laboratories use terminal reheat year round, they become excellent sinks for low-temperature energy sources (with temperatures as low as 87°F to 69°F [31°C to 21°C]). Figure 18-10 shows a diagram with a low-temperature loop that permits heat recovery of many sources.

ENERGY EFFICIENCY—SOURCE

Options for producing the required energy for laboratory buildings, such as heat recovery chiller/heat pumps, ground-source heat pumps, chillers, and other sources such as condensing boilers and direct-contact heaters, are discussed in this section.

Heat Recovery Chillers/Heat Pumps

In a heating-driven climate, all internal heat gains should be recovered (from computer rooms, electrical rooms, equipment rooms, etc.) by a chilled-water loop and transferred to perimeter spaces to supply the heating needs of a building. Recovered heat should be considered for all heating needs before free cooling is considered.

Ground-Source Heat Pumps

If internal heat gains are not sufficient to fulfill the heating loads of perimeter spaces, the potential offered by ground-source heat pumps should be analyzed.

Chillers

Whenever possible, use a high-chilled-water-temperature-difference chiller to reduce flow rate and pumping horsepower. Variable-speed chillers should also be considered. It is important to verify the performance of a chiller to make sure it can deliver high temperatures on the condenser side while running at part load. This is not the case for most centrifugal chillers.

Other Sources

Condensing boilers and direct-contact heaters have an efficiency of 95%+ when supplied with a return water temperature of ±90°F (±32°C). Figure 18-11 shows a schematic for a direct-contact boiler.

Figure 18-11
Direct-Contact Boiler (Natural Gas)

(Courtesy of Pageau Morel)

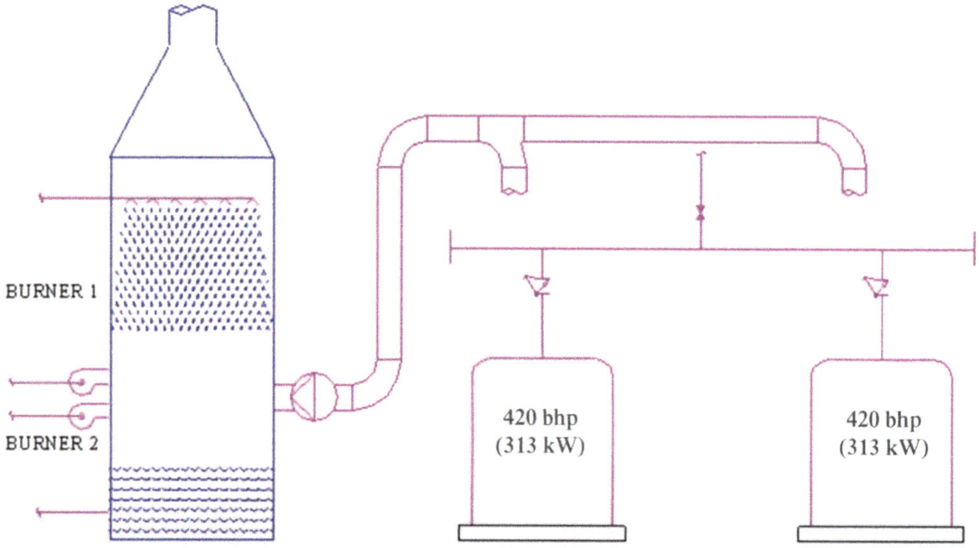

ENERGY EFFICIENCY—CONTROLS

As described in Chapter 11, there are many potential control strategies for optimizing a laboratory building's energy efficiency. Those strategies should attempt to reduce the amount of air exhausted from the building while maintaining a safe laboratory environment. Control strategies that enable this include VAV control of room and fume hood exhaust airflows, demand-based control or active contaminant sensing of the laboratory environment, and hydronic room sensible-only cooling equipment such as active or passive beams and fan-coil units.

Static Pressure Setpoint Reset

A control strategy should be developed to reset the static pressure setpoint in supply and exhaust duct networks as the flow reduces while maintaining the required air supply to all terminal units. A duct pressure model must be developed to predict the required setpoints at various flow conditions (Taylor 2007; Murphy 2011).

Exhaust Airflow Reset

The discharge velocity in the fume exhaust stacks can be reduced if wind condition and direction do not create a risk of reentrainment. Also, IAQ sensors can be used to assess the quality of the fume exhaust air, and when it is evaluated as sufficiently "clean" the discharge velocity can be reset (Carter et al. 2011).

Other Control Strategies

There are additional opportunities for improving the operational efficiency of an HVAC system. Following is a list of potential measures that should be considered.

- Demand-controlled ventilation
- Static pressure reset
    - Requires an algorithm to evaluate and determine minimum and maximum VAV box damper positions, which are used to reset the static pressure. If the most critical box airflow is greater than 95%, then static pressure can be increased; if less than 90%, then it should be reduced.
    - Notes:
        - Ignore rogue boxes that are "out of control."
        - Specify how often to run the algorithm.

        - Define static pressure limits.
        - Specify the location of the static pressure sensor.
- Supply air temperature reset
    - VAV systems and some constant-volume systems operate with a constant temperature discharge.
    - Additional control sequences should be added to provide for good humidity control.
    - In VAV systems, boxes will throttle down if less cooling is needed.
    - Under certain circumstances, this system may not be the most efficient system, and at minimum flow certain zones may subcool under low load.
- Discharge temperature dynamic reset
    - Adjust discharge temperature based on the most critical space temperature.
- Chiller plant (Taylor 2011a, 2011b, 2011c)
    - Include a chiller sequencing program in the building automation system (BAS).
    - Optimize chiller machine sequencing.
    - Optimize cooling tower operation.
    - Program a chilled-water temperature reset.

ON-SITE ENERGY PRODUCTION

Energy could be produced on site by a cogeneration plant that generates electricity and thermal energy, which in turn can be used to produce chilled water using absorption chillers.

On-site energy production options also include solar photovoltaic and wind turbine electricity production, solar thermal technologies, fuel cells, and steam- or gas-fired turbine generators.

ONGOING COMMISSIONING

Laboratory systems are complex and need occasional recommissioning. Because exhausted air is the major driver of the energy consumption of a laboratory building, continuous tracking of total air intake and exhaust provides a good indication of how well the systems are functioning. Installing flow monitors in air intake and exhaust fan inlets is simple and provides very good information on the operation of the systems. Figure 18-12 shows a continuous flow tracking graph of a laboratory system.

OPERATION/DESCRIPTION MANUAL

The operation/description manual contains the Owner's Project Requirements (OPR), Basis of Design (BOD), and conceptual design criteria for the building. It is normally used as a learning and troubleshooting tool for the facility operator.

LABORATORY SUSTAINABILITY CHECKLIST

The laboratory sustainability checklist presented as Table 18-4 can help designers review potential opportunities to include in the building design for enhancing the sustainability of a laboratory. Because this checklist is an authorized adaptation of one firm's

Figure 18-12
Airflow Tracking

(Courtesy of Pageau Morel)

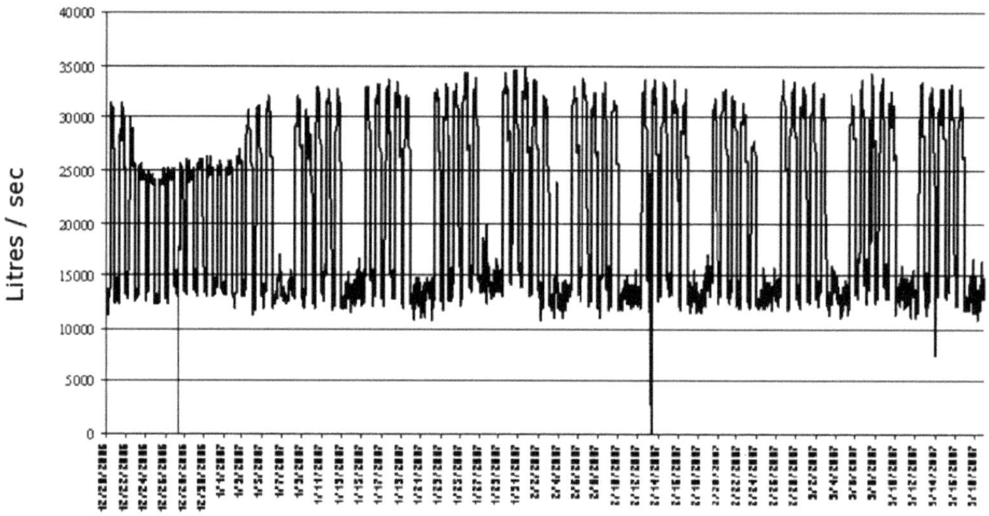

Table 18-4
Laboratory Sustainability Checklist
(Adapted with permission from Pageau Morel)

	Yes	No	+/–
RESOURCES CONSERVATION			
Siting laboratory building to recover heat from an adjacent building			
Shared services within the building			
Sharing of rooms			
Sharing of research equipment			
Building thermal envelope			
Create an energy simulation model			
Optimize wall and roof thermal performance			
Optimize glazing and SHGCs			
Use the highest ΔT possible on air and water?			
Round or flat oval ductwork			
Diversity factor on exhaust (air makeup) needs			
Fume hood exhaust manifolding			
No ceiling or partial ceiling in laboratories to take advantage of thermal inertia			
Natural ventilation of atriums and public spaces			
Vertical shafts to reduce horizontal ductwork			
Carpool-dedicated parking			
Construction waste management			
Use of low-volatile-organic-compound (low-VOC) materials and caulking			
Flexible open laboratory approach			
WATER EFFICIENCY			
Rainwater collection cistern			
Process-cooling water loop			
Central or local vacuum pumps to reduce potable water use			

Table 18-4
Laboratory Sustainability Checklist
(Adapted with permission from Pageau Morel)

	Yes	No	+/–
Cooling coil condensate recovery			
To feed cooling towers			
To flush toilets			
To cool exhaust with evaporative cooling and improve heat recovery efficiency			
Low-flow plumbing fixtures			
Dual-flush toilets			
Electronically activated faucets and flush valves			
Waterless urinals			
Reverse osmosis rejection water recovery			
Bioswale for gray water treatment			
ENERGY EFFICIENCY—REDUCTION			
Building thermal envelope optimization with energy modeling			
Reduction to minimum ventilation rate			
Zone presence sensors/night setback			
IAQ sensors/demand-based control of laboratory air change rates			
Local sensible cooling/airflow decoupling: fan-coils, active or passive beams			
Plug-load reduction and diversification			
Interior/exterior light shelves			
Daylighting with artificial lighting controls			
Dimming ballast			
High-efficiency lighting (fixtures and sources)			
White ceiling/pale-color walls and floor			
Zone presence sensors for lighting control and setback			
High-efficiency electrical transformers			
Premium efficiency motors			
Exhaust fans arrangements: paralleling, grouping, etc.			
High-efficiency fan selection			
Select for maximal efficiency at most probable operating condition			
High-efficiency pump selection			
Select for maximal efficiency at most probable operating condition			
Static pressure reduction			
Low-velocity HVAC units			
High-efficiency duct fittings (low pressure losses)			
High-efficiency equipment			
Direct-driven fans			
VAV systems with VFDs			
More vertical shafts and fewer horizontal ducts			
Dedicated fans for high static pressure zones			

Table 18-4
Laboratory Sustainability Checklist
(Adapted with permission from Pageau Morel)

	Yes	No	+/–
Free cooling (water-side economizer)			
Outdoor air preheating			
Cooling tower free cooling			
Recycle nonlaboratory air as laboratory makeup air?			
Separate low- and high-temperature cooling loops?			
Locate air returns above heat-producing equipment?			
High-performance, low-flow fume hoods			
Horizontal-sash fume hoods			
Limited-height vertical-sash opening (sash stops)			
Radiant hydronic heating/cooling concrete slabs			
Variable exhaust stack geometry to maintain constant velocity exhaust			
Booster fan on high static equipment exhaust			
Carbon dioxide (CO_2) sensors and demand-controlled ventilation in densely occupied spaces			
Optimized design criteria (temperature, relative humidity, noise levels)			
Ventilated cage racks in animal rooms			
Tightly sealed ductwork			
Low-pressure-loss filters			
Timers on canopy hood exhaust			
Variable exhaust fan exit velocity control (wind speed/direction sensing)			
ENERGY EFFICIENCY—RECOVERY			
Separate fume hood exhaust and general exhaust?			
Enthalpy "total energy" heat recovery on general exhaust			
Condensing boilers			
Chillers' condensers heat recovery			
Exhaust air heat recovery			
Heat pump chillers			
Chilled-water energy recovery			
From electrical rooms			
From equipment rooms			
From refrigerator rooms			
From cold-room compressors			
From computer rooms			
From other sources			
Adiabatic humidification			
High-efficiency frost-free alternate-flow heat recovery on general exhaust			

Table 18-4
Laboratory Sustainability Checklist
(Adapted with permission from Pageau Morel)

	Yes	No	+/–
ENERGY EFFICIENCY—SOURCE			
Recover excess heat rejected from an adjacent building?			
Low-temperature water-heating loop			
Heat recovery chillers/heat pumps			
Condensing boilers			
Direct-contact boilers			
Low-temperature water-heating loop			
For preheating of outdoor air			
For adiabatic humidification			
For preheating of domestic hot water			
Peak load management			
Thermal energy storage			
Full modulation of boilers' burners			
Variable-speed chillers			
Lake-source heat pump			
Variable primary flow			
Condenser-water temperature reset			
Off-peak energy storage			
Ground-source heat pump			
ENERGY EFFICIENCY—CONTROLS			
Night airflow reduction/setback in controlled- or limited-access laboratory spaces			
Vary exhaust stack discharge velocity based on wind direction?			
Local alarm reminder to close the fume hood sashes			
Peak load management			
Static pressure setpoint reset			
Local timers on canopy hoods and snorkels			
Watchdogs from EMCS for all exhausts			
Vary exhaust stack discharge velocity based on exhaust plenum contaminants?			
Demand-based control of laboratory and animal holding room ventilation levels			
ON-SITE ENERGY PRODUCTION			
Wind turbines			
Solar photovoltaic panels			
Combined heat and power			
Solar thermal collection panels			
Fuel cells			

Table 18-4
Laboratory Sustainability Checklist
(Adapted with permission from Pageau Morel)

	Yes	No	+/–
ONGOING COMMISSIONING			
Continuous airflow tracking			
Postoccupancy performance verification			
Annual verification of pressure setpoints			
Recalibration of control sensors			
ENVIRONMENT PROTECTION			
Adsorption filters on air exhaust			
Air scrubber on exhaust			
Used chemicals recovery management plan			

internal checklist, the presentation of the items does not correspond exactly to discussions in this chapter. A version of this checklist in Microsoft® Excel® format is available with this Guide at www.ashrae.org/LabDG.

REFERENCES

ASHRAE. 2015. Chapter 40, Computer applications. In *ASHRAE handbook—HVAC applications*. Atlanta: ASHRAE.

ASHRAE. 2013a. *ASHRAE greenguide: Design, construction, and operation of sustainable buildings*, 4th ed. Atlanta: ASHRAE.

ASHRAE. 2013b. *ASHRAE handbook—Fundamentals*, pp. 35.6–35.7. Atlanta: ASHRAE.

ASHRAE. 2013c. Normative appendix G, Performance rating method. In ANSI/ASHRAE/IES Standard 90.1-2013, *Energy standard for buildings except low-rise residential buildings*. Atlanta: ASHRAE.

ASHRAE. 2013d. ANSI/ASHRAE Standard 62.1-2013, *Ventilation for acceptable indoor air quality*. Atlanta: ASHRAE.

ASHRAE. 2014. ANSI/ASHRAE Standard 140, *Standard method of test for the evaluation of building energy analysis computer programs*. Atlanta: ASHRAE.

CAD/BIM Technology Center. n.d. The CAD/BIM Technology Center website. Vicksburg, MS: The CAD/BIM Technology Center for Facilities, Infrastructure and Environment, U.S. Army Engineer Research and Development Center. https:cadbim.usace.army.mil.

Carter, J.J., B.C. Cochran, and J.D. Reifschneider. 2011. Saving energy in lab exhaust systems. *ASHRAE Journal* 53(6):26.

EERE. n.d. Building life cycle cost programs. Washington, DC: Office of Energy Efficiency and Renewable Energy. www1.eere.energy.gov/femp/information/download_blcc.html.

EERE. 2012. When should inverter-duty motors be specified? Energy tips: Motor systems. Motor systems tip sheet #14. Washington, DC: Advanced Manufacturing Office, Energy Efficiency and Renewable Energy, U.S. Department of Energy. www1.eere.energy.gov/manufacturing/tech_assistance/pdfs/motor_tip_sheet14.pdf.

EPA. n.d. What is energy use intensity (EUI)? Washington, DC: U.S. Environmental Protection Agency. www.energystar.gov/buildings/facility-owners-and-managers/existing-buildings/use-portfolio-manager/understand-metrics/what-energy.

EPA. 2015. ENERGY STAR website. Washington, DC: U.S. Environmental Protection Agency. www.energystar.gov.

Fuller, S.K., and S.R. Petersen. 1995. *Life-cycle costing manual for the federal energy management program*. NIST Handbook 135. Gaithersburg, MD: National Institute of Standards and Technology.

GSA. 2015. 3D-4D-BIM overview. Washington, DC: U.S. General Services Administration. www.gsa.gov/portal/content/102276.

Hayter, S.J., P.A. Torcellini, R. Judkoff, B.C. Snead, and R.B. Hayter. 2000. Designing for sustainability. Presented at the 20/20 Vision CIBSE/ASHRAE Joint Conference, Dublin, Ireland. www.nrel.gov/docs/fy00osti/27797.pdf.

I^{2}SL. n.d. Laboratory energy efficiency profiler tool. Arlington, VA: International Institute for Sustainable Laboratories. http://i2sl.org/resources/toolkit/leep.html.

ICC. 2018. IgCC/189.1, *International green construction code®* powered by ANSI/ASHRAE/ICC/USGBC/IES Standard 189.1. Washington, DC: International Code Council.

Itard, L. 2005. Analysis of the exergy-consumption of four types of buildings. Presented at World Sustainable Building Conference, September.

Keeler, M., and B. Burke. 2009. *Fundamentals of integrated design for sustainable building*. NJ: Wiley.

Labs21. 2005a. *Laboratories for the 21st Century: Best practice guide—Right-sizing laboratory equipment loads.* Washington, DC: Laboratories for the 21st Century. www.i2sl.org/documents/toolkit/bp_rightsizing_508.pdf.

Labs21. 2005b. *Laboratories for the 21st Century: Best practices—Low-pressure-drop HVAC design for laboratories*. Washington, DC: Laboratories for the 21st Century. http://labs21.lbl.gov/DPM/Assets/01-18-05_Lab21_Low-Pres.pdf.

Labs21. 2009. *Laboratories for the 21st Century: Best practice guide—Chilled beams in laboratories: Key strategies to ensure effective design, construction, and operation.* Washington, DC: Laboratories for the 21st Century. http://i2sl.org/documents/toolkit/bp_chilled-beam_508.pdf.

May, M.E. 2009. *In pursuit of elegance: Why the best ideas have something missing*. NY: Broadway Books.

Mills, E., G. Bell, D. Sartor, A. Chen, D. Avery, M. Siminovitch, S. Greenberg, G. Marton, A. de Almeida, and L.E. Lock. 1996. *Energy efficiency in California laboratory-type facilities*. Report LBNL-39061. Berkeley, CA: Lawrence Berkeley National Laboratory.

Moran, M.J., and H.N. Shapiro. 2000. *Fundamentals of engineering thermodynamics*, p. 314. NY: Wiley.

Murphy, J. 2011. High performance VAV systems. *ASHRAE Journal* 53(10):18.

Sartor, D., M.A. Piette, and W. Tschudi. 2000. Strategies for energy benchmarking in cleanrooms and laboratory-type facilities. *Proceedings of the 2000 ACEEE Summer Study on Energy Efficiency in Buildings*, August 20–25.

Smith, D.K., and A. Edgar. 2015. Building information modeling (BIM). *Whole Building Design Guide.* Washington, DC: National Institute of Building Sciences. www.wbdg.org/bim/bim.php.

Taylor, S.T. 2007. Increasing efficiency with VAV system statue pressure setpoint reset. *ASHRAE Journal* 49(6):24.

Taylor, S.T. 2011a. Optimizing design and control of chilled water plants, Part 1: Chilled water distribution system selection. *ASHRAE Journal* 53(7):14.

Taylor, S.T. 2011b. Optimizing design and control of chilled water plants, Part 2: Condenser water system design. *ASHRAE Journal* 53(9):26.

Taylor, S.T. 2011c. Optimizing design and control of chilled water plants, Part 3: Pipe sizing and optimizing delta T. *ASHRAE Journal* 53(12):22.

Index

A

B

C

D

E

F

G

H

I

L

M

N

O

P

Q

S

T

U

V

W

Z